Studies in Space Policy

Studies in Space Policy
Volume 11

Edited by the
European Space Policy Institute
Director: Peter Hulsroj

More information about this series at
http://www.springer.com/series/8167

Marco Aliberti

When China Goes to the Moon...

Springer

Marco Aliberti
European Space Policy Institute
Vienna
Austria

ISSN 1868-5307 ISSN 1868-5315 (electronic)
Studies in Space Policy
ISBN 978-3-319-19472-1 ISBN 978-3-319-19473-8 (eBook)
DOI 10.1007/978-3-319-19473-8

Library of Congress Control Number: 2015943788

Springer Cham Heidelberg New York Dordrecht London

Cover Picture Credit: ESA/NASA, The Antarctic aurora, photographed by ESA astronaut Alexander Gerst

Printed on acid-free paper

Springer International Publishing AG Switzerland is part of Springer Science+Business Media (www.springer.com)

Foreword

The world is in flux. Twenty-five years of relative hegemonic stability seem to be coming to an end. And with this comes challenges to the dominating ideology of liberal democracy and to its proponents. ISIS offers a return to the Middle Ages as an approach, Singapore trumpets a unique authoritarian and benevolent model, and Latin America is revisiting various left-wing governance concepts.

China, never much impressed with liberal democracy, has completed the transition to the fifth leadership generation. The new leadership has to deal with the consequences of the relentless growth model introduced by Deng Xiaoping while balancing carefully party rule and the cry for more freedoms. As opposed to Mao era's fervent proselyting, it is doubtful that China will embark on a new journey of ideological warfare now. In fact, this has never been China's path, apart from Mao's time. Chinese exceptionalism is culturally non-expansionist—Chinese exceptionalism lies in the alleged possession of the heavenly mandate, and that cannot be shared. American exceptionalism is founded on its Constitution and the values embodied therein, and there is an inherent ideological expansionism involved. Books like Francis Fukuyama's *The End of History* illustrate the mind-set. This asymmetry in exceptionalism is not well recognised, particularly in the United States, but might provide hope for a future without superpower confrontation. A rising power without interest in evangelising might be able to establish a constructive relationship with the established hegemon so deeply attached to its "universal" values. However, this requires that the hegemon understands the opportunity for non-confrontation, and in this respect there is some way to go in the United States.

This brings us to the specific topic of Marco Aliberti's book, namely, the possible role of space as a bridge builder between actors that see a divide but find few tools to bridge it. Space has frequently been a harbinger of things to come. Space has often been used as a geopolitical tool, not only in times of confrontation but also as a symbol and instrument of cooperation. Marco Aliberti's book explores the possibilities of using China's likely quest to go to the Moon as a tool to create trust and cooperation with a reluctant American partner and Europe's possibilities to be a facilitator and participant. But the book also seeks to assess the situation if cooperation in the reconquest of the Moon cannot be achieved. The book draws attention to the cultural fallout for the existing major space powers if China goes to the Moon on its own and highlights how a possible new space race is likely to lead to embarrassment for the United States and its space allies.

When China Goes to the Moon seeks to reach far beyond the traditional space community—to the geostrategists, overall policymakers, and interested general public. In doing so, it makes a rather introvert field generally accessible by providing comprehensive and easy-to-digest overviews of China's space programmes and space organisations. In a similar fashion, it zooms in on the current state of play of China's efforts in human spaceflight and the rationale for China possibly going to the Moon and the technical challenges in this respect. However, the special merit of Marco Aliberti's work is that it puts China's space endeavours into the broader political and societal setting, something few other books, if any, have done.

When analysing history it is easy to see how it contains a number of watershed points. Identifying such points without the benefit of hindsight is not so straightforward, yet it is relatively safe to say that the global community is currently in front of one. Many forces need to be aligned to make a positive outcome possible. It is my hope that *When China Goes to the Moon* will demonstrate to a wide audience that space can be a potent tool for such an alignment. The global community is not involved in a zero-sum game. Humankind has a unique possibility to continue the path of prosperity and relative peace on Earth. Is it not a beautiful thought that by going back to the Moon together a contribution could be made to a splendid common future on Earth?

Peter Hulsroj
Director, European Space Policy Institute
Vienna, Austria

Acknowledgements

My sincerest gratitude goes to the many people who have supported and contributed to the completion of this book, each in their own way. My greatest debt is owed to Peter Hulsroj, Director of ESPI, who gave me the opportunity to work on this project and guided me throughout the writing process with advice and support that simply proved invaluable. Without our thought-provoking and challenging dialogue, many of the ideas contained in this book would not have come to light.

I would also like to mention the useful inputs provided by all the befriended colleagues of ESPI throughout the research and drafting period. Likewise, I am grateful to many former peers and professors at the University of Naples "L'Orientale" and in particular to Giuseppe Scarpa, who provided invaluable research material and many interesting suggestions.

My thanks are also extended to former ESA Director General Roy Gibson, for his review of the manuscript and for his valuable advice, and to Frances Brown and Jaque Grinberg, whose fine touch and critical input further enhanced the quality of this work.

It remains, of course, that the responsibility for errors, inaccuracies, and infelicities rests with me, despite all the help and inspiration given by so many.

In conclusion, I want to thank and dedicate this book to my family and in particular to my son Ascanio, my most treasured source of strength and inspiration.

Contents

List of Figures

List of Tables

Chapter 1
Introduction

> La nation la plus sage et la plus policée de tout l'univers.
>
> Voltaire
>
> Le gouvernement chinois, comme celui de tous les peuples esclaves, est trop vicieux pour se rendre respectable par ses propres forces. . .
>
> Montesquieu

When European travellers and missionaries retraced the steps of Marco Polo back to China at the beginning of the sixteenth century, they had heard the stories about its wealth and exquisite culture. Yet, they marvelled. In scale and sophistication, China was in a league of its own. There were similarities to Europe, and yet the many differences were striking.

It was a world unto itself, the cradle of an ancient civilisation which, incomprehensibly to the Europeans, blossomed *outside* and *before* the Biblical order. Its culture, language and sociopolitical institutions were all symbols of a refined and millennia-long tradition, which did not simply assert the status of a great civilisation but claimed to be civilisation itself. To emphasise this superiority over the non-Chinese world, China called itself the "Central Kingdom"[1]; a potentially universal empire from which values radiated and whose borders were only set by cultural isobars. This elevated perception of its status was matched and supported by a level of scientific and technological sophistication that often outshone that of Europe. At least initially, Europeans were also surprised by the presence of a prosperous and ordered society that was ably administered by a highly educated class of literati selected on a meritocratic basis. As for the presiding Ming dynasty, it resonated with grandiosity.

To Europeans, affluence and virtue no longer appeared to be the natural monopoly of Europe. The *Great Encounter* with the Chinese civilisation understandably

[1] This book utilises the term "Central Kingdom", rather than "Middle Kingdom", to designate China. Although the term "Middle Kingdom" finds a broader application within the scholarly production and the Chinese ideograms 中国 "zhong-guo" denominating China comprise both meanings, the word "central" better grasps the concept of Sino-centrism in China's *Weltanschauung*. Indeed, while the word "middle" appears to have only a geographical and political connotation, the term "central" also expresses the "civilisational" aspect of China's centrality (thus their cultural superiority) within the *Tianxia* (what is under the heaven, the world).

M. Aliberti, *When China Goes to the Moon...*, Studies in Space Policy 11,
DOI 10.1007/978-3-319-19473-8_1

had a profound impact on the Occident. It simultaneously fascinated and frightened Europeans, generating powerful and contradictory feelings that in some sense have resurfaced dramatically today. Indeed, China's astonishing rise—or more properly, resurgence—as a great power on the international stage, combined with an awareness of its sociocultural "singularity" and multidimensional immensity (in territorial, demographic, and historical terms), has increasingly captured the world's attention in recent years. It should, therefore, come as no surprise that, according to the tracking by the Global Language Monitor of more than 50,000 media sources worldwide, China's rise figured as the most read-about news in the first decade of the twenty-first century.[2]

Interestingly, the current and ever-growing attention paid by the general public and global leaders to China seems to perpetuate the debate that arose in the aftermath of Europe's rediscovery of the Central Kingdom. Just like the contradictory views expressed by Voltaire and Montesquieu, our time is simultaneously generating divergent, and even conflicting interpretations. Portmanteau words such as "coopetitive relations" and "congagement" have, for instance, made their entry into current academic and political debates, demonstrating the inherent difficulty of finding a fixed consensus on what China's resurgence means for the world.

These conceptual and analytical ambiguities are also dramatically mirrored in the space arena, where Beijing's ambitious space programme has increasingly seized the imagination of the global space community, generating as much positive expectation as apprehension and angst.

Of course, all the leading space powers are fully aware that China's ascendancy as a space power represents a significant and potentially disruptive occurrence that can no longer be ignored. If a large part of the debate has so far focused on the geopolitical implications of its ascendancy and on the perils this might hold for the sustainability of space activities, the impressive achievements of this relatively new space actor have also dramatically raised the question of the ensuing cooperation possibilities. Particularly at a time when the undisputed leadership of the USA seems to be faltering or at least face a serious "crisis of identity", and all the traditional space powers are undergoing a period of prolonged austerity, much thought is going into whether China could also be an auspicious partner in the costly and demanding area of space exploration. A comprehensive reflection on how to best deal with (and benefit from) Beijing's arrival on the international space scene has thus become a necessity.

This book is about China's ambitions in its most complex and internationally visible space endeavour, namely, its human space exploration programme. It will provide a comprehensive reflection on China's strategic direction and objectives in space, including in particular those set forth in its human spaceflight programme, and will analyse the key endogenous and exogenous factors that are bound to affect the country's presumed manned lunar ambitions.

[2] "Top News Story of the Decade". Global Language Monitor. 9 December 2009. Web. http://www.languagemonitor.com/top-words-2/top-news-stories-of-the-decade/

However, the focus will not be on China's space exploration programme as such. While it is essential to provide a better understanding of China in order to avoid reductive and potentially misleading interpretations and hence have tools to better engage with the country, the objective is to disentangle the opportunities and challenges China's space ambitions are creating for other spacefaring nations and for Europe in particular. The book will therefore include an in-depth analysis of possible European postures towards China in space exploration and will attempt to stimulate a debate on future space strategies in a broader geopolitical context.

The book is comprised of eight chapters. The next chapter will provide an introductory overview of the fast-developing and increasingly complex Chinese space programme. Attention is paid in particular to its organisational set-up, budgetary allocation, and technological capabilities, as well as to its policies and long-term strategies. China's space programme appears to be one of the most complex and opaque in the world and the difficulties encountered when navigating the ocean of its organisational and bureaucratic structures have often raised fears and fuelled speculation. Providing new tools and perspectives to reach behind the public facade of China's space programme represents the underlying objective of the chapter.

Chapter 3, "Why the Moon?", provides a detailed investigation of the rationales and objectives guiding China's leaders towards a possible manned lunar exploration programme. The analysis seeks to provide a better understanding of the underlying philosophy of China's space programme and, more broadly, China's sociopolitical behaviour, besides the pervasive but too reductive interpretation of a strategic confrontation between a *fast-rising power* and a *declining hegemon.* In fact, overemphasis on an inevitable confrontation between the two juggernauts, China and the USA, can only encourage a simplistic interpretation that would hinder understanding of the multifaceted purposes of China's space programme, many of which are historically and cultural derived behaviours. The intent of the analysis is to provide a window in understanding China's plans and intentions from *their* perspective and thus to permit better engagement with the country. Indeed, in considering China's motivations to send its taikonauts to the Moon, the possibilities for international cooperation in this pursuit may become more visible.

The objective of Chap. 4 is to assess China's long-term ambitions for a manned lunar landing. The analysis is comprised of two main sections. In the first, an extensive review of the precursor functional programmes for embarking upon a lunar endeavour—in particular of the manned spaceflight programme and of the lunar exploration programme—is provided. This will in turn be used as a basis for discussing the current state of play of Chinese lunar plans. More specifically, the second part of the chapter will set out considerations of the skills and hardware development required for the implementation of the programme and an assessment of how the overall organisation of this programme might be managed and structured. Some reflections on the potential mission configuration will also be provided.

Chapter 5 shifts the focus to an examination of what can be regarded as the "conditioning factors" for securing Chinese success in the reconquest of the Moon. It is in fact quite evident that concrete plans and strong motivations for reaching the Moon are not, on their own, sufficient for the country to send its taikonauts there. The high complexity of a manned lunar exploration programme involves a number of conditioning factors and prerequisites that must be fulfilled in order to succeed. By considering "China going to the Moon" as a dependent variable, we can identify the series of independent variables that could ultimately affect China's capacity to carry out its manned lunar exploration programme. The chapter identifies four macro-variables influencing the country's space ambitions—socio-economic, political, technological, and international. Chapter 5 focuses on what can be regarded as endogenous conditioning factors, while Chap. 6 will assess the international ones. Rather than predicting the future of China in each of these domains, the different sections of the chapter aim to assess why, how, and to what extent the variables considered could affect a manned lunar exploration programme. These conditioning factors will eventually be summarised in the last section, which will also try to answer the question of whether China can or cannot go to the Moon on its own and discuss why it might not be willing to embark upon a solo mission.

International variables are then assessed in Chap. 6, "China, the Moon and the World". The main aim of this chapter is to investigate how a Chinese determination to go to the Moon would affect the rest of the international institutional landscape in the period leading up to the country getting there. At the start the chapter will reflect on the *nature* and the *extent* of China's impact on the global space community and hence provide an account of the posture the leading space powers could adopt vis-à-vis its ambitions in space. In doing so, the chapter will in particular elaborate on the much-discussed scenario of an intra-Asian space race (between China, Japan, and India) and of a Sino-American space race. The various sections will, however, also seek to accompany the analysis with suggestions of a limited amount of scenario alternatives at the various junctures, where more cooperative pathways for space exploration might eventually become possible.

The final chapter, "Europe and China in Space: Constraints, Opportunities and Options", will specifically elaborate on the opportunities and challenges China's possible lunar ambitions are raising for Europe and will provide an assessment of the different strategies available to European stakeholders in this regard.[3]

Given the inherent geopolitical dimension of space activities, the chapter will first provide an assessment of the most recent evolution in the broader political relationship between Europe and China. An account of the long-standing framework of cooperation in space activities between China and different European institutions will subsequently be provided. The two analyses will in turn be used as a basis for a strengths, weaknesses, opportunities, and threats (SWOT) analysis of potential Sino-European cooperation with regard to human space exploration and to identify a set of policy options for Europe. Finally, a qualitative assessment of the various options and a series of recommended actions for European stakeholders will be provided.

[3] Within this study Europe is regarded and examined as a unified, though sui generis, internationally acting body, whose *space actorness* results from the complex interplay of three main constituencies (ESA, EU, and their member states).

The book closes with an epilogue reflecting on the potential contribution that a major European initiative in space exploration could bring to the contemporary quest for a new global order.

It is the author's hope that this study will contribute to promoting a better understanding of China's posture in the international space arena and stimulate further reflections on this complex and exceedingly relevant topic.

Chapter 2
China's Space Programme: An Overview

This chapter provides an introductory overview of China's fast developing and increasingly complex space programme. The analysis is performed according to a categorisation created by Jim Dator,[1] who developed a framework to understand the process of technology advancement. In his view, all technological areas of development—including space programmes—can be understood as a product of three components: *hardware*, *software*, and *orgware*.

The term *hardware* in this categorisation refers to the material resources and technological capabilities of a space programme. It basically makes up the national capacities in terms of space systems (e.g. launchers, satellites, and ground facilities) and budgetary expenditures. *Orgware*, on the other hand, comprises the organisational structures set up to develop and run the hardware. The *software* of the space programme denotes the norms and rules applied to use the technological capabilities for specific purposes. These are captured in the national space policies and strategies.

In line with this taxonomy, particular attention will be paid to the organisational set-up of China's space programme, to the budgetary allocation, and to the space policies and long-term strategies adopted by Beijing. Specific consideration of China's technological capabilities will be provided in Chap. 4.

[1] Dator, Jim (1983). "Loose Connections: A Vision of Transformational Society". In: Masini, Eleonora (ed). *Visions of Desirable Societies*. Pergamon Press, Oxford. Dator's categorization has been successfully adopted and applied to the analysis of space programmes also by space policy analyst Stacey Solomone. See Solomone, Stacey (2013). *China's Strategy in Space*. Springer, New York: pp. 17–22.

M. Aliberti, *When China Goes to the Moon...*, Studies in Space Policy 11,
DOI 10.1007/978-3-319-19473-8_2

2.1 Organisation of Space Activities in China

China's space programme is one of the most complicated and non-transparent in the world, and understanding its organisational and bureaucratic structures can involve significant difficulties.

These difficulties are not just a result of the high level of secrecy surrounding the programme; rather, they are determined by the combination of secrecy with other four main features, which are (a) the existence of a "Byzantine maze" of bureaucratic structures that involve a myriad of organisations, as well as countless organisations within organisations[2]; (b) the general complexity of the inner workings of China's power structures and hierarchies; (c) the multiple restructurings, renaming, and relocation of bureaucratic offices and institutes that have occurred through the past 50 years in the Chinese space organisation; and (d) the continuous expansion of space governance in terms of the creation of new administrative entities designed to respond to the needs of new programmes and missions.

The combination of these multiple factors not only confuses any attempt to correctly pair the various institutions, and eventually to peer into the inner workings of the Chinese system, but also raises many fears and fuels speculation. It has even been noted that often "the renaming, relocation, and lack of transparency within organisations has left employees themselves unaware" of their office's position within the overall organisational structure.[3]

The following section can thus only be an attempt to assess the functions and responsibilities of the most important, large, and central organisations currently involved in the governance of China's space programme.

2.1.1 *A Leading Small Group on Space?*

In order to reach behind the public facade of the governance of China's space programme, an insight into the structures of power and working relationships of the leadership system is provided first of all.

The first point to note is that the governance regime of the People's Republic of China (PRC) consists of three major vertical systems (*xitong*): the Chinese Communist Party (CCP), the government, and the military.[4] The three systems operate

[2] Johnson-Freese, Joan (1998). *The Chinese Space Program. A Mystery Within a Maze*. Krieger Publishing Company, Malabar, FL.

[3] Cheng, Dean, and Kerry Murray (2001). "Orbital Dragons: Implications of Chinese Access to Dual-Purpose Space Technologies". In: Williamson, Ray A. *Dual-Purposes Technologies: Opportunities and Challenges for US Policymaking*. Space Policy Institute. Washington DC: p.72.

[4] Ning, Lu (2001). "The Central Leadership, Supraministry Coordinating Bodies, State Council Ministries, and Party Departments". In: Lampton, David M. (ed). *The Making of Chinese Foreign and Security Policy in the Era of Reform*. Stanford University Press, Stanford: pp. 45–49.

in a symbiotic relationship, but the role and power of the CCP—and of its Central Committee in particular—are ultimately the most prominent, and its overwhelming presence continues to overshadow the entire system. For this reason, China's leadership system has been correctly described as centred on a party-based, oligarchic, consensus-driven structure that reflects a balance among the institutional interests of its three organisational pillars.[5]

In order to build consensus on issues that cut across the government, party, and military systems and to develop rational, coherent, and balanced decision-making, high-level coordinating and consulting bodies have regularly been set up. These bodies, usually labelled Leading Small Group (LSG, *lingdao xiaozu* in Chinese), provide a mechanism for top decision-makers to exchange views on sensitive issues, build consensus, and create a framework for the general direction in which the subordinate bureaucracies should move. As noted by the US scholar Alice Miller, because these groups deal with sensitive leadership processes, they are never incorporated into publicly available charts or explanations of party/government/military institutions, but their existence has to be nonetheless acknowledged and their role ultimately considered crucial in any coherent policymaking elaboration on sensitive issues.[6]

LSGs do not generally formulate concrete policies, but create—through the provision of recommendations and guiding principles—the framework for their development. As noted by several scholars, these recommendations are likely to exert considerable influence on the policymaking process because they are an expression of the consensus reached by the leading members of the relevant government, party, and military agencies. In some cases, the Chinese leadership will adopt an LSG's recommendations with little or no modification.

An important feature of these high-level coordinating bodies is that they can be formed not only to build consensus on issues that cut across the government, party, and military systems but also on sensitive issues involving different interests within one of these three systems. In short, the State Council, the Central Committee of the CCP, and the People's Liberation Army (PLA)—respectively, the highest ranking organs of the government, the party, and the military—often create their own leading groups to coordinate policies.

LSGs are formed in regard to a broad range of issues; examples include foreign affairs, finance and economic affairs, national energy resources, environmental protection, and agricultural affairs. Sometimes, these groups are also formed with regard to specific issues, such as the LSG for the 2008 Olympics set up by the State

[5] Swaine, Michael D. (2012). "China's Assertive Behavior Part Three: The Role of the Military in Foreign Policy". China Leadership Monitor No. 36. Hoover Institution.

[6] The practice of creating Leading Small Groups has become so relevant for China's policymaking processes, that these groups are now considered the most important national coordinating bodies and the centres of cross-ministry negotiation and consultation. Miller, Alice (2008). "The CCP Central Committee's Leading Small Groups". China Leadership Monitor No. 26. Hoover Institution.

Council, or the LSG for the Lunar Probe Project, jointly established by the State Council and the Central Military Commission of the CCP in February 2004.[7]

Considering the widespread utilisation of LSGs for the management of sensitive issues and the political, economic, and strategic significance that space activities have for China, it is highly plausible to also envisage the existence of a high-level LSG for the overall coordination of space activities.

Notwithstanding the absence of official documents and the dearth of extensive analysis in this regard,[8] the necessity and plausibility of a "Space Leading Group" (SLG) is reinforced in particular by the simultaneous involvement of different key stakeholders in the management of the space programme.[9]

Such an SLG would not only be intended to serve as an oversight body and arena for consensus building among the leading members of the relevant government, party, and military agencies; it would also form the core programmatic leadership of China's space programme. The members of the SLG would be senior officials of the CCP, the PLA, and the government, including the prime minister and high-level representatives of the different ministries involved in the programme (e.g. the Ministry of Foreign Affairs, the Ministry of Industry and Information Technology, and the Ministry of Finance).

Like the other LSGs, the SLG is unlikely to formulate concrete policies, but more likely provides the various stakeholders with a series of recommendations and guidelines about the general direction, which the various stakeholders have to respect.

2.1.2 The State Council and SASTIND

Among the major stakeholders under the shadow of an SLG, a primary role would be played by the State Council, which is the highest ranking government organ. The State Council mainly exercises its authority over national space affairs through its ministries and by having the final word on funding decisions for programmes. In addition, the State Council issues the five-year space plan—in the form of a government White Paper—defining the medium-term national strategy in space.

---

[7] *Ibid.*

[8] Only little analysis in the literature has so far acknowledged the possible existence and role of a Space Leading Group. One of the first is provided by the Chinese scholar Yanping Chen in an article published by Space Policy in 1993 ("China's space commercialisation effort. Organisation, policy and strategy". Space Policy Vol. 9 (1). 1993: 45–53). The SLG is also mentioned, although not extensively explained in the books of Joan Johnson-Freese (*The Chinese Space Program. A Mystery Within a Maze*. Krieger Publishing Company, Malabar, 1998) and Brian Harvey (*China in Space. The Great Leap Forward.* Springer, New York, 2013).

[9] The likelihood of an SLG is also reinforced by the acknowledged creation of an ad hoc LSG for the management of specific highly sensitive space projects like Shenzhou and Chang'e.

The State Administration on Science, Technology and Industry for National Defence (SASTIND) is the main administrative body under the State Council tasked with coordinating and managing the country's space activities. It was created through the March 2008 reforms of the State Council that "consolidated and rearranged a number of existing government bodies into larger 'super-ministries'".[10] These reforms dismantled the Commission on Science, Technology and Industry for National Defence (COSTIND) and shifted most of its responsibilities and personnel to the newly established SASTIND.

Unlike COSTIND, SASTIND is no longer an organisation under the direct authority of the State Council, but has become part of the super-Ministry of Industry and Information Technology (MIIT). Its main role is to act as the administrative and regulatory hub for the general aspects of China's defence and aerospace industry (in particular development, procurement, and supply). Concretely, SASTIND issues space and defence industry regulations and monitors their implementation, allocates R&D funds through research programmes—which are supervised in collaboration with the Ministry of Science and Technology (MOST) and presumably also with the Ministry of Finance (MOF)—and determines which enterprises may or may not engage in the research and production of aerospace technologies and systems.[11] Specifically, with regard to space activity administration, SASTIND also plays an important role in terms of coordinating space policy and plans for the State Council; it is in charge of executing the main space-related regulations, including the "Measures for the Administration of Registration of Objects Launched into Outer Space".[12]

2.1.3 The China National Space Administration

Under SASTIND in the hierarchy, the China National Space Administration (CNSA) formally holds responsibility for "defin[ing] the national space policies, administer[ing] the civilian space programme and manag[ing] the development of

[10] Francis, Ed, and Susan M. Puska (2010). "Contemporary Chinese Defense Industry Reforms and Civil-Military Integration in Three Key Organizations". Study of Innovation and Technology in China. Policy Brief No. 5. Web. http://igcc.ucsd.edu/assets/001/500870.pdf. Accessed 18 January 2014.

[11] *Ibid.*

[12] In the measures, it is for instance specified that COSTIND (SASTIND) is in charge—together with Ministry of Foreign Affairs—of the national registration of space objects (art. 4). SASTIND is also responsible for maintaining the National Register. See "Measures for the Administration of Registration of Objects Launched into Outer Space". Unofficial translation by the Faculty of International Law of China University of Science and Law. 8 February 2001. Available at: http://www.spacelaw.olemiss.edu/library/space/China/Laws/JSL_33.2_China%20Law.pdf. For a commentary, see Ling, Yan (2008). "Comments on the Chinese Space Regulations". Chinese Journal of International Law. Vol. 7 (3). Web. http://chinesejil.oxfordjournals.org/content/7/3/681.full.pdf.

national space science, technology and industry".[13] Although CNSA appears on paper to be a fully fledged national space agency, it would be erroneous to consider it as such. In spite of having a name similar to that of its better-known US counterpart, the CNSA in fact is not an all-encompassing space agency, tasked with similar responsibilities and functions to those exercised by the space agencies of the major spacefaring nations.

Rather, the CNSA appears to be, in essence, a clearing house carrying out only a few tasks, namely, serving as the public international face of China's space programme and, second, acting as the liaison office between SASTIND and the aerospace industries. It should be recalled that CNSA was established in 1993 along with the China Aerospace Corporation (CAC) to replace the dismantled Ministry of Aerospace Industry. The underlying intention was to provide the country's space programme with a visible governmental face and apparently to separate space-related governmental functions (theoretically to be assigned to the CNSA) from industrial ones (assigned to CAC). In fact, many of the administrative and managerial responsibilities and functions of this defunct ministry have remained inside CAC. As a result, CNSA's role has remained rather narrow: it has ended up operating as a liaison office between SASTIND and CAC, besides serving as the public face of China's space programme internationally, working with foreign national space agencies.

In sum, while CNSA can be seen as China's external space policy organisation, carrying out China's international obligations and representing the country in international organisations and events (e.g. the ISECG), CAC can be seen as a more powerful internal complement, wielding real power over national space programme matters.[14] Perhaps, these two organisations should really be viewed as one large agency which, not by chance, shares both personnel and management, as well as a very similar logo. A more detailed description of CAC (now restructured as CASC and CASIC) and the aerospace industry's role is provided later in this section.

Confirmation of CNSA's limited role comes from the fact that CNSA is not responsible for the elaboration of the Five-Year Guidelines on space activities, these Guidelines falling within the same framework as China's overall national economic development plans and being decided at the highest political level. Even the derived document, the "White Paper on China's space activities", is not issued by the CNSA but by the State Council on the basis of the targets envisaged in the Five-Year Plan and subsequently released by its Information Office.

---

[13] See "Organisation and Functions". China National Space Administration. 20 February 2013. Web. http://www.cnsa.gov.cn/n615709/n620681/n771918/index.html.

[14] Cheng, Dean, and Kerry Murray (2001). "Orbital Dragons: Implications of Chinese Access to Dual-Purpose Space Technologies". In: Williamson, Ray A. *Dual-Purposes Technologies: Opportunities and Challenges for US Policymaking*. Space Policy Institute. Washington DC.: p.74.

2.1.4 The China Satellite Launch and Tracking Control General

Compared to CNSA, a more substance-orientated organisation under the authority of SASTIND is the China Satellite Launch and Tracking Control General (CLTC). This organisation, headquartered in Beijing, directly controls and oversees the country's space missions and projects, including its launch infrastructure (thus the three launch sites of Xichang, Jiuquan, and Taiyuan and the forthcoming launch centre of Wenchang), as well as the hub of China's telemetry, tracking, and control (TT&C) network, the Xi'an Satellite Control Centre (XSCC).[15] Although the CLTC falls under the civilian authority of SASTIND, it is run by the General Armament Department (GAD) of the PLA for both the military and civil space programmes. This civil–military mixture in the governance of the CLTC can ultimately be regarded as evidence of the aforementioned intricate web of functions and responsibilities surrounding the Chinese space programme. It clearly shows how the different dimensions (civil, military, commercial, and academic) of the programme—although not fully integrated—are hardly distinguishable.[16] Additional information on China's TT&C network, control centres, and launch sites will be provided in Sect. 4.2.

2.1.5 The General Armaments Department of the PLA

The General Armaments Department (GAD) is one of the four departments of the PLA operating under the control of the Central Military Commission (CMC). [17] It is primarily in charge of managing the procurement and acquisition of weapon systems for the PLA and ensuring defence industry core capabilities. These essential tasks, however, give GAD a broad portfolio of administrative functions and responsibilities. Besides acting as the defence industry's main customer, GAD has also widely engaged with the defence and aerospace industry as regulator, in particular in terms of R&D and production programme management. This role is exercised together with SASTIND on a complementary and peer-to-peer basis. It should be noted, however, that, although GAD and COSTIND were once of equal bureaucratic rank, since the March 2008 reforms and the subsequent subordination

---

[15] "China Satellite Launch and Tracking Control General". Nuclear Threat Initiative. 20 January 2014. Web. http://www.nti.org/facilities/124/.

[16] Solomone, Stacey (2013). *China's Strategy in Space*. Springer, New York: p. 21.

[17] The other three departments are the General Staff Department, the General Political Department, and the General Logistics Department.

of SASTIND to the MIIT, the new protocol parity is no longer between GAD and SASTIND, but between GAD and MIIT.[18]

In collaboration with SASTIND, GAD issues defence industry regulations and monitors their implementation; allocates R&D funds through research programmes, such as the 863 programme, supervised in collaboration with the Ministry of Science and Technology (MOST); and determines which enterprise may or may not engage in the research and production of space technologies and systems.[19]

Besides sharing responsibility for the R&D and production programmes of China's aerospace sector and for the administration of space-related infrastructure with SASTIND, GAD is directly responsible for the development of military space capabilities. It also takes part in the management of sensitive space programmes, like human spaceflight. The China Manned Space Engineering (CMSE) Office, which is the bureau of an ad hoc LSG established to manage the Shenzhou manned spaceflight programme, is not by accident headed by a representative of GAD.

This active involvement of the PLA in the management and execution of China's space programme has obviously raised serious concerns and led many Western analysts to assert that the role of the PLA is ultimately the overwhelming one. Reports produced by the *US–China Economic and Security Review Commission* issued for the US Congress have repeatedly emphasised this aspect.[20]

This claim can be considered accurate insofar as the key infrastructural elements (like launch and tracking facilities) are run and staffed by the military, and a highly visible endeavour such as human spaceflight sees its direct involvement. Affirming that projects are run by the PLA, however, does not automatically imply that they are ultimately decided on and controlled by the military. In fact, not only are core responsibilities shared with other leading stakeholders (e.g. SASTIND, the MOST, and CAS), but key decisions on the implementation of space policies and the overall direction of the programme ultimately reside in the hands of the high-level decision makers of the Party.[21]

In this regard, it must be emphasised that the PLA is far from being an autonomous and independent player within the power structures of the PRC. As mentioned, the GAD is one of the four departments of the PLA operating under the control of the Central Military Commission (CMC) of the CCP, which is the leading organ of the armed forces within the Communist Party. The fact that Xi Jinping,

[18] Francis, Ed, and Susan M. Puska. (2010) "Contemporary Chinese Defense Industry Reforms and Civil-Military Integration in Three Key Organisation". Study of Innovation and Technology in China. Policy Brief No. 5. Web. http://igcc.ucsd.edu/assets/001/500870.pdf. Accessed 18 January 2014.

[19] *Ibid*: pp. 2–3.

[20] US–China Economic and Security Review Commission. 2011 Annual Report to Congress. US Government Printing Office, Washington DC, United States. November 2011.

[21] For this interpretation, see also Kulacki, Gregory, and Jeffrey Lewis. (2009). *A Place for One's Mat: China's Space Program, 1956–2003*. American Academy of Arts and Sciences, Cambridge, MA. See also Harvey, Brian (2013). *China in Space. The Great Leap Forward*. Springer, New York.

president of the PRC and chairman of the CCP, also holds the position of chairman of the CMC not only shows the close interconnection between the Army and the Party but is also a significant piece of evidence of the overall subordination of the military to the party political leadership.

In addition, the orientation given by the CMC on the development of space competences does not necessarily indicate that there is a clear priority given to the military component of the space programme.[22] GAD participation in a number of relevant space endeavours concretely serves as a catalyst for faster and broader innovation throughout the PLA, as well as for spurring civil–military integration in the Chinese defence, science, technology, and industry system,[23] but cannot be portrayed as the specific, ultimate goal of China's space programme.

Furthermore, it should be noted that possible cooperation with China in human spaceflight does not automatically imply cooperation with its military: indeed, scientists of the China's Academy of Science have for instance already cooperated with German researchers on SIMBOX, a package of biological and medical experiments launched on board the Shenzhou-8 spacecraft in 2011. In addition, the scientific community is planning joint cooperative undertakings in the forthcoming second space laboratory (e.g. the POLAR experiment with Switzerland, France, and Poland and the SVOM mission with CNES).[24]

In conclusion, it would seem more appropriate to reverse the perspective proposed by the scholar Dean Cheng by affirming that one of the myths surrounding China's space programme is that "it is military in nature".[25]

2.1.6 The Aerospace Industry

In an almost symbiotic relationship with SASTIND, GAD, and the central government, China's aerospace industries occupy a key position in the overall organisational structure of China's space activities. As mentioned, not only do they make up the backbone of China's space programme through the provision of space technologies and systems; they also act as proactive and fully fledged players, wielding real power in the administration of space activities and exerting primary

---

[22] Rathgeber, Wolfgang (2007). "China's Posture in Space. Implications for Europe". ESPI Report 3. European Space Policy Institute, Vienna: p. 32.

[23] Francis, Ed, and Susan M. Puska. (2010) "Contemporary Chinese Defense Industry Reforms and Civil-Military Integration in Three Key Organisation". Study of Innovation and Technology in China. Policy Brief No. 5. Web. http://igcc.ucsd.edu/assets/001/500870.pdf. Accessed 18 January 2014.

[24] Harvey, Brian (2013). *China in Space. The Great Leap Forward.* Springer, New York: p. 18 and p. 359.

[25] Cheng, Dean. "Five Myths about China's Space Program". The Heritage Foundation. 29 September 2011. Web. http://www.heritage.org/research/reports/2011/09/five-myths-about-chinas-space-program.

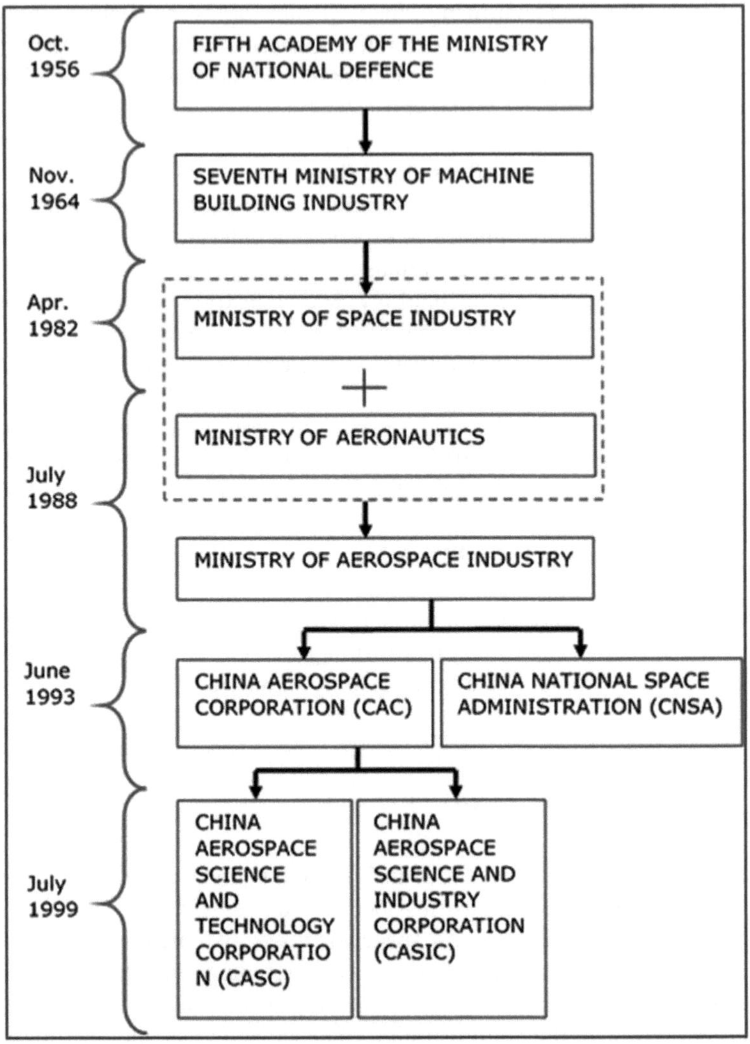

Fig. 2.1 Evolution of China's aerospace industry (author's visualisation)

control over the execution of the space programme in terms of day-to-day operations. As State-Owned Enterprises (SOEs), it should be recalled that they originated as governmental entities that were eventually transformed into actual enterprises, but a large part of their former administrative responsibilities and functions has nonetheless remained in their hands (see Fig. 2.1).

Currently, there are two huge state-owned industrial groups that are actively involved in the administration and execution of China's space programme:

- The China Aerospace Science and Technology Corporation (CASC)
- The China Aerospace Science and Industry Corporation (CASIC)

Both CASC and CASIC were officially set up in July 1999, when the CAC, established in 1993, was once again restructured and split into these two large conglomerates. It merits note, however, that CAC too was the result of a long-standing process of transformation started in October 1956 with the creation of the Fifth Academy of the Ministry of National Defence. In 1965, this academy became an autonomous ministry (the Ministry of the Seventh Machinery Industry) and subsequently went through numerous organisational and name changes that include the Ministry of Space Industry (in 1982), the Ministry of Aerospace Industry (in 1988), and the final transformation into a real corporation in 1993 when the ministry was split into CNSA and CAC.[26] This evolutionary path deserves attention, as it captures well the reason for arguing that CASC acts as the actual "operating agency" of the country's space programme.

It is quite likely that the driver behind the restructuring approved in July 1999 was the twofold effort of China's "New Right" policymakers to create a more organic partition between the defence industry (CASIC) [27] and the space industry (CASC) and to loosen the state's control over the running of enterprises in order to spur innovation and inject some degree of competition into the aerospace and defence industry, thus strengthening overall procurement for space technologies and systems. The ultimate outcome, which still appears to be a work in progress, can nevertheless be seen as a half success for the "New Right": although both the CASC and CASIC aerospace industries are no longer directly government-managed companies, they nonetheless remain government owned and controlled. Indeed, the two corporations still need to report directly to the Central Government, which exercises control through three agencies: the State-owned Assets Supervisions and Administration Commission (SASAC) of the State Council[28] and the previously described SASTIND and GAD.

2.1.6.1 CASC

CASC is a large-scale conglomerate of more than 130 companies and industrial plants scattered nationwide and employing more than 140,000 staff.[29] As the main contractor of China's space programme, CASC is primarily engaged in the research, design, manufacture, and supply of space technologies and systems, as well as in the provision of international commercial satellite launch services. The conglomerate comprises eight major R&D and production complexes—each of

[26] "History of CASC". China Aerospace Industry Corporation. Web. http://english.spacechina.com/n16421/n17138/n382513/c386575/content.html. Accessed 20 February 2014.

[27] For more information on the so-called New Right policymakers, see Sect. 5.2.

[28] For a description of the role of SASAC, see "Main functions and responsibilities of SASAC". State-owned Assets Supervisions and Administration Commission. Web. http://www.sasac.gov.cn/n2963340/n2963393/2965120.html. Accessed 20 February 2014.

[29] "Company profile". China Aerospace Science and Technology Corporation. Web. http://english.spacechina.com/n16421/n17138/n17229/c127066/content.html. Accessed 20 February 2014.

them having their own research institutes, manufacturing plants, and commercial enterprises—14 specialised companies, nine listed companies, and a number of subordinate units.[30] The eight major complexes, which form the backbone of CASC and are often referred to as academies, thanks to their close connection with the China Academy of Science (CAS), are:

- China Academy of Launch Vehicle Technology (CALT)
- China Academy of Space Technology (CAST)
- Shanghai Academy of Spaceflight Technology (SAST)
- Academy of Aerospace Solid Propulsion Technology (AASPT)
- Academy of Aerospace Liquid Propulsion Technology (AALPT)
- Sichuan Academy of Aerospace Technology (SAAT)
- China Academy of Aerospace Electronics Technology (CAAET)
- China Academy of Aerospace Aerodynamics (CAA)

Among these academies, CAST, CALT, and SAST are the most prominent. CAST is the primary R&D and production complex that designs and manufactures scientific and applications satellites. Like the other academies, CAST has a significant infrastructure, with a number of subordinate institutes, centres, and factories.[31] CALT and SAST are mainly involved in the overall research, design, development, manufacturing, and testing of the Long March (LM) Launch Vehicles, manned spacecraft, and related products. Usually referred to as the Beijing and Shanghai "bureaus" of China's space programme, the two academies are sometimes regarded as competing organisations. In order to have a harmonised and balanced "distribution of work and responsibilities", competences have been carefully distributed between the two. With regard to the development of the next generation of LM vehicles, it can be noted that, whereas SAST has been assigned the development of the LM6 and LM7, the Beijing-based CALT acts as the primary stakeholder for development of the future heavy and super-heavy rockets: the LM5 and the LM9, respectively (see Sect. 4.2).

Besides these eight large R&D and production complexes, CASC exercises control over a number of specialised companies. Among them, a significant role is played by the China Great Wall Industry Corporation (CGWIC). Established in 1980, CGWIC "is the sole company authorised by the government to provide commercial satellite launch services and space technology to international clients".[32] As a large corporation, CGWIC also has a number of subsidiary companies. Other specialised companies of CASC include the China Satellite Communication

[30] *Ibid.*

[31] For detailed examination of CAST, see Harvey, Brian (2013). *China in Space. The Great Leap Forward*. Springer, New York: pp. 52–53.

[32] *Cit.* "Company profile". China Great Wall Industry Corporation. Web. http://www.cgwic.com/About/index.html. Accessed 20 February 2014. See also "China Great Wall Industry Corporation (CGWIC)." Nuclear Threat Initiative. Web. http://www.nti.org/facilities/50/. http://www.cgwic.com/Partner/. Accessed 20 February 2014.

Corporation, the China Aerospace Engineering Consultation Centre, and the China Aerospace Electronics Corporation.

Overall, the organisational structure of CASC reveals a high degree of complexity; a graphical representation can be found in Appendix C, which reinforces the aforementioned argument about CASC acting largely as a national space agency.

2.1.6.2 CASIC

Like CASC, CASIC is a huge conglomerate of more than 140 companies, factories, and R&D institutes scattered nationwide. These are comprised of seven main academies, two research and production bases, and six publicly listed companies, which employ 135,000 people, 40 % of whom are specialists and technicians.[33]

Although CASIC acts as the main contractor of China's aerospace defence programme and is particularly focused on the production of short- and medium-range ballistic missiles and cruise missiles, it also plays an active part in many space endeavours. Through its subsidiaries, a large number of stand-alone technologies and products, industrial basic parts, electronic components, and software testing and evaluation services have been provided for the Chang'e lunar exploration and the BeiDou satellite navigation programmes. In addition, CASIC has provided ground-to-space integrated support for each of the ten spaceflights undertaken by the Shenzhou missions, including technical support for their accurate injection into orbit, precise docking, stable operation, and safe return. CASIC radar equipment and technology are also used in meteorological observations and for the ground receiving system.[34] Compared to CASC, however, CASIC's overall role remains less substantial and not so completely entrenched in the overall governance of the Chinese space programme.

2.1.7 The Chinese Academy of Sciences

The Chinese Academy of Sciences (CAS) is an additional and very relevant player in the scheme of China's space activities. Established immediately after the CCP takeover of China (on 1 November 1949), from the outset CAS has been assigned with the responsibility for providing S&T consultations for the nation's decision-making and leading the nation's S&T development. CAS is a prestigious and vast institution, numbering roughly 60,000 regular staff, 79.9 % of whom are

[33] "Introduction to CASIC". China Aerospace Science and Industry Corporation. Web. http://english.casic.cn/n189298/n189314/index.html. Accessed 21 February 2014.

[34] "Footprints in Space". China Aerospace Science and Industry Corporation. Web. http://english.casic.cn/n189300/n1547687/index.html. Accessed 21 February 2014.

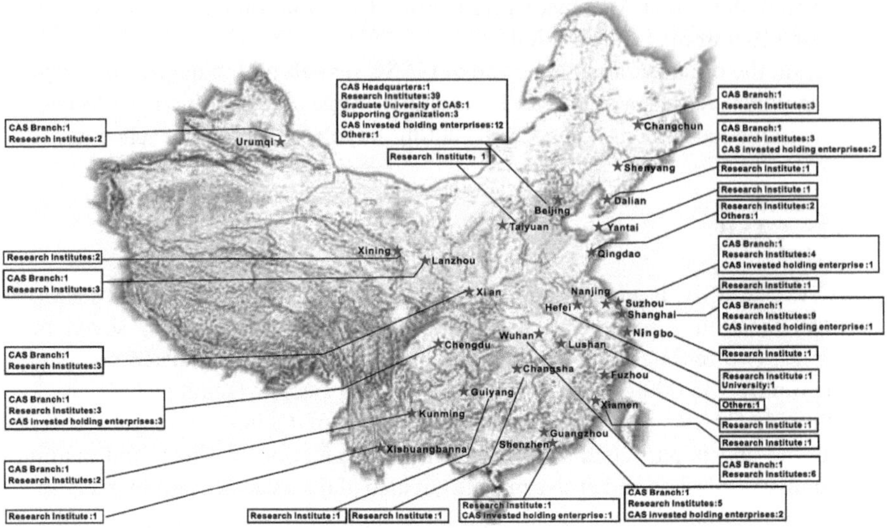

Fig. 2.2 Distribution of CAS institutions (*Source*: CAS)

professional and technical employees.[35] There are 124 institutions directly under its authority, with 104 research institutes, five universities and supporting organisations, 13 management organisations featuring headquarters and branches in China's major cities, and three other units. Moreover, there are 25 legal entities affiliated with it and 22 holding companies with CAS investment,[36] the most well-known being Lenovo.[37] Research is organised along six divisions: mathematics and physics, chemistry, life sciences and medical sciences, Earth sciences, information technology sciences, and technological sciences. A map illustrating the distribution of CAS institutions is presented in Fig. 2.2.

As the highest academic institution for S&T and the linchpin of China's overall S&T planning, CAS has been playing a crucial role also in regard to the space programme. This role has manifested itself both in terms of influence exercised on the decision-making process and in terms of concrete management of a number of space-related programmes. As the analysis of the Shenzhou manned spaceflight and Chang'e lunar exploration programmes will show (see Sect. 4.2), CAS scientists

[35] "Directory of the CAS Subordinate Institutions". Chinese Academy of Sciences. Web. http://english.cas.cn/CASI/In/200909/P020120813569850110467.pdf. Accessed 21 February 2014.

[36] "CAS Institutions". Chinese Academy of Sciences. Web. http://english.cas.cn/CASI/. Accessed 21 February 2014.

[37] The CAS holds a 36 % stake in Legend Holdings Ltd., which is Lenovo's largest shareholder with a 32.5 % stake. Osawa, Juro and Lorrainne Luk. "How Lenovo Built a Chinese Tech Giant". Wall Street Journal. 30 January 2014. Web. http://online.wsj.com/news/articles/SB10001424052702303973704579352263128996836. Accessed 22 February 2014.

have often turned into policy entrepreneurs, proposing ideas, setting agendas, and implementing policy within the national space programme.[38]

This influence is primarily wielded through the prestige CAS holds at national level and is reinforced by its independence from the ministries and its direct connections to the State/Party authorities. In fact, it is worth-noting that CAS does not report back to the MIIT or the Ministry of Science and Technology, the Academy being an institution under the direct authority of the State Council.

Through the work of its research institutes, CAS also exercises a substantial role in space programme management. Among the numerous institutes under CAS authority that are more visibly involved in the overall execution of China's space programme are:

- The Institute of Remote Sensing and Digital Earth (RADI), which was established in November 2012, through the merging of two CAS institutes: the Institute of Remote Sensing Applications (IRSA) and the Centre for Earth Observation and Digital Earth (CEODE). RADI focuses on the construction and operation of major earth observation infrastructure and the air–space–ground integrated earth observation technology system.[39]
- The National Space Science Centre (NSSC), which is the key national institute responsible for planning, selecting, developing, and managing the operation of China's space science satellite missions.[40]
- The National Astronomical Observatories of the Chinese Academy of Sciences (NAOC): this institution was officially founded in April 2001 through the merger of four observatories, three observing stations, and one research centre, all of which were suborganisations of CAS. NAOC is headquartered in Beijing, with four subordinate units distributed across the country: Yunnan Observatory, Nanjing Institute of Astronomical Optics and Technology, Urumqi Observatory (now called Xinjiang Astronomical Observatory), and Changchun Observatory. Purple Mountain Observatory (PMO) and Shanghai Astronomical Observatory (SHAO) are separate institutes of CAS, but are subject to the same academic strategies and research policies as NAOC.[41]
- The Shanghai Institute of Micro-systems and Information Technology, which had major responsibility for the development of microsatellites, including the *Chung Xin* (literally "Innovation") store-and-forward communications

[38] Even the decision to create the BeiDou national satellite navigation system resulted from the efforts of a core group of scientists of CAS. Besha, Patrick (2010). "Policy making in China's space program: A history and analysis of the Chang'e lunar orbiter project". Space Policy Vol. 26 (4): 214–221.

[39] "Institute of Remote Sensing and Digital Earth-Chinese Academy of Sciences". Chinese Academy of Sciences. Web. http://english.irsa.cas.cn/. Accessed 18 February 2014.

[40] "National Space Science Centre-Chinese Academy of Sciences". Chinese Academy of Sciences. Web. http://english.nssc.cas.cn/au/mfdg/. Accessed 18 February 2014.

[41] "National Astronomical Observatories Chinese Academy of Sciences". Chinese Academy of Sciences. Web. http://english.nao.cas.cn/au/history/. Accessed 21 February 2014.

microsatellite, is also focused on systematic core technology innovation and microsatellite integration innovation.

The existence of a large number of institutes under CAS that are tasked with concrete management responsibilities creates additional complexities in the attempt to draw a clear picture of "who is in charge" of China's space activities. The institutes and R&D centres do not report to SASTIND via the CNSA, but only to CAS management organisations.

2.1.8 Other Space-Related Organisations

Finally, there are a number of separate administrative entities and organisations that do not directly fall within the frame of the principal space players presented so far, but are nonetheless involved in the overall management of China's space activities. Noteworthy among them are:

- The China Meteorological Administration (CMA),[42] which is inter alia responsible for the procurement and operation of China's meteorological satellites and for the organisation of meteorological research projects. CMA reports directly to the State Council.
- The China Satellite Navigation Project Centre, which is in charge of the design, development, and operation of the BeiDou/COMPASS navigation system and comprises two main departments: the project management department and the general technology department.[43]
- The National Satellite Oceanic Application Center (NSOAS), which operates under the jurisdiction of the State Oceanic Administration and is mainly responsible for development and data processing of the *Hai Yang* oceanographic satellite series, as well as for the development and provision of satellite oceanographic applications.[44]
- The National Remote Sensing Centre of China (NRSCC), an entity under the MOST, which is in charge of planning and overall policy decisions on remote sensing technology and its industrialisation.[45]

---

[42] CMA is a large-scale organisation employing 52,988 staff members and accounting for 2300 bureaus nationwide. "China Meteorological Administration". China Meteorological Administration Web. http://www.cma.gov.cn/en/aboutcma/introduction/201203/t20120319_166488.html. Accessed 18 February 2014.

[43] "BeiDou Navigation Satellite System". Web. http://en.beidou.gov.cn. Accessed 22 February 2014.

[44] "National Satellite Oceanic Application Center". Web. http://www.nsoas.gov.cn/NSOAS_En/index.html. Accessed 21 February 2014.

[45] "National Remote Sensing Centre of China". Web. http://www.nrscc.gov.cn/nrscc/en/functions/. Accessed 23 February 2014.

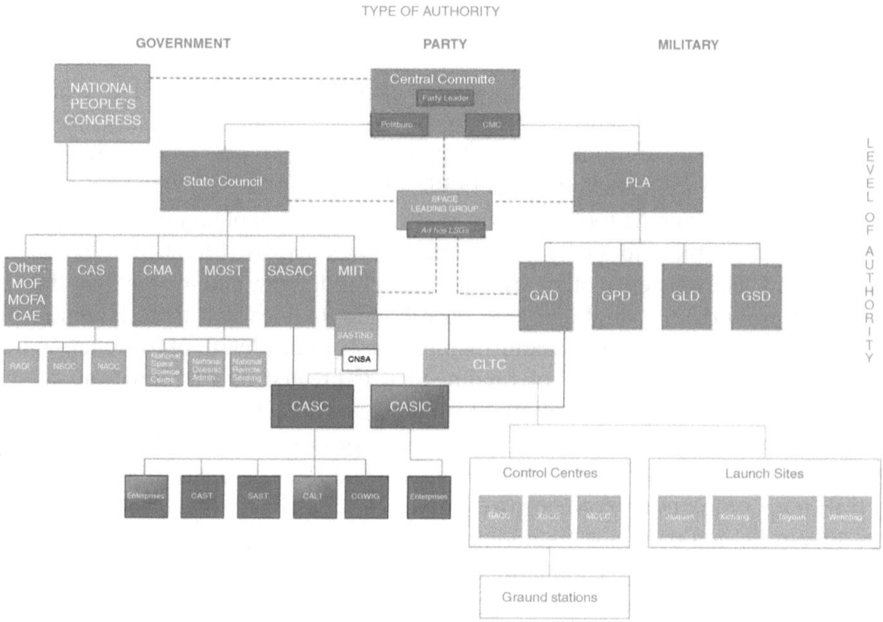

Fig. 2.3 Organisation of Chinese space activities (author's visualisation)

2.1.9 *An Overview*

As shown, there are many players involved in the governance of Chinese space activities. Their interplay creates an intricate web in policymaking processes of programmes, R&D, and related activities and hampers any attempt to separate the different dimensions (civil, military, academic, commercial) of China's space programme. Furthermore, it should be noted that whereas this section has only provided an assessment of the functions and responsibilities of the most important, large, and directly involved entities, other actors are presumably playing important roles, and new bodies will doubtless be set up.

In order to shed some light on this maze, a graphical representation showing the interconnections and inter-responsibilities between the major players in the organisational structure of China's space programme is presented in Fig. 2.3. Notwithstanding its possible inaccuracy, this organigram can nonetheless be of some help in understanding China's space programmes; it should thus contribute to eliminating some of the frequent and prevailing misinterpretations in the majority of Western assessments.

Hopefully, the ever-present reticence among Chinese policymakers about divulging information to foreigners will in the future decrease, as organisational secrecy does not support the Chinese goal of increasing international cooperation.[46]

[46] See Johnson-Freese, Joan (1998). *The Chinese Space Program. A Mystery Within a Maze.* Krieger Publishing Company, Malabar, FL.

2.2 Space Activities Budget

China does not publish official figures on its overall space spending. This dearth of information, combined with the previously described opacity in structures and organisation, makes obtaining reliable data on the expenditure and budget allocation for space-related activities difficult.

It would, however, be too simplistic to attribute the lack of an official budget for space activities solely to a desire for secrecy motivated by the sensitive nature of this domain. The existence of an official budget for defence and the fact that figures are often provided for single projects corroborate the idea that there is more to it than this. [47] Perhaps, one of the contributing causes of the dearth of information lies in the "multichannel system" of funding that is created by the numerous actors involved in the management of space projects and R&D tasks (the Government, CAS, GAD, the SOEs, etc.). The explanation is probably not one dimensional.

In order to have an idea of the total annual spending on space activities in China, it is thus necessary to rely on estimates. A variety of estimates have been offered in the literature, and different methodologies have been proposed. One of the most recognised approaches to estimating the Chinese space budget is that proposed by the Space Foundation in its Space Report, which suggests comparing China to its "peers". On average, the major spacefaring nations—excluding the United States and Russia, where spending is significantly higher than in *any* other country—devote approximately 0.042 % of their current-price Gross Domestic Product (GDP) to civil space activities. [48]

Using this method and China's 2012 current-price GDP of 51,894.20 trillion yuan (US$ 8227 billion),[49] the country's 2013 space spending can be estimated at 21.80 billion yuan ($3.50 billion) (see Table 2.1 for the estimated budget for China's space programme over the past few years). [50]

Similar estimates have been provided through research conducted by other institutions, such as Euroconsult and the European Space Directory, thus confirming a baseline of at least $3 billion for China's total space budget.

[47] The cost of project 921, for instance was given as ¥18 billion (about 1.5 billion €), of which ¥8 billion covered new facilities and ¥10 billion the development of Shenzhou. Later they quoted costs for an unmanned Shenzhou launch of ¥800 million and manned at ¥1 billion (80 million € and 100 million €, respectively). The cost of Chang'e up to 2012 was given as ¥2.3 billion (230 million €). Harvey, Brian (2013). *China in Space. The Great Leap Forward.* Springer, New York.

[48] The Space Foundation (2013). The Space Report 2013 | The Authoritative Guide to Global Space Activity. The Space Foundation, Colorado Springs. See also ESD Partners (2013). European Space Directory. ESD Partners Publication, Paris.

[49] Source: "GDP (current US$)". The World Bank (2014). Web. http://data.worldbank.org/indicator/NY.GDP.MKTP.CD. Accessed 13 February 2014.

[50] "World Economic Outlook Database." International Monetary Fund (2014). Web. http://www.imf.org/external/pubs/ft/weo/2013/02/weodata/index.aspx. Accessed 24 February 2014.

Table 2.1 China's space budget (in current US$ billion)

Year	2010	2011	2012	2013
GDP growth (%)[a]	10.4	9.3	7.7	7.7
GDP (current $ billion)[b]	5930.5	7322.0	8227.1	8860.6
Space budget (0.042 % of the previous year GDP, in $ billion)	2.1	2.5	3.1	3.5

[a]"Regional Outlooks. East Asia and the Pacific." The World Bank (2014). Web. http://www.worldbank.org/en/publication/global-economic-prospects/regional-outlooks/eap. Accessed 13 February 2014
[b]"GDP Annual Growth Rate | Forecast." The World Bank (2014). Web. http://www.tradingeconomics.com/forecast/gdp-annual-growth-rate. Accessed 13 February 2014

The table relates budgetary growth to overall economic performance. It shows that the budget has consistently increased in the past few years, moving from $2.1 billion in 2010 to $3.5 billion in 2013 and thus accounting for a growth of $1.4 billion over only 3 years. In line with the overall economic growth rates, the rate of expansion of China's space programme is considerable; however, if compared to the size of its national economy, the overall investment appears quite modest.

It bears noting in this regard that the above methodology could present several pitfalls. First, it should be asked whether China can be treated as a "normal spacefaring nation" with a space budget that represents only 0.042 % of its GDP and whether the space budget is merely linked to the growth of the economy or must be assumed to represent a deeper political—and thus financial—involvement. In addition, it is not clear whether the estimates provided using this method are inclusive of infrastructure-related expenditures,[51] development costs, or military programmes. Finally, and more importantly, these estimates do not take into account the difficulties related to currency exchange rates, cost of living/cost of labour, and specific market prices.

Since labour and manufacturing costs are rather low in China, as are market prices, it thus appears to be more useful to convert Chinese expenditures utilising Purchasing Power Parity (PPP) rates. Notwithstanding that the consistency of this method can also be questioned, the utilisation of a current international dollar—a hypothetical currency with the same purchasing power of goods that the US dollar had in the United States at a given point in time—gives a better idea of the actual size of China's space programme (see Table 2.2).

[51] Many analysts do not include infrastructure-related expenditures in the overall budget—i.e. the expenditures for the launching facilities, the tracking systems, and the testing facilities.

Table 2.2 China's space budget (billion $)

Year	2010	2011	2012	2013
GDP (current $)	5930.4	7322.0	8221.0	8854.0
PPP GDP (international $)[a]	10,039.9	11,189.1	12,261.3	13,205.4
China's space budget	2.1[b]	2.5	3.1	3.5
China's PPP space budget	3.8[c]	4.2	4.7	5.1

[a]"Regional Outlooks. East Asia and the Pacific." The World Bank (2014). Web. http://www.worldbank.org/en/publication/global-economic-prospects/regional-outlooks/eap. Accessed 13 February 2014

[b]0.042 % of the previous year GDP, in $ billion

[c]0.042 % of the previous year PPP GDP, in $ billion

2.2.1 Budget Breakdown

Assessing the budget breakdown is also quite problematic, given that few figures are available. In addition, these figures are similarly complicated by currency rates, different market prices, and how inclusive they are of military programmes, development costs, and commercial revenues. In light of these obstacles, the figures should be treated with extreme caution.

According to the analysis provided by *Euroconsult*, one of the most authoritative sources in this regard, the expenditure split between civil and defence space programmes can be estimated at 58 % and 42 % (US$ 2.022 billion and 1.410 billion, respectively, out of a total budget of US$ 3.432 billion).[52]

Although the percentage dedicated to military-related space activities appears impressive, it has to be underlined that the supposed budget managed by the PLA includes expenditure for the human spaceflight and launcher programmes, which represent the first and third largest items of China's total space budget.[53] Expenditure specifically dedicated to space security (mainly Space Situational Awareness) is, on the other hand, estimated to represent less than the 0.3 %. At the same time, programmes financed by the civil budget (e.g. earth observation and navigation) are dual-use systems benefitting also the PLA.

Separating military programmes from civil ones can thus be quite difficult: budget composition by application might be more meaningful. A graphical representation is therefore provided in Fig. 2.4.

With $790 million allocated in 2012, human spaceflight represents the largest budget item, roughly one quarter of China's total space budget.[54] It can be anticipated that budget growth in this domain will continue in order to ensure proper investment for the planned launch of a second space laboratory in 2015 and the

[52] For figures elaborated on the data provided by Euroconsult, see Euroconsult (2013). *Government Space Markets—World Prospects to 2022*. Fourth Edition. A Euroconsult Research Report, Paris.

[53] *Ibid.*

[54] *Ibid.*

Budget Breakdown 2012	*USD million*	*Percentage*
Civil	**2022**	**58%**
Defence	**1410**	**42%**
Manned Spaceflight	790	23.02%
Earth Observation	769	22.41%
Launcher	620	18.07%
Science & Exploration	479	13.96%
SatNav	416	12.12%
Other	216	6.29%
SatCom	132	3.85%
Space Security	10	0.29%
Total	**3432**	**100.00%**

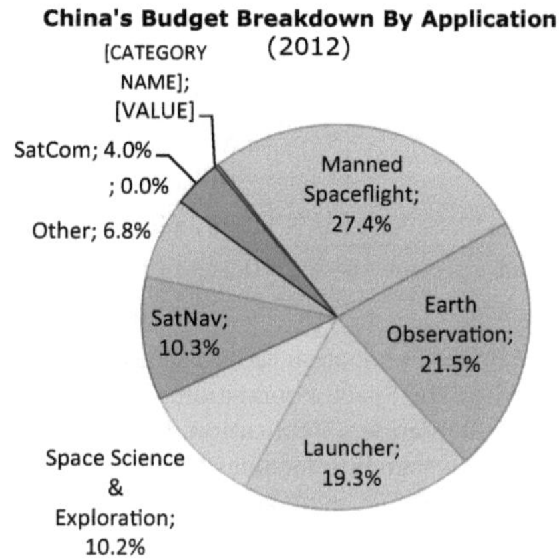

Fig. 2.4 Budget breakdown 2012

construction of a large space station by 2022.[55] The second largest budget is for the earth observation programme ($769 million allocated in 2012). There are currently four systems in operation and a bilateral programme with Brazil (see Sect. 3.1.3). The size of the EO programme expenditure will likely continue to expand, considering that roughly 30 EO satellites are expected to be launched by 2022.

Accounting for 18 % of the total budget, the launcher segment is the third largest budgetary item in the programme ($620 million); a budgetary augmentation can be also expected in this domain, given the ongoing development of a new launcher family (see Sect. 3.3.1).

2.2.2 *China's Space Budget in Comparative Perspective*

Beside the *absolute* numbers and the budget breakdown of China's space programme, the *relative* position China has when setting its budget in a comparative international perspective is also quite significant. Table 2.3 depicts the space-related expenditures of the major spacefaring nations and their respective global ranking.

[55] Jones, Morris. "The Next Tiangong". Space Daily. 3 March 2014. Web. http://www.spacedaily.com/reports/The_Next_Tiangong_999.html. Accessed 3 March 2014. See also Lan, Chen, and Jacqueline Myrrhe (eds). *Go Taikonauts. All about China's space programme.* Issue 10. December 2013: pp. 31–33.

Table 2.3 Estimated government space budget (2012)

Country	Rank	Space report and ESD[a] (€)	Space report and ESD ($)	Euroconsult ($)
United States[b]	1	37.160 billion €	$47.911 billion	$42.689 billion
Europe[c]	2	6.193 billion €	$7.985 billion	$9.606 billion
Russia	3	3.596 billion €	$4.636 billion	$8.597 billion
Japan	4	2.616 billion €	$3.373 billion	$3.699 billion
China	5	2.397 billion €	$3.090 billion	$3.432 billion
India	6	0.938 billion €	$1.210 billion	$1.259 billion
Canada	7	0.318 billion €	$0.411 billion	$0.618 billion

[a]The Space Foundation (2013). The Space Report 2013 I The Authoritative Guide to Global Space Activity. The Space Foundation, Colorado Springs. ESD Partners (2013). European Space Directory. ESD Partners Publication, Paris
[b]US space spending includes the budget of NASA and that of the Department of Defence
[c]Europe's budget includes all contributions to ESA (EU, EUMETSAT, and Cooperating States, but Canada excluded) and all national space expenditures of ESA Member States

Table 2.4 Number of launches 1957–2013[a]

Country	1957–2013	2007–2010	2011	2012	2013
Russia	2998	119	30	24	32
United States	1439	71	17	13	19
Europe	221	25	5	10	7
China	188	41	18	19	15
Japan	84	8	3	2	3
India	33	9	3	2	3
Total	4965	272	77	70	79

[a]Elaboration of the data provided by the US Federal Aviation Administration (FAA) for the launches performed between 2007 and 2013 and by Brian Harvey for the launches performed during the period 1957–2011; see Harvey, Brian. China in Space. The Great Leap Forward, Springer, New York, 2013

As the table shows, China is estimated to be the fifth largest space spender in the world, lagging far behind not only the United States, but also Russia and Europe, although in a catching-up position with Japan and well ahead of India and Canada. If compared to China's ranking in the global economy (second), the investment appears relatively modest.

As already mentioned, the figures should be treated with extreme caution. Considering the several complications involved in this type of comparison, additional comparative methods may be relevant for capturing China's relative position among the leading spacefaring nations. Particularly meaningful are, for instance, statistics comparing the number of launches. Table 2.4 depicts the launches performed by the major spacefaring nations over the period 1957–2013, with a special focus on the period 2007–2013.

As the table shows, China accounts for only a tiny portion of world launches since 1957. The country lags far behind the leading spacefaring nations, in particular Russia and the United States. However, given that China's ascendancy in space is relatively recent, it is more indicative to look at the last decade, while during the period 1970–2003 the average launch rate was just over two launches a year, since the early 2000s the number has exponentially increased, pushing China into the top-three league of launching states: in 2007, the country outstripped Europe as the third leading nation in terms of launches and in 2011 and 2012 overtook the United States. Considering the impressive number of satellites that China plans to launch in the next few years (see the next section), the number of Russian launches is also likely to be matched. Overall, these statistics place China at the pinnacle of the international space hierarchy, alongside the United States and Russia, although they should not be taken as indicators of the level of operational and technological capabilities compared with other countries. Especially in regard to the United States, parity in terms of financial resources and technological expertise still appears some way off.

2.2.3 China's Space Budget: A Forecast

Finally, it may be relevant to try to forecast the evolution of China's space budget over the next decade. A projection can be made by directly linking the increase in the space budget to the forecasted growth of China's economy, in particular to the estimated GDP growth rate. As before, the figures neither reflect the possible deeper political involvement in space activities nor account for inflation, but only display the pace of growth of China's space budget in relation to its forecasted economic performance to 2030. Both nominal and PPP space budget evolution are considered in Table 2.5.

A graphical representation of Table 2.5 can be found in Fig. 2.5.

Notwithstanding the possible pitfalls, Fig. 2.5 gives quite a good indication of the likely pace of growth of China's space programme. Like its economic growth rate, the space budget can be expected to increase at a Compound Annual Growth rate (CAGR) of 7.5 % between 2013 and 2029, when it will reach the level of roughly US$10 billion (international $14 billion in PPP terms). It can be anticipated that such a growth rate will make China the largest space spender in the world, after the United States, by the end of the next decade.

Table 2.5 China's future economic performance and space activities budget (in current US$ billion)

Year	2012(e)	2013(e)	2014 (f)	2015 (f)	2016 (f)	2020 (f)[a]	2030 (f)[a]
GDP growth (%)[b]	7.7	7.7	7.7	7.5	7.5	5.0	4.0
GDP	8221.0	8854.0	9535.8	10,251.0	11,019.8	14,374.4	23,191.3
PPP GDP[c]	12,261.3	13,205.4	14,222.2	15,288.9	16,435.5	21,438.7	34,588.8
Space budget (0.042 % of GDP)	3.1	3.5	3.7	4.0	4.3	5.7	9.4
PPP space budget (0.042 % of PPP GDP)	4.7	5.1	5.5	6.0	6.4	8.6	14.0

e estimated, *f* forecasted

[a]2017–2019: 7.5 % GDP growth rate; 2020–2029: 5.0 % GDP growth rate; 2030: 4.0 % GDP growth rate

[b]"Regional Outlooks. East Asia and the Pacific." The World Bank (2014). Web. http://www.worldbank.org/en/publication/global-economic-prospects/regional-outlooks/eap. Accessed 13 February 2014. See also "GDP Annual Growth Rate | Forecast." The World Bank (2014). Web. http://www.tradingeconomics.com/forecast/gdp-annual-growth-rate. Accessed 13 February 2014

[c]"Regional Outlooks. East Asia and the Pacific." The World Bank (2014). Web. http://www.worldbank.org/en/publication/global-economic-prospects/regional-outlooks/eap. Accessed 13 February 2014

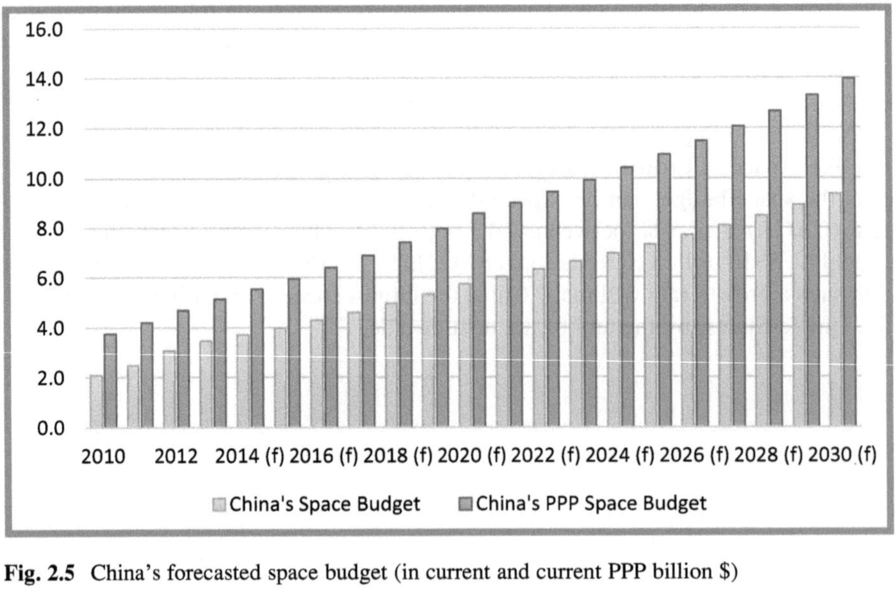

Fig. 2.5 China's forecasted space budget (in current and current PPP billion $)

2.3 Space Policy and Targets

Compared to the high level of secrecy that Beijing maintains in its policy processes and the dearth of information provided on its capabilities and budgetary allocation, the overall direction of China's space policy agenda is intentionally more transparent, to a degree.[56]

Traditionally, China's space policy has been articulated in a series of broader socio-economic, S&T, industrial or defence-related development plans, policy papers, and guidelines (e.g. "the National Medium- and Long-Term Plan for the Development of Science and Technology (2006–2020)" issued by the State Council in February 2006, which provides guidelines for leapfrogging China into a leadership role in science-based industry by 2020).[57]

More generally, the space programme has been planned and executed within the framework of China's Five-Year Plans (5YP), the broad coordinating mechanisms of national social and economic planning that have been issued since 1953. Discussed and adopted by the Central Committee of the CCP and subsequently ratified by the National People's Congress, the 5YP provides the grand blueprint of the overall objectives and goals related to social and economic growth and industrial planning in key sectors (e.g. strategic emerging industries such as biotech, information technology, advanced materials, aerospace, etc.) and regions.[58] Currently, China is in its 12th Five-Year Programme, governing the period from 2011 to 2015.

The specific 5YP for the space sector was announced by China's State Council in the form of a government White Paper, entitled "China Space Activities in 2011".[59] The paper, released on 29 December 2011, is the third space policy document of this kind, the other two having been issued in 2001 and 2006, in conjunction with the 10th and the 11th Five-Year Plans, respectively.

---

[56] As brilliantly explained by Stacey Solomone, for China it is more important to show intentions than capabilities. "It is a sign of weakness to show one's capabilities, while using secrecy to hide them is a means to retain harmony and balance. Revealing capabilities would create an imbalance. This might seem counter-intuitive to a Western policy maker, but not from a Chinese perspective". For them, mutual dependence among actors in the global space community maintains balance. Because the Chinese clearly state their intention in the Space White Paper, then there is no need to reveal their hardware capabilities. Solomone, Stacey (2013). *China's Strategy in Space*. Springer, New York: p.59.

[57] McGregor, James (2010). "China's Drive for Indigenous Innovation. A Web of Industrial Policies". Global Regulatory Cooperation Project—US Chamber of Commerce. APCO Worldwide. A more detailed analysis of these guidelines is also provided in Sect. 4.3.

[58] For a detailed description of the 12th Five-Year Plan's policy processes, see Gilligan, Greg. "China's 12th Five-Year Plan. How it actually works and what's in store for the next five years". APCO worldwide. 10 December 2010.

[59] Government of the People's Republic of China. "China's Space Activities in 2011". White Papers of the Government of China. Beijing, China. 29 December 2011. Web. http://www.china.org.cn/government/whitepaper/node_7145648.htm.

Considering that the space programme was not previously subject to a national policy statement in its own right, the release of these documents merits attention, as it clearly signals both the increased relevance attached to space activities and the growing efforts made by Beijing policymakers in terms of making their space agenda more open to the scrutiny of the international community.

Like the previous versions, the most recent White Paper first highlights the most important achievements and breakthroughs realised by the national programme during the previous 5YP,[60] subsequently enunciates the plans and key priorities for the following years, and finally discusses the policy measures to undertake as well as China's international space-related policies.

2.3.1 China's Key Space Policy Targets

A specific feature of the latest space policy document is that it provides an unprecedented level of technical and operational information,[61] demonstrating China's increased level of confidence and pride in the country's space capabilities. It has even been noted that in this regard, "the White Paper provides much more information than similar US documents on its space programme".[62] Where the document is, however, much lacking compared to other national space strategies is—according to most analysts—in providing a clear picture of the underlying intentions regarding the pursuit of the space goals listed and the mechanisms through which policies will be implemented.

Although the statements might seem rather bland and the aspirations quite vague, the document is nonetheless far from a mere compendium of China's intended space activities for the next 5 years. By reading between the lines, a precise set of building blocks and priorities for the space policy comes into focus.

In designing its space policy, China has specified that the space programme "is subject to and serves the national overall development strategy and adheres to the principles of independent, peaceful, innovative, and open development".[63] In line with these four main principles, the principal axes for the development of space activities in China are identified as:

---

[60] In the case of the 2001 Space White Paper, the document describes the progress and achievements realised by China since 1956, thus filling an information gap regarding the development of the Chinese space programme during the previous 45 years. For a comparative analysis of the three White Papers, see Lele, Ajey, and Gunjan Singh (2012). "China's White Paper on Space: An Analysis". Institute for Defence Studies and Analyses Issue Brief.

[61] See Pollpeter, Kevin (2012). "China's Space White Paper: Increasing Transparency…to a degree". China Brief Vol. 12: (3).

[62] *Ibid.*

[63] Government of the People's Republic of China. "China's Space Activities in 2011". White Papers of the Government of China. Beijing, China. 29 December 2011. Web. http://www.china.org.cn/government/whitepaper/node_7145648.htm.

- Keeping the path of technological and operational self-reliance and independence in the development of the space programme/industry
- Adhering to the peaceful utilisation of outer space and opposing its *weaponisation* or any arms race in space
- Spurring the creation of a broader genuine innovation system through the implementation of space science and technology programmes
- Balancing independence and self-reliance with the adoption of an open and constructive attitude to international space cooperation on the basis of equality and mutual benefit

All these guiding principles eventually relate space activities to achieving the general objective of enhancing China's comprehensive national power (CNP): in short, space activities are intended as a tool designed to concurrently meet the demands of economic growth, scientific and technological development, national security, social progress, and increased international influence.

Beside these general principles, the documents provide a clear and comprehensive description of China's programmatic intentions for its space activities in the next 5 years. By combining the unprecedented level of operational technological details offered by the document with a reading of the long-term strategies and objectives identified by the CAS in its R*oadmap for Space Science and Technology,*[64] the specific targets of the current Chinese space policy 5YP can be spelled out and described in detail. These can be encapsulated in the five mission areas of space transportation, satellite development, orbital spaceflight, applications, and infrastructure building.

2.3.2 Space Transportation

In the area of space transportation, which is one of the largest budget items in the programme, China has clearly announced its intention to keep improving its launch vehicle series by enhancing the reliability and adaptability of the vehicles in service and developing a new generation of launchers in order to meet the country's future launch requirements. The *Chang Zheng* (Long March—LM) launch vehicle has so far been developed in four main configurations, of which three are still active: the LM-2, the LM-3, and the LM-4. Since 1970, these rockets have conducted 186 launches, of which 178 were successful.[65]

In the next few years, China will continue to enhance launcher reliability and adaptability, while in the meantime focusing on the development of a new launcher

---

[64] Guo, Huadong, Ji Wu (eds) (2010). *Space Science and Technology in China: A Roadmap to 2050*. Chinese Academy of Sciences, Science Press (Springer), Beijing.

[65] See Harvey, Brian (2013). *China in Space. The Great Leap Forward*. Springer, New York, and the data provided by the annual compendiums of commercial space transportation of the Federal Aviation Administration (FAA).

fleet, using more efficient engines and an entirely new upper stage. The fleet will comprise three different configurations:

- The *Long March 5 (LM5)*, which will be the workhorse of the new fleet, is designed to be a heavy-lift launcher having a payload capacity of 25 tons to LEO and 14 tons to GEO. LM-5 has primarily been conceived for launching the large modules for the country's future space station, but up to six variants will be developed for different purposes.
- The *Long March 6 (LM6)*, described as a "high-speed launch vehicle",[66] will be a light launcher, similar in appearance and capability to the European Vega, with a payload capacity of 1 ton to LEO. This launcher will "provide China an operational response capability for the first time, with obvious national security and commercial applications".[67]
- The *Long March 7 (LM7)*, a medium-lift launcher able to carry from 3 to 10 tons to LEO and from 1.6 to 6 tons to GEO.[68] The LM-7 will be developed from the LM-2 F and built with a variety of booster combinations in five different versions.

These new launch vehicles, as clearly set out in the White Paper, will be less toxic and more reliable than the ones currently in use. They will also be based on a "modular approach" in order to maximise commonalities and efficiency and, in parallel, provide a high level of adaptability to the diversity of launch requirements. The realisation of this programme clearly underscores China's readiness to attain comprehensive and flexible access to space matching, and competing with, that of the other major spacefaring nations. For instance, the LM-5D version, while primarily intended to launch large space station modules, has been identified by the Chinese as a potential rival to Ariane 5, "able to put two satellites into 24-h orbit simultaneously, compared to Ariane's one large and one medium".[69] As for the LM6, it will also compete in the expanding market of lightweight satellites.

In addition, the 2011-released policy document for the first time officially announces China's intention to conduct pre-research on key technologies for a new super-heavy-lift launcher, dubbed by the media and analysts as Long March 9 (LM9). More information on this new launcher fleet, and in particular about the "Moon rockets", will be provided in Sect. 4.3.

---

[66] Government of the People's Republic of China. "China's Space Activities in 2011". White Papers of the Government of China. Beijing, China. 29 December 2011. Web. http://www.china.org.cn/government/whitepaper/node_7145648.htm.

[67] *Ibid.*

[68] More detailed information on the new launcher's fleet is provided in a number of analysis documents. See in particular Harvey, Brian. *China in Space. The Great Leap Forward*, Springer, New York, 2013: pp. 361–366.

[69] *Ibid.* p. 364.

2.3.3 Satellite Development

In the area of satellite development, the full spectrum of programmes is considered. The new policy calls for the development of an entirely new earth observation and navigation satellite series and for enhanced meteorological and communication satellites:

- *Earth Observation.* EO satellites are one of the most important lines of development of China's space programme, and more than forty EO satellites have been launched since 2001. In this field, China plans to continue the development of four main satellite series:

  (a) The *ZiYuan* series for natural resources: this series was initially developed in collaboration with Brazil as CBERS[70] and five satellites had been launched as at 2012. One of the main purposes of the latest generation of this series (inaugurated in January 2012) is to obtain indigenous access to high-resolution geographical information and replace foreign commercial sources for imaging China.[71]

  (b) The *HaiYang* series for oceanography and maritime observation: two generations of these satellites have been deployed, and the third is now under development. While the HY1 series (comprising 2 satellites) concentrated on ocean colour monitoring and the HY2 series (two satellites) used microwaves to monitor ocean dynamics, the third series includes three satellites that will be equipped with Synthetic Aperture Radar (SAR) for ocean surveillance and monitoring. In addition, in the field of EO oceanographic missions, China is planning a joint mission with France in 2015.[72]

  (c) The *HuanJing* series for environmental surveillance: these satellites are intended to form a constellation known as the Environmental Disaster Monitoring Constellation aimed at monitoring a variety of disasters, from

[70] China–Brazil Earth Resources Satellites (CBERS) is a joint China–Brazil programme in the field of earth observation developed under an agreement signed in July 1988. Three satellites were launched between 1999 and 2007. In November 2008, the two governments agreed to jointly continue the development of CBERS: CBERS-5 and CBERS-6 are expected to launch in 2017 and 2020, respectively.

[71] Sensibly, China continues to rely on the Dragon programme with ESA for obtaining high-resolution geographical information of China. For more information on the Dragon programme, see "ESA-MOST Dragon Cooperation Programme". European Space Agency. 21 August 2011. Web. http://earth.esa.int/dragon/. Accessed 19 February 2014. See also Desnos, Yves-Louis, and Li Zengyuan. "EO Science and Applications development in China". In: *Dragon Programme Mid-term results*. Proceedings of the 2005 Dragon Symposium. European Space Agency. January 2006.

[72] The French–Chinese Oceanic Satellite (CFOSAT) aims at monitoring wind and waves globally for the purpose of marine meteorology, ocean dynamics, climate variability, and surface processes. See Harvey, Brian (2013). *China in Space. The Great Leap Forward.* Springer, New York: p. 209.

floods and earthquake to forest fires and pollution. The constellation, to be launched by 2019, will comprise six optical and five radar satellites.[73]

(d) The *Yaogan* series for disaster management and surveillance: Yaogan are dual-use satellites utilised both for land survey, crop yield assessment, and disaster monitoring and for security-related observations.[74] Indeed, it is believed that their development is primarily aimed at providing China with a comprehensive military surveillance system combining optical, radar, and electronic intelligence (Elint). Nineteen Yaogan satellites were launched between 2006 and 2013, and improved versions can be expected to be developed and launched over the next few years.

Besides these four specific series, China has in addition expressed its intention to carry out development of a high-resolution all-weather, 24-h, multispectral EO system.[75]

- *Communication Satellites*. The new guidelines, in addition to reiterating China's goal of developing enhanced communication satellites for fixed communication services as well as television, radio, and mobile communications, emphasise the country's interest in developing a new "satellite platform of higher capacity and higher power for the next generation of GEO communications and broadcasting satellites".[76] In effect, along with two more powerful variants of the *Dongfanhong-4* (*The East is Red, DFH*-4) satellite platform, China is now focusing on the development of the new DGH-5 platform.

  While the DFH-4 was primarily developed for domestic missions and with the aim of reducing dependence on foreign technologies, the new platforms (the DFH-4S, the DFH-4E, and the DFH-5) are specifically intended to strengthen China's position in the global market of telecommunications platforms. China aims to seize 10 % of the international commercial satellite market by 2015, thanks to the development/export sales of the new DFH-4 variants. For its part, the DFH-5 will in the future help China gain an important share of the large spacecraft platform market, particularly thanks to its improvement in performance and reliability and its low cost compared to those offered by the United States, Russia, and Europe.[77]

---

[73] Euroconsult (2013). *Government Space Markets—World Prospect to 2022*. Fourth Edition. A Euroconsult Research Report, Paris.

[74] *Yaogan* satellites specifically used for military reconnaissance are named *Jianbing*.

[75] Government of the People's Republic of China. "China's Space Activities in 2011". White Papers of the Government of China. Beijing, China. 29 December 2011. Web. http://www.china.org.cn/government/whitepaper/node_7145648.htm.

[76] *Ibid*.

[77] "Reaching for the stars: China's commercial space industry." Editorial: China Brain. Web. http://www.china-brain.com/inner.php?eid=0082&goback=%2Egde_131398_member_2479741$32#13#.UvPYjfldUj4. Accessed 10 February 2014. See also "Chinese DFH-4 Platform Product Line Improvement." 2012. International Astronautical Federation. http://www.iafastro.net/iac/archive/browse/IAC-12/B2/4/14084/. Accessed 16 February 2014.

Not explicitly mentioned in the White Paper are the communications satellites for the PLA, in particular the *FengHuo* ("fire and smoke"[78]) and *ShenTong* series, which started to be launched in the early 2000s. The precise performances are not known, but ShenTong is believed to provide Ku-band communication and FengHuo only C-band and UHF communication.[79] Further development in the field of military satellite communication systems can be anticipated.

- *Meteorological Satellites.* For a large country that is highly dependent on agriculture but often subject to damaging storms and floods, the development of reliable meteorological satellites has become a key priority, clearly reflected in the 2011 space policy. China has developed a meteorological satellite system, called *FengYun* (wind and cloud, FY), currently comprising two LEO and three GEO operational satellites.[80] A new generation of both LEO and GEO satellites (the FY-3 and FY-4, respectively) is now under development within the frame of the 12th Five-Year Guidelines (2011–2015) and is planned for launch around 2015. While the four LEO satellites of the FY-3 series that are yet to be launched will be equipped with instruments for 3D atmospheric detection,[81] the FY-4 series will comprise four optical satellites and two microwave satellites to be launched by 2020.[82]
- *Satellite Navigation.* In this field, Chinese policy reaffirms the "three-step" development plan followed for the development of its BeiDou/COMPASS Satellite Navigation System,[83] and it clearly emphasises China's ultimate goal of providing a global satellite navigation capability. In accordance with the construction plan, the initial satellite navigation system has been enhanced and is now providing coverage in the Asia-Pacific region with positioning, navigation, timing, and short-message communication service capabilities. Since December 2012, the service has been made available to foreign customers, such as Pakistan. Needless to say, like the other navigation systems, BeiDou has also been designed to provide encrypted signals to the PLA. The system is

[78] The "fire and smoke" satellites are named after an ancient system of communicating utilising beacons along the Great Wall which were lit all along the wall in case of barbaric invasions. See Harvey, Brian (2013). *China in Space. The Great Leap Forward.* Springer, New York: p. 146.

[79] *Ibid.* p. 146.

[80] Euroconsult (2014). *Profiles of Government Space Programs, Analysis of Over 80 Countries & Agencies.* Euroconsult Profiles Series, Paris.

[81] Three satellites of the FY-3 series have been launched so far, the last of which on 23 September 2013.

[82] Euroconsult (2014). *Profiles of Government Space Programs, Analysis of Over 80 Countries & Agencies.* Euroconsult Profiles Series, Paris.

[83] The first phase of the programme, started in 1994 and completed in 2007, envisaged the creation of an experimental system. The second phase (2008–2012) was aimed at creating a regional system. The current and third phase (2013–2020) aims at building a global satellite navigation capability.

scheduled to complete the deployment of its 35 satellite constellation and to provide global coverage by around 2020.[84]

- *Small and Microsatellites*. The development of small (less than 500 kg) and micro (less than 100 kg) satellites is not explicitly identified as one of the main targets for the 12th Five-Year Guidelines period (2011–2015), but advances are also expected in this domain. The main series of microsatellites that are currently in development are the China Seismo-Electromagnetic Satellites (CSES) and *Chuangxin* (literally "creation") series. The CSES form a constellation of microsatellites aimed at detecting electromagnetic anomalies in the atmosphere and are part of China's earthquake monitoring network. Launches of these satellites are slated for 2014 and 2017 and will enable China to have an operational earthquake prediction system by 2020. As for the *Chuangxin* satellites, they are being developed by CAS in collaboration with the Shanghai Engineering Centre for Microsatellites and are designed to store and forward communications in case of disasters.

The above list of China's main targets in the field of satellite development is on its own impressive and is indicative of the confidence and importance China's leadership attaches to the development of the full spectrum of satellite capabilities. However, there is more.

2.3.4 Space Exploration and Human Spaceflight

In the area of space exploration and orbital spacecraft development, the White Paper identifies robotic lunar exploration and manned spaceflight as the two key priorities that will help boost the comprehensive development of space science in China. For the lunar exploration programme (CLEP), the document specifies once again the "three-step strategy" of orbiting, landing, and returning a sample and extends this model to other exploration missions. However, a new exploration mission to Mars (the *Yinghuo* programme) is not mentioned.[85]

As for human spaceflight, the new policy reiterates China's ambition to develop all technologies necessary for enabling human spaceflight and for maintaining a permanent human presence in orbit. The construction of a space station appears the main longer-term goal envisaged in the document in regard to human spaceflight. An in-depth analysis of both the CLEP and manned spaceflight programme will be presented in the following chapter (Sect. 4.2).

---

[84] The constellation comprises 5 GEO satellites and 30 non-geo satellites. "BeiDou Navigation Satellite System". Web. http://www.beidou.gov.cn/2012/12/14/2012121481ba700d7ca84dfc9ab2$32#ab9ff33d2772.html. Accessed 20 February 2014.

[85] Evans, Ben. "A Red Flag on the Red Planet: China's Mars Ambitions". AmericaSpace. 15 October 2013. Web. http://www.americaspace.com/?p=43535. Accessed 15 March 2014.

Other relevant projects in the area of space science planned within the next 5 years include the following:

- *The KuaFu mission*, which represents one of the key missions within the broader Sun–Earth Connection (SEC) programme,[86] is specifically intended to establish a space weather forecasting system composed of a constellation of three satellites. One of these satellites will be placed at the Sun–Earth L1 Lagrangian point, with the other two in polar orbits. Initially slated to launch between 2012 and 2014, the mission has been postponed to the beginning of the 13th Five-Year Plan period (2016–2020).
- *The Space Solar Telescope* (SST), another major programme initiated in 1992 by CAS, is aimed at studying the solar magnetic field, solar activities, and Sun–Earth interactions.[87] Its launch has not been scheduled yet.
- *The Black Holes Probe (BHP) programme,* which comprises a series of projects and experiments, is intended to study high-energy processes of cosmic objects and black hole physics.[88]

2.3.5 Applications Development

The development of applications is specifically identified by the 2011 White Paper as the chief objective in the practical implementation of the Chinese space policy during the 12th Five-Year Guidelines period. The document consistently underlines the importance of exploring ways of making space profitable, by pushing the country's emerging space capabilities to become industrialised, commercialised, and economically viable.

Three key areas are identified for the development and full utilisation (and commercialisation) of satellite-based applications and services, namely, earth observation, communications, and navigation. For each of these areas, the document envisages an extensive market-oriented provision of services.

This insistence on the development and commercialisation of the applications side of the satellite industry not only reflects the need to update and extend the industrial and commercial scope of China's space industry; it also subtly discloses Beijing's long-term ambitions to become a global provider of commercial satellite services alongside the provision commercial space launches.

---

[86] For more information on the SEC programme, see Guo, Huadong, Ji Wu (eds) (2010). *Space Science and Technology in China: A Roadmap to 2050.* Chinese Academy of Sciences, Science Press (Springer), Beijing: pp.64–65.

[87] *Ibid:* pp. 59–64.

[88] The main missions of the BHP programme include Hard X-ray Modulation Telescope (HXMT) satellite, Space Variable Object Monitor (SVOM) satellite, and Gamma-ray Burst Polarisation (POLAR) experiment on board China's spacelab. *Ibid.* p.60.

At the same time, this emphasis is indicative of the underlying intention to utilise space assets to meet the broader demands of national economic and social development targets identified in the 12th 5YP. In this regard, it should be noted that the Five-Year Plan in which the 2011 space policy is positioned has triggered a significant shift in the nation's economic growth model. Indeed, as emphasised by many analysts, the 12th 5YP is the first commercialisation-driven and pro-internal consumption plan in the macroeconomic planning of modern China.[89] The development of the satellite applications industry and the commercialisation of space-based services are thus particularly intended to support social and economic development targets.

2.3.6 Infrastructure Development

Lastly, the new Chinese space policy pays special attention to the need to develop an infrastructure for space activities that is capable of supporting the achievement of the aforementioned goals. Specifically, the policy document pledges concrete efforts to enhance the reliability and automation levels of the three existing launch sites and to proceed with the construction of a new launch site on the island of Hainan, southwest China. The new Wenchang cosmodrome, which is currently under construction and planned to be operational before the end of the 12th Five-Year Plan period (2015), is specifically designed to enable future manned and space exploration programmes. The new launch fleet, and the LM-5 in particular, will be launched exclusively from this site.

In terms of infrastructural development, attention is also given to the need to improve China's space TT&C network, to build Deep Space Network stations and to develop advanced TT&C technologies.

2.3.7 Policy Measures and International Cooperation

The list of goals China has set in its White Paper is as impressive as it is ambitious and underpins an increased level of confidence and pride in the country's space capabilities that was previously lacking. At the same time, the new Chinese space policy seems aware that the space industry might be taking on too many programmes and pursuing too many diverse objectives at the same time. In an attempt to prioritise goals, the White Paper specified that space applications take priority over space science and exploration in the implementation of the country's

[89] Roach, Stephen S (2011). "China's 12th Five-Year Plan: Strategy vs. Tactics". Morgan Stanley Asia. Web. http://www.law.yale.edu/documents/pdf/cbl/China_12th_Five_Year_Plan.pdf. Accessed 20 March 2014.

space policy. Yet the central government is directing the space industry to simultaneously develop a lot of space programmes in space sciences and exploration. As underlined by Stacey Solomone, while the government is guaranteeing a steady investment, with the mounting resources demand by the plethora of space programmes, the aerospace leadership will likely find difficulties in balancing national security needs, developing highly technical space programmes, and opening space to commercial use.[90] In addition to the programmes already underway, China's space policy—as noted above—has for the 2011–2015 period also planned a myriad of projects: it simply does not appear likely that China will be able to accomplish all these goals simultaneously, and delays—especially in terms of launcher development, human spaceflight and space science development—can thus be anticipated.

Doubtless aware of these potential pitfalls, the fourth section of the White Paper articulates a set of what it calls "development policies" designed to assist the realisation of the aforementioned programmes and objectives. Far from providing detailed mechanisms for the implementation of policies, these should rather be seen as a grand blueprint mapping out the overall strategy. What emerges is nonetheless a coherent and well-thought-out set of policy measures to effectively sustain the appropriate development of the space industry and its activities.

These measures envisage plans to promote a mutually beneficial interplay and integration between the space industry, academia, and the research community, which will gradually spur the creation of an ecosystem of innovation beneficial to space as well as to China's economy in general. In addition, they recommend a broad restructuring of the space-related industrial and R&D base, including the creation of new research facilities and engineering centres and, more importantly, the renewal of the country's skilled "social" capital relevant to space activities.

Interestingly, the document also highlights the fact that the emergence of new industrial capabilities is equally to be accompanied by steady and suitable financial investment and by a legislative framework creating a favourable environment for "the development of space entrepreneurship and market-oriented satellite utilisation schemes".[91] The adoption of a national space law is explicitly envisaged as a necessary step for creating a favourable environment for space activities.

Finally, a specific section is dedicated to the role of international cooperation, also regarded as an important policy measure, and intended as both a means to carry out the country's national objectives in space and to garner tangible recognition of its rising space status.[92] Besides providing a very detailed account of China's principal bilateral and multilateral cooperation agreements signed up to 2011, the

[90] Solomone, Stacey (2013). *China's Strategy in Space*. Springer, New York: pp. 94–95.

[91] Government of the People's Republic of China. "China's Space Activities in 2011". White Papers of the Government of China. Beijing, China. 29 December 2011. Web. http://www.china.org.cn/government/whitepaper/node_7145648.htm.

[92] Al-Ekabi, Cenan (2014). "European Space Activities in the Global Context". In: Al-Ekabi, Cenan, Blandina Banares, Peter Hulsroj, Arne Lahcen (eds). *ESPI Yearbook on Space Policy 2011/2012. Space in Times of Financial Crisis*. Springer, Vienna: pp. 42–25.

policy document emphasises the requirements of a global space regime and gives the impression that the United Nations (UN) should have a major role to play in this regard. At the same time, it indirectly identifies regional cooperation as an important building block to consolidate global space governance and thus promises consistent advances within the frame of the China-led Asia-Pacific Space Cooperation Organisation (APSCO) (see Box 2.1).

Box 2.1: Asia-Pacific Space Cooperation Organisation

APSCO is a regional international organisation established in 2005 and operational since 2008. Currently, nine countries are signatories of the convention: Bangladesh, China, Indonesia, Iran, Mongolia, Pakistan, Peru, Thailand, and Turkey, while Kazakhstan, Malaysia, and Tajikistan are expected to accede to the Convention soon.

The main purpose of this intergovernmental organisation, set out in Article 4 of its Convention, is "to promote and strengthen the development of collaborative space programmes between Member States, to assist Member States, to promote cooperation, joint development, and to share achievements among the Member States". The fields of cooperation identified by Article 6 of the Convention are space technology and applications, earth observation, space science research, education and training, space law, policy, and regulations. The implementation of a number of projects in each of these fields has already been initiated, together with a progressive consolidation of the institutional structure.

Within APSCO, China acts as a primus inter pares, providing the direction of the organisation and having very significant decision-making powers, while it also bears the major part of the financial cost. Overall, APSCO plays an important role in China's broader space diplomacy, which aims to provide the country with a leadership role in Asia and among developing countries.

Key cooperation areas for future space endeavours are identified by the document. Particular interest is shown in the field of space science, deep-space exploration (including the TT&C), and applications development, the last being the focal point of the ongoing 5YP for space activities.

Interestingly, the need for cooperative undertakings with regard to space debris monitoring and mitigation is also raised. The document clearly commits China to working together with the international community to maintain a peaceful and clean outer space, presumably both in terms of concrete joint undertakings and the elaboration of intergovernmental "best practices" and guidelines.

More importantly, the document also highlights the desire to engage extensively in international cooperation within the framework of China's human spaceflight programme, including technological and scientific cooperation for the future

Chinese Space Station (CSS). Since the publication of the White Paper, this plea has been increasingly reiterated by China's space officials. No occasion has been missed to underline the fact that China's space policy is open to the world. Indeed, China's commitment to cooperation has become the buzzword used by Chinese space officials in all the international fora with Chinese participation: for instance, the declarations released at the occasion of the International Astronautical Congress (IAC) meeting held in Beijing in September 2013 and at the International Space Exploration Forum (ISEF), held in Washington in January 2014, can be taken as consistent evidence of this approach.[93]

[93] Lan, Chen; Jacqueline Myrrhe (eds). *Go Taikonauts. All about China's space programme*. Issue 10. December 2013. See also Lan, Chen, and Jacqueline Myrrhe (eds). *Go Taikonauts. All about China's space programme*. Issue 11. February 2014.

Chapter 3
Why the Moon?

An assessment of Chinese plans to land its taikonauts on the Moon first requires a detailed investigation of the rationales possibly guiding China's leadership towards this ambitious endeavour. Such an analysis helps us better understand the underlying philosophy of China's space programme and, more broadly, China's sociopolitical behaviour, beyond the pervasive but too reductive interpretation of a strategic confrontation between a *fast-rising power* and a *declining hegemon.*

In fact, viewing Chinese ambitions towards the Moon through the lens of US–Chinese relations, with its overemphasis on the inevitable confrontation between the two juggernauts, can only encourage a simplistic—not to say misleading—interpretation of China's space programme that would hinder understanding of its multifaceted purposes, many of which are historically and cultural derived behaviours. The goal of this analysis is thus to provide "insider access"—an attempt to understand China's plans and intentions from a Chinese perspective and thus to provide the tools to better engage with the country. Indeed, an assessment of China's rationales for shooting its taikonauts Moon-wards will eventually help define the possibilities for international cooperation which the implementation of a lunar endeavour could open.

Furthermore, the analysis shows the type of variables or conditioning factors that could affect the realisation of this ambitious endeavour. The following sections reveal, alongside a multitude of rationales and motivations, the existence of a variety of influencing factors, whose relative weight may vary in the future according to the domestic and international environment within which China operates.

M. Aliberti, *When China Goes to the Moon...*, Studies in Space Policy 11,
DOI 10.1007/978-3-319-19473-8_3

3.1 Backgrounder: The Path of "Cultural Grandiosity"

At first glance it could be argued that China's great leap into space forms part of its recently demonstrated ambitions of *grandeur*. China has increasingly impressed the "West" with its ability to realise colossal projects, to rapidly allocate and move huge human, intellectual, and financial resources with determination, efficacy, and far fewer concerns over consensus than in any other country. The construction of the Three Gorges Dam, the extremely rapid erection of brand new cities every year, and a number of technologically avant-garde projects (e.g. the Maglev Train) well exemplify this unique capability.

The explanation for this is that China's type of government and political leadership allows it to freely and powerfully implement its policies, overcoming the complex—at least for Western countries—"consensus procedures" and concerting its impressive financial, human, and intellectual resources in every targeted sector. In this light, the much rumoured intention to send humans to the Moon presents itself as the ultimate target of the CCP regime's *grandeur* ambition.

However, this unique ability of the Central Kingdom is only partially a factor derived from the current historical juncture; it constitutes much more a constant and abiding propensity towards grandeur made possible by a millennia-long practice. More properly, it represents one of the striking sociocultural features of Chinese civilisation.[1] Of course, other civilisations also created magnificent works (suffice it to think about the Egyptian pyramids or the Roman Coliseum), but in no other cultural experience has the realisation of ambitious undertakings—combining grandiosity, beauty, and technical complexity—been such a constant and almost orthogenetic factor and that for more than 2000 years! Sinologists have noted that the origin of this phenomenon took shape in a precise era: with the creation of China's first centralised empire by the Qin dynasty (206–221 BC).[2] In fact, after the unification of China, Qin Shi Huang Di, the first celestial emperor, initiated the realisation of a series of titanic projects bound to symbolise not only his glory and power but also the unity finally attained by China and its people: the *Tianxia* (all under heaven). Besides the well-known Mausoleum with its 10,000 terracotta statues, whose construction involved 700,000 workers, the construction of the capital city Xianyang, a city of gigantic dimensions (25 km^2), as well as the construction of the first 2200 km (out of the 8800) of the Great Wall, should be recalled.[3]

These large undertakings required enormous levies of manpower and resources and combined a high level of technical complexity with the aesthetic of a refined grandiosity. More importantly, they soon became the premise of a cultural and political tradition that has continued throughout the centuries. Indeed, the

[1] See Calza, Giancarlo (2006). "Cina. Grandiosità e bellezza". In Lionello Lanciotti, Maurizio Scarpari (eds). Cina. Nascita di un impero, Skira editore, Milano

[2] *Ibid.*

[3] Fahr-Becker, Gabriele (ed) (1999). *The Art of East Asia*. Könneman, Köln: p. 71

undertakings of Qin Shi Huang Di were turned into a model of social and political intervention to which subsequent dynasties could not avoid referring.

A simple look at Chinese history reveals that its development is constantly seised by the pursuit of similar grandiosities. Just a few examples give an idea of this peculiar feature. Among the myriad of candidates: the royal garden of Wu Di of the Han dynasty (206 BC–220 AD), which contained hundreds of architectural elements and more than 3000 species of rare plants from every corner of China, aiming to offer a reproduction of the emperor's kingdom[4]; the imposing Yungang and Longmen Grottoes of the following Wei dynasty (386–535), also designed to celebrate technical capacity together with religious grandiosity; Chang'an, the majestic capital city of the Tang emperors (618–907); the 980 buildings forming the Forbidden City, built through the mobilisation of one million workers immediately after the establishment of the Ming dynasty (1368–1644); and the Hebei Imperial Tombs of the Qing emperors (1644–1912).

All these undertakings were characterised by the magnitude, complexity, and rapidity of their realisation. Of course, they bear distinctive witness to the precise era in which they were realised, but all of them were executed to express—and thus to become an expression of—the political, economic, and technical strength possessed by the ruling dynasty, a sign of its glory and ability to govern: in short an expression of the *Tianming*, the celestial mandate.

Even following the fall of the imperial system and the establishment of a republic, such grandiosity has been pursued by the new generations of *Tianzi* (the Son of Heaven), first under the leitmotif of Maoist ideology and later under that of economic development: witness the above-mentioned project of the Three Gorges Dam—the biggest hydrological project in the world, whose completion required the evacuation of over two million people—or the highly symbolic Tiananmen square, with its imposing Mao Mausoleum, built in 2 years by 700,000 workers with materials coming from every part of China.

To sum up, it can be argued that the strength China nowadays represents and its ambition to achieve a demanding endeavour such as a manned lunar landing do not stem from current historical circumstances (at either international or national level) but are an expression of a specific sociocultural feature. Indeed, there appears to be a continuous thread between the construction of Xianyang and a future manned lunar exploration programme. The lunar endeavour, like the erection of China's ancient capital 2000 years ago, may eventually become another grand project and a symbol of collective self-esteem, thus representing the nation, its unity, and its continuity.

To be sure, the intention of these introductory words is not to suggest a cultural determinism or a teleological interpretation of Chinese history and imply that China's culture is inevitably pushing the CCP policymakers to go to the Moon. Rather, these few observations aim to show that it is the "grandiosity path" of Chinese history and culture, more than ideological motives or political calculations,

---

[4] *Ibid.* p. 83

which is creating the "narrative substratum" and offering Chinese leaders the possibility of pursuing the lunar endeavour.

As brilliantly explained by Zheng Wang, in China cultural roots permeate nationalism more than any ethnic, religious, or ideological element[5]; and it is the awareness of this cultural and historical grandiosity that boosts the Chinese people's identity, sense of unity, and pride in their country. After suffering a humiliating decline during the "century of shame" (1839–1949),[6] the Chinese people are unwavering in their commitment to helping their country reclaim its ancient glory and splendour. Space achievements can clearly be perceived—and used by the regime—as highly symbolic but concrete expressions of this long-awaited renaissance.

In this light, what eventually becomes more evident is the nationalism and prestige-related rationales guiding China's policymakers in their great leap into space.

3.2 The Domestic Political Goals: National Pride and Propaganda Themes

As with most spacefaring nations, China's space endeavours, and human spaceflight in particular, respond to a variety of political objectives. A key driver is the enhancement of national prestige, utilised as a springboard for the realisation of broader political objectives in relation to both the domestic and the international arena.

Many scholars have argued that China's pursuit of prestige in space has an overarching international dimension, with similar objectives to those expressed by the two original space antagonists of the first space age (namely, demonstration of technological and military capabilities, reflecting positively on their political systems). In fact, an analysis of the broader dynamics surrounding the country's space endeavours shows that the pursuit of prestige is primarily driven by domestic, rather than international, considerations. As also underlined by Michael Sheehan, one of the specific features of China's space programme lies in the fact that it is "aimed overwhelmingly at influencing the perceptions of the Chinese people, rather than those of the international community"[7] and thus first of all at supporting political

[5] Wang, Zheng (2013). "Il nazionalismo cinese tra mito e trauma". Istituto Affari Internazionali, Orizzonte Cina July-August 2013. See also: Wang, Zheng. "Not Rising But Rejuvenating: The Chinese Dream". The Diplomat. 5 February 2013. Web. http://thediplomat.com/2013/02/chinese-dream-draft/?allpages=yes. Accessed 20 March 2014

[6] The first Opium war and the takeover of China by the Communist Party are identified as the demarcating events of the "century of shame". See Mazzei, Franco, Vittorio Volpi (2006). *Asia al Centro*. Universita' Bocconi Editore, Milano

[7] *Cit.* Sheehan, Michael (2013). "Did you see that, grandpa Mao? The prestige and propaganda rationales of the Chinese space program". Space Policy Vol. 29 (2): 89

needs within the domestic realm. Obviously, this is not to downplay the relevant role in terms of foreign policy objectives assigned to human spaceflight and eventually to a manned lunar landing; rather it is to emphasise that in a vast and complex reality like China's, the primary political focus has always been inwards, not outwards.

Indeed, the country's past and present experiences show that the constant preoccupation of every ruling class has predominantly been the preservation of national unity, social stability, and cohesiveness. These goals are not only essential for retaining the legitimacy to govern; they are also perceived as indispensable preconditions for safeguarding the country against the everlasting perils of the "outside world". As usefully documented by several sinologists, there is in fact a strong link between these two political dimensions (the striving for unity internally and the sense of vulnerability towards the outside world), which finds its expression in the ancient formula *nei luan, wai huan* (translated as "internal disorder, external danger"), generally used during periods of dynastic crisis.[8] For the Chinese government, the preservation of social harmony (*he*) is thus a necessity for both its internal and international protection.

By the early 1990s, at a time of increasing irrelevance of the communist ideology and of rampant dissatisfaction among the people towards the ruling class, the CCP would have seen prestigious endeavours like the space programme as able potentially to act as a unifying cause and a source of pride for the Chinese people and their governing regime.[9] To be sure, China's leadership has not from then on based its survival strategy merely on the space programme. Yet it is evident that, since the outset, both the Shenzhou manned spaceflight and the lunar exploration programmes have been used as powerful instruments to pursue these domestic political objectives: in particular, they were conceived as a means for reinforcing national unity, pride, and cohesiveness as well as a "crucial validator of China's political system and ruling party".[10] Even President Hu Jintao publicly recognised the significant political benefits brought to the country by its space endeavours and declared: "we should broadly publicise the spaceflight virtues across society, in order to increase national pride and confidence and reinforce national solidarity".[11]

[8] According to the sinologist David Bachman, China's foreign policy can be seen merely as an extension of its domestic one. See Bachman, David. "Domestic Sources of Chinese Foreign Policy". In Kim, S.S. (ed) (1994). *China and the World.* Westview Press, Boulder: p. 44.

[9] For some scholars, the initiation of the Shenzhou programme in 1992 may have been intended in particular to create a positive focal point for national pride to counter the negative 1989 Tiananmen Square images. See Johnson-Freese, Joan (2005). "Space Wei Qi. The Launch of Shenzhou V". Naval War College Review Vol. 57 (2): p. 124.

[10] *Cit.* Handberg, Roger, and Zhen Li (2007). *Chinese Space Policy. A study in domestic and international politics.* Routledge, New York

[11] President Hu Jintao on 26 November 2005. Quoted from: Sheehan, Michael (2013). "Did you see that, grandpa Mao? The prestige and propaganda rationales of the Chinese space program". Space Policy Vol. 29 (2): 89–166

Fully aware of the valuable role played by cultural elements in boosting patriotic sentiments, a specific instrument used to promote the space programme among the Chinese people can be seen in the pervasive practice of linking key elements of the programme with historical and cultural themes. In 1994, for instance, when the manned spaceflight programme took off, the name *Shenzhou* (divine or spiritual vessel) was applied to the Soyuz-derived capsule as a tribute to China's ancient culture.[12] More significantly, the Long March vehicle as well, named after the Mao Zedong's historical march from Jiangxi province to Shaanxi province, was renamed *Shenjian* on the occasion of Shenzhou's first flight. The word *Shenjian* has its origins in the name of an ancient Chinese rocket that was made of a long arrow with a small bamboo container of gunpowder attached to the back of the arrow and lit for launch.[13] This decision could not but indicate a particular sensitivity to China's cultural heritage, which is profoundly rooted among the Chinese people. In addition media reports noted that President Jiang Zemin wrote by hand the two ideograms that were transposed onto the side of the rocket body[14]: a poetic recall of the highest form of art among the mandarin literati—calligraphy! In a perfect mix of old and new elements, Shenzhou-1 was scheduled to launch on the 50th anniversary of the CCP's takeover of China (1 October 1999), although it was eventually postponed to 19 November.

Another significant example of the relevant role played by cultural elements in China's space endeavours is offered by China's lunar exploration programme (CLEP). The lunar orbiters are named Chang'e, after the ancient legend of a Chinese goddess residing in the Moon. Besides highlighting the fact that reaching the Moon has been a cherished desire in the Chinese nation since ancient times, this decision had a special meaning for Chinese people. Chang'e is one of the most celebrated figures of Chinese mythology and popular culture, and every year her story is remembered at the well-attended Moon Festival.[15] Several variations of the legend exist, but according to the most popular version, Chang'e was a beautiful woman who, after stealing the elixir of immortality from her husband, the great hero Yi, became immortal and flew to the Moon, where she lives forever with Yutu, a Jade Rabbit.[16] Together with Chang'e, the Jade Rabbit also constitutes a central figure in Chinese poetry and popular culture. Little surprise, then, that the online poll set up by the CLEP authorities to decide the name of China's first lunar rover eventually came up with the largest number of votes for Yutu. When announcing the name of the rover at a press conference, Li Benzheng, CLEP deputy chief

[12] The meaning of the word comes from an ancient name for China and has the same phonetic sound, although the ideogram for *zhou* is different in meaning.

[13] Solomone, Stacey (2013). *China's Strategy in Space*. Springer, New York: p. 65

[14] Harvey, Brian (2004). *China's Space Program. From Conception to Manned Spaceflight*. Springer, New York

[15] Even Mao Zedong's most famous poem, "Reply to Li Shuyi", has as a central tenet the figure of the "lonely moon goddess", which is utilised to recall his murdered wife Yang Kaihui.

[16] Yang, Lihui, *et al* (2005). *Handbook of Chinese Mythology*. Oxford University Press, New York: p. 86

commander, not only remarked on this cultural reference but also that Yutu was "a symbol of kindness and purity reflect[ing] China's peaceful use of space".[17] Worth mentioning, en passant, is the fact that before the launch of Chang'e 3, Xinhua—the PRC's news agency—brilliantly recalled the importance of the Moon in Chinese culture by publishing some of the most representative poems (from Li Bai of the Tang period to Mao Zedong) on the theme.[18]

The practice of linking Chinese cultural elements to space endeavours is not limited to manned spaceflight or CLEP; other examples are offered by the naming of the *Great Wall Industry Corporation*, the *BeiDou* (the North Container) navigation satellites, and the *Kuafu* space weather programme.[19] It is anticipated that the names of the future interplanetary exploration missions will be also based on those of ancient Chinese astronomy. In sum, there is no shortage of examples showing how historical and cultural heritage has permeated China's space programme. This practice has been very effective in terms of outcomes, since it has increasingly allowed the Chinese people to associate the programme with the country's path to grandiosity, while at the same time helping the regime to popularise science and technology among the public.

Besides this recurrent utilisation of historical–cultural themes, confirmation of the relevant domestic policy goals assigned to the space programme is highlighted by the continuous and active participation of political leaders at key events of China's space endeavours. From the Jiang Zemin era to that of Xi Jinping, prime ministers and presidents of the PRC have attended the "sacred and glorious missions" of Shenzhou and Chang'e.[20] This involvement—which has so far proved to exceed that of any other spacefaring nation—has been promptly covered by the national media and utilised to explicitly associate the programme's successes with the CCP's political legacy.[21] The clear objective has been to enhance the regime's legitimacy and prove its ability to respond to the "lofty aspirations" of the Chinese people. Public commentaries released by government ministries throughout the

[17] "China names moon rover 'Yutu'". Xinhua News. 27 November 2013. Web. http://www.spacedaily.com/reports/China_names_moon_rover_Yutu_999.html. Accessed 20 March 2014

[18] Qiang, Hou. "The Moon in Chinese poetry". Xinhua News. 2 December 2013. Web. http://news.xinhuanet.com/english/china/2013-12/02/c_132932871.htm. Accessed 2 April 2014

[19] Kua Fu is another important figure in Chinese mythology: according to the legend, he was a giant who chased the sun and died while getting too close.

[20] "Sacred and glorious mission" are the words utilised by both China's Premier Wen Jiabao on occasion of the Shenzhou 6 mission and by President Xi Jinping at the launching operations of the *Shenzhou* 10.

[21] A highly indicative demonstration of the relevance attributed to the human spaceflight in terms of identification with the ruling class is quite ironically shown by the fact that the first manned mission in 2003 was not broadcast live on television to avoid embarrassment in case of failure. As explained by Joan Johnson-Freese, President Jiang Zemin made it very clear that if space successes are spectacular, failure would be devastating for the regime. Live broadcasting thus started with the *Shenzhou*-6 missions—that is when a higher level of confidence was gained. See Johnson-Freese, Joan (2005). "Space Wei Qi. The Launch of Shenzhou V". Naval War College Review Vol. 57 (2).

duration of the various missions have also been intended to improve the party's image and reiterate a number of propaganda themes.

Noteworthy is that, following their re-entry to Earth, taikonauts have been celebrated by the media as national heroes[22] and transformed by the government into "internal ambassadors" of its policies: as again noted by Michael Sheehan, "whereas the USSR used to send its cosmonauts abroad to bolster Soviet international prestige, the Chinese practise is to send them to cities around China itself [including the former colonial enclaves of Hong Kong and Macao[23]], in order to encourage patriotism, scientific awareness and national unity".[24] Judging from the eager crowds of citizens gathered to welcome them and the enthusiastic demonstrations of patriotism recorded in every city they visited, the move has been highly effective. According to several scholars, not only has this type of promotional activity succeeded in stirring up the people's pride in their country, it has also succeeded socially by "minimising the divide between the rich and the poor, because all Chinese people can share in the successes of the manned spaceflights".[25]

All these dynamics are highly indicative of the ways Beijing policymakers intend to utilise ongoing and future space endeavours—and in particular a possible manned lunar landing programme—to bring domestic political gains. There is still, however, a much more interesting indicator of the use the Chinese regime will make of the lunar endeavour in terms of domestic political goals: this aspect can be seen in the well-thought-out integration of China's increasing space ambitions and the broader political theme of the national "great rejuvenation".

As previously mentioned, rejuvenation (*fuxing*) is a deeply rooted motive within the national ethos, which has been explicitly "invoked by almost every modern leader from Chiang Kai-Shek to Jiang Zemin and Hu Jintao"[26] to mobilise the population and get their commitment in helping the country reclaim its ancient glory and splendour. After taking office in November 2012, President Xi Jinping

---

[22] Besides China's first astronaut Yang Liwei, who was promoted to Major General and became a national icon, it should be mentioned that after her flight, Liu Wang, China's first space woman, soon became one of the most celebrated women in contemporary China.

[23] The Chinese overseas diaspora has been also targeted by promotion activities related to the human spaceflight endeavours.

[24] Sheehan, Michael (2013). "Did you see that, grandpa Mao? The prestige and propaganda rationales of the Chinese space program". Space Policy Vol. 29 (2): 110

[25] Solomone, Stacey (2006). "The Culture of China's Space Program: A Peking Opera in Space". Journal of Future Studies. Vol. 11(1): 43–58

[26] *Cit.* Wang, Zheng. "Not Rising But Rejuvenating: The Chinese Dream". The Diplomat. 5 February 2013. Web. http://thediplomat.com/2013/02/chinese-dream-draft/?allpages=yes. Accessed 20 March 2014

has similarly and repeatedly emphasised that "realizing the *great renewal* of the Chinese nation is the greatest dream for the Chinese nation in modern history"[27] and has thus pledged to continue targeting this long-awaited national renewal as his government's main objective.

Given the enormous reach this catchword has in terms of inspiring, unifying, and encouraging individuals to make personal sacrifices in order to contribute to the greater national good, consistent efforts have also been made in the direction of transforming China's space endeavours into an integral part of this collective, grand undertaking. When looking at public statements, official documents, and media messages, it becomes evident that they all now emphasise the relevant role played by space achievements in both accompanying and embodying the quest for China's "great rejuvenation".

In December 2013, for instance, when the Chang'e 3 mission was launched, the media promptly announced that "the space dream, a source of national pride and inspiration for further development, is part of the dream to make China stronger and will surely help realize the broader Chinese dream of national rejuvenation".[28] Similar words were also pronounced directly by President Xi in his congratulatory message for the Shenzhou-10 mission and by astronaut Nie Hiasheng,[29] as well as being regularly reiterated by—among others—the CAS in its forward-looking report "Space Science and Technology in China: a Roadmap to 2050".[30]

The self-reinforcing pattern of this strategy is remarkable: the rejuvenation motive helps space endeavours permeate and reach out to society, while at the same time space endeavours, given their high visibility and undisputed technical difficulty, help substantiate with concrete evidence the process of achievement of the *Zhongguo meng*: the Chinese dream of a great national rejuvenation.

---

[27] Speech by Xi Jinping at the exhibition "the Road toward Renewal" on 29 November 2012. (Emphasis added).Quoted from: "Xi pledges "great renewal of Chinese nation". Xinhua News. 29 November 2012. Web. http://news.xinhuanet.com/english/china/2012-11/29/c_132008231.htm. Accessed 10 April 2014

[28] "Lunar probe boosts Chinese Dream". Space Daily. 3 December 2013. Web. http://www.spacedaily.com/reports/Commentary_Lunar_probe_boosts_Chinese_dream_999.html. Accessed 10 April 2014

[29] For instance, on that occasion President Xi Jinping affirmed: "China will take bigger steps in space exploration in pursuit of its space dream... the space dream is part of the dream to make China stronger. The mission's crew members carry a space dream of the Chinese nation, and represent the lofty aspirations of the Chinese people to explore space ...." Quoted from Yi, Yang. "Exploration part of the China dream". Xinhua News. 23 June 2013. Web. http://news.xinhuanet.com/english/china/2013-06/25/c_132483952_2.htm. See also: "Xi wows bigger strides in space exploration. Space Daily. 25 June 2013. Web. http://www.spacedaily.com/reports/Xi_vows_bigger_stride_in_space_exploration_999.html.

[30] In its report CAS stressed "China fell from a world economic power into a poverty-stricken country, subject to insult and humiliation by other powers". Science and technology are offering a way forward for the great rejuvenation. Guo, Huadong, Ji Wu (eds) (2010). *Space Science and Technology in China: A Roadmap to 2050*. Chinese Academy of Sciences, Science Press (Springer), Beijing.

It can thus be expected that a human lunar landing programme will be powerfully presented as an indispensable constituent for the realisation of the Chinese dream and—after the landing—as the final concrete manifestation of the long-awaited rebirth of China as the "Central Kingdom".

In sum, for China, going to the Moon will be a powerful unifying symbol that will boost national pride and reinforce cohesiveness and confidence among the Chinese people. It will also allow the ruling class to pursue the double imperative of gaining the commitment of the masses and retaining—at least in their eyes—the legitimacy to govern a revived "Celestial Kingdom". Flying to the Moon will eventually act as a source of inspiration and imagination for an entire nation, which will witness a most cherished dream since ancient times become true. In this regard it is plausible to expect that, to maximise the social and political effects of this ambitious undertaking, the government will pursue the ultimate logic: after sending robotic orbiters named after the ancient mythical goddess who flew to the Moon, a woman would be tapped to be the first Chinese to set foot in her realm. Clearly, a daughter of *Chang'e* flying to the Moon will not only have a huge impact in the imagination of over one billion Chinese but is also bound to produce a huge and far-reaching effect on the cultural perception of the international community, not least with that half of the world population who saw 12 American men go—but none of their own gender.

3.3 The International Dimension: Status and Geopolitics

As mentioned, international considerations strongly influence the motives and directions of China in space in a complex interface with domestic politics and culture.

Apart from being a highly visible demonstration to the rest of the world that China can produce more than cheap clothing, human spaceflight achievements are important "status markers" that put the country at the forefront in the international arena and contribute to its image as a major player. With the historic flight of Jiang Liwei in October 2003, for instance, China not only dissipated the widespread belief that it was a backward country in terms of technological prowess; it projected the image of being a prominent space technology leader in Asia, as evidenced by the counteractions of India and Japan (see Sect. 6.2).

Although a status-based explanation for China's manned space programme is clearly not a point of contention, status is itself driven by different motivations—motivations that in the case of China are hard to disentangle.

3.3.1 Deciphering Chinese Intentions

For the last several years, many analysts have warned that behind China's strides in this highly visible and prestigious technological arena lies not only the intention to move up the international space pecking order; the real underlying ambition is to catch up and even surpass what the USA has done in space and ultimately to directly challenge US leadership with a new space race, one that the USA would be in danger of losing this time.[31]

In particular, many leading members of the US space community have insisted on this interpretation, warning that the Chinese have the ability to beat the USA back to the Moon. Among them, former NASA administrator Michael Griffin has persistently expressed concerns about America's fading dominance in space by reason of China's astonishing rise. In a communication intended for a Congress Committee Griffin pointed out that "if China were to achieve this before the return of a manned American spacecraft to the Moon for the first time since 1972, the bare fact of accomplishment will have enormous, and not fully predictable, effects on global perceptions of US leadership in the world".[32]

As China is projected to outstrip the USA in economic terms by 2030, sending taikonauts Moon-wards would also be intended by Beijing policymakers to symbolise the "passage of the relay baton" to China in the technological and geopolitical world hierarchy. The logic, simple and compelling, is the same as that aptly described more than 40 years ago by the then US Vice President Lyndon Johnson: "In the eyes of the world, first in space means first, period; second in space is second in everything!"[33]

Historical analogies are often highly tempting, yet the idea of a "new space race" with the USA appears simplistic, despite its currency with the media and a significant number of scholars. For one thing, China's manned spaceflight programme has followed a "marathon approach", rather than a race proper. Initiated more than 20 years ago, it never appeared to be in a hurry to accomplish any of its key objectives by a specific deadline or before other countries. The USA and Russia accomplished many of the same objectives decades ago, and regional powers like Japan and India—which it was initially thought would soon become involved in this

---

[31] See, for example, Seedhouse, Erich (2010). *The New Space Race. China vs the United States*, Springer—Praxis Publishing, Chichester, UK. Handberg, Roger, and Zhen Li (2007). *Chinese Space Policy. A study in domestic and international politics*. Routledge, New York. Moltz, James Clay (2011). *Asia's Space Race: National Motivations, Regional Rivalries, and International Risks*. Columbia University Press, New York.

[32] Quoted from Kaufman, Marc. "NASA Star Is Fading, Its Chief Says". Washington Post. 14 September 2008. Web. http://www.washingtonpost.com/wp-dyn/content/article/2008/09/13/AR2008091302142.html. Accessed 2 April 2014

[33] Quoted from Griffin, Michael D. "To Explore Strange New Worlds". Remarks to the 39th Lunar and Planetary Science Conference. 10 March 2008. Web. http://www.nasa.gov/pdf/216616main_LPSC_10_Mar_08.pdf. Accessed 10 April 2014

domain—remain well behind China.[34] As a result, Chinese political leaders have never seen themselves involved in a space race nor have they felt compelled—like the original space antagonists—by the need to prove the superiority of their own system or ideology in a one-on-one contest.

Even though in the long run the possibility that China might strive for a dominant international position through its space programme should not a priori be excluded, the primary international rationale for Chinese human spaceflight is to regain the status of great power that in Chinese eyes was lost with the emergence of the so-called great divergence,[35] but which remained nonetheless deeply rooted in their identity. As explained by Fiona Cunningham, this sense of victimisation and entitlement has driven China to emulate existing great powers by acquiring "markers" that identify great power status in the contemporary international system.[36] Put simply, contemporary Chinese leaders have invested in space, because it is the ultimate expression of what being a scientifically and technologically advanced nation means. Having a space programme, manned spaceflight in particular, has been a means for China to express to itself and the world its entitlement to join the club of great powers.[37]

From the beginning, and throughout the development of the human spaceflight programme, the goal was never to catch up or surpass the leading space powers, but to avoid falling far behind and to come to the table as an "equal".[38] Thus, politically, China has understood its space efforts to be a measure of national

---

[34] For this interpreation see in particular Kulacki, Gregory, and Jeffrey Lewis. (2009). A Place for One's Mat: China's Space Program, 1956–2003. American Academy of Arts and Sciences, Cambridge, MA

[35] The term was coined by the scholar Kenneth Pommeranz. In China's narrative, the emergence of this divergence was due to the Chinese purported inability to embrace modern science and develop technology.

[36] See Cunningham, Fiona (2009). "The Stellar Status Symbol: True Motives for China's Manned Space Program". China Security Vol. 5 (3): 73–88.

[37] Kulacki, Gregory, and Jeffrey Lewis. (2009). *A Place for One's Mat: China's Space Program, 1956–2003*. American Academy of Arts and Sciences, Cambridge, MA

[38] Indeed, as noted by some scholars, when in 1992 the 30-year plan to send humans into space and build a national space station was approved, Chinese space professionals believed that they would have been latecomers to an expanding human presence in low Earth orbit. Establishing a sustainable human presence in LEO was perceived by some leading members of the Chinese space community as a technological benchmark that would be reached by many of the nations that China saw as developmental role models or peers. Not only were the USA and Russia implementing ambitious plans to develop a space station, but Japan, Canada, and Europe were also making consistent investments in this domain. Chinese space experts believed that these efforts would continue well into the early decades of the following century and that Korea and India would invest in manned spaceflight and ultimately either develop space station plans of their own or join the USA and Russia as partners. China decided to move forward to keep pace with the development supposed to be taking place in other nations, in order to avoid losing its voice at international level. See Kulacki, Gregory (2012). "Why China is building a space station". Union of Concerned Scientists: pp. 6–7.

accomplishment necessary to qualify for inclusion among the great powers that set the rules.[39]

As documented by Gregory Kulacki and Jeffrey Lewis, who have analysed original sources from China itself, Chinese space professionals have in fact used a particular phrase to describe why China made significant investments in space programmes. They explain their goal of making China a strong spacefaring nation with the phrase *yi xi zhi di*: "a place for one's mat", equivalent to the English "a seat at the table", the difference explained by the fact that people in ancient China sat on mats on the floor, not in chairs. "The fundamental idea is that China deserves a place among spacefaring nations. Throughout the history of the Chinese space programme that has meant taking technical cues from the leading space programmes—usually the US",[40] in order to gain a seat that both the Americans and the Russians have nonetheless always been reluctant to concede.

For instance, this seat was denied when China was excluded from participation in the US-dominated International Space Station (ISS) programme: US congressmen were concerned about having Chinese spies running around *their* space station and taking significant advantage of *their* technology![41] To Chinese politicians this exclusion appeared to be a definition of their lack of status, underscored by the accusation of "stealing and copycatting" spacefaring nations' capabilities. As they often insist by quoting the opposite example of China's car industry, this isolation eventually acted as a powerful boost for the development of indigenous space capabilities.

Even so, China has continued to look for membership, respect, and equality in the world space community, but neither in competition with nor isolation from it.[42] Indicative of this is the fact that, even in 2006, when the second phase of the human spaceflight programme was already underway, former CNSA Director Sun Laiyan stated to reporters that China might not need to construct and operate its own space station if it were allowed to participate in the ISS.[43] Rather than the ambition to compete with US leadership in maintaining a human presence in LEO, this posture unveils the desire to "join the club" of great powers and to be accepted as such by them.

---

[39] Kulacki, Gregory, and Jeffrey Lewis. (2009). A Place for One's Mat: China's Space Program, 1956–2003. American Academy of Arts and Sciences, Cambridge, MA: p. vii

[40] *Ibid*. p. 4

[41] It bears noting that still in 2012 similar words were pronounced by Senator Wolf: "we don't want to give [China] the opportunity to take advantage of our technology, and we have nothing to gain from dealing with them" (see Sect. 6.3).

[42] Harvey, Brian (2013). *China in Space. The Great Leap Forward*. Springer, New York.

[43] Kulacki, Gregory (2012). "Why China is building a space station". Union of Concerned Scientists: pp. 6–7.

3.3.2 China's Foreign Policy and the Lunar Endeavour

The idea that China might not intend to use its space endeavours to race the USA also finds confirmation in China's broader foreign policy behaviour.

Compared to Deng's *taoguang yanghui* policy (literally, "cultivate the shadow, hide the brightness", i.e. keep a low profile in international politics and security in favour of concentrating on building up the economy as a source of strength), Beijing's external policy behaviour has certainly become more assertive and less accommodating of Washington's preferences. While it no longer denies its rising power status per se and the ambition to play a greater role in shaping global rules, norms, and institutions, Beijing does not, however, act as a revisionist power trying to wreak havoc on the structure of the international system or striving for a dominant role.

Indeed, notwithstanding the existence of an increasingly animated debate among Chinese elites on the most appropriate international posture for China (see Chap. 5 for a more detailed analysis), Chinese strategists do not counsel challenging the USA as the predominant global power in the foreseeable future. Despite what they perceive as initial signs of decline, they are fully aware that the USA is bound to remain the global hegemon for several decades: the formula used to describe their view of the international system is "one superpower, many major powers" (*yichao duoqiang*).

More importantly, Beijing sees more opportunities than constraints in using the current system to advance its interests. As extensively documented by Evan Medeiros, among others, China's leaders do not want to displace the USA as the global superpower.[44] "They view their domestic challenges as too great to assume the burdens associated with such a role, and they recognize that they still lack the material resources to be able to project and sustain economic and military power across the world. They also fear that playing such role could deplete much needed resources and foster a backlash against China".[45]

Obviously, in the long term Chinese leaders are working for a "multipolar world, one in which multilateralism reigns and US power is constrained".[46] The guiding principle, however, is to avoid direct confrontation with the USA. Not only would this hamper the credibility of China's peaceful rise and its efforts to dampen threat perceptions; it could eventually act as a catalyst for the emergence of a broader coalition aimed at containing its rise.[47]

---

[44] Medeiros, Evan S (2009). *China's International Behaviour. Activism, Opportunism and Diversification*. RAND Corporation, Santa Monica, CA: p. 208.

[45] *Cit. Ibid.* p. 209.

[46] *Cit. Ibid.* p. xxii.

[47] As underlined by many sinologists, one of the most constant obsessions of China's foreign policy is that of becoming contained and encircled by a hegemonic state and its allies. See, among the analyses, Kissinger, Henry (2011). *On China*. Penguin Books, New York. See also: Mazzei, Franco; Vittorio Volpi (2006). *Asia al Centro*. Universita' Bocconi Editore, Milano.

Not by chance the harmonious world policy *(hexie shijie)* inaugurated by Hu Jintao and further pursued by Xi Jinping under the catchphrase "a new type of great power relationship" comprises as its core features: "no conflict or confrontation", "mutual respect", and "win-win cooperation".[48]

This behaviour—which is in a sense diametrically opposed to the relations of mutual hedging, mutual deterrence, and zero-summing between the USA and the USSR during the Cold War—is thought by China to be the most appropriate measure to favour its rise and promote greater multipolarity (and multilateralism), while at the same time lessening US influence on the world stage. On the one hand, it creates an environment in which states do not view China only as a threat they must try to contain, while on the other it enables the adoption of mechanisms of "soft balancing" that may indirectly constrain US actions and advance China's diplomatic interests.

In the light of this foreign policy, utilising the space programme to directly and visibly compete with the USA would simply be nonsensical. To the contrary, in pressing ahead with its human spaceflight programme, China will probably increasingly attempt to bring the cooperative balance into play and gradually set up an alternative space environment that is more attuned to a multipolar rather than a unipolar world system.[49] The future Chinese Space Station (CSS)—which the authorities have already announced will be at the disposal of other spacefaring and non-spacefaring nations and not a closed club like the ISS—will constitute a political response to this objective. In particular, the future international joint experiments, joint manned missions, docking of foreign visiting vehicles, and additional pressurised modules built and launched by various partners will all contribute to establishing a "more democratic" space environment and differentiate China's role in space from that of the USA. In this regard, it should be noted that a prominent role is assigned to the UN to promote international cooperation on the CSS. As announced at the 55th plenary session of the UN Committee on the Peaceful Uses of Outer Space (COPUOS), the Human Space Technology Initiative (HSTI), launched by the United Nations Office for Outer Space Affairs (UNOOSA) in 2010,[50] will in fact work with the China Manned Space Agency to review

[48] Zhao, Suisheng. (2013). "A new model of Great Power Relationship and China-US Competition in the Asia Pacific". ISPI Analysis No. 211. Istituto per gli Studi di Politica Internazionale, Milano.

[49] Dellios, Rosita (2005). "China's space program: A strategic and political analysis", *Culture Mandala: The Bulletin of the Centre for East–west Cultural and Economic Studies* Vol. 7 (1). Web. http://epublications.bond.edu.au/cm/vol7/iss1/1

[50] The core purpose of the HSTI programme is to build space skills in countries currently without a space programme. In particular it aims at involving more countries in activities related to human spaceflight and space exploration and at increasing the benefit from the outcome of such activities through international cooperation, to make space exploration a truly international effort. The role of the initiative in these efforts consists of providing a platform to exchange information, to foster collaboration between partners from spacefaring and non-spacefaring countries, and to encourage emerging and developing countries to take part in space research and to benefit from space applications. More information on the HSTI is available at the UNOOSA website: http://www.oosa.unvienna.org/oosa/en/SAP/hsti/index.html.

possible collaboration in utilising the CSS. At the 64th IAC it was also said that international cooperation on the CSS will be included in the framework of HSTI and that UNOOSA will invite international partners to join cooperation on the CSS.[51]

Such initiatives support the argument that in the long run "space may not only become an attractive *indicator* of Beijing's commitment to multilateralism; it could also act as an *instigator* by leading to deeper international cooperation".[52] Indeed, if Chinese leaders eventually decide to embark upon a human lunar landing, it can be expected that they will use (and cleverly present) this endeavour as the *antithesis of a space race*: instead of being an arena of competition, the programme will be intended as a catalyst for international cooperation.

Although some might claim that China would prefer to go it alone as an "imperial demonstration" of capability, for the Beijing policymakers investing in a costly, complex, and highly ambitious programme like human lunar exploration will require far more than the mere pursuit of status in the international hierarchy. Status considerations will admittedly continue to weigh on their minds. But it is clear that such considerations will be pragmatically accompanied by the pursuit of more tangible political objectives.

The drivers for pursuing international cooperation in space endeavours are myriad.[53] In terms of political objectives, a set of two self-reinforcing considerations can be identified. These are, respectively, linked to the benefits China will gain through the implementation of such endeavours and to those resulting from their accomplishment.

In terms of benefits to be gained through the implementation of the programme, cooperation will not only allay the fears of the international community about the nature of China's rise and favour the emergence of a more amicable environment; it will also serve as the means through which bridges among nations can be created; bridges that might eventually become broader political axes intended to reinforce rather than isolate China's position on the international chessboard. In short, cooperation in space will ultimately suit China's interests on Earth.

The configurations that might develop as a result of China's call for cooperation depend on the attitude of other spacefaring nations. An in-depth analysis of their position vis-à-vis China's space ambitions will be presented in Chap. 6. For now suffice it to say that the cooperation pursued on board the forthcoming CSS will act as a test-bed for verifying and experimenting with the potential partnership configurations to be deployed to achieve the ambitious target of a manned lunar landing.

---

[51] Lan, Chen. "Quarterly Report on the Chinese Space Programme. July-September 2013". In Lan, Chen, and Jacqueline Myrrhe (eds). *Go Taikonauts. All about the Chinese space programme*. Issue 10. December 2013: p. 6.

[52] *Cit*. Dellios, Rosita (2005). "China's space program: A strategic and political analysis", *Culture Mandala: The Bulletin of the Centre for East–west Cultural and Economic Studies* Vol. 7 (1). Web. http://epublications.bond.edu.au/cm/vol7/iss1/1

[53] For a more detailed assessment on the benefits China expects to harvest from international space cooperation, see Sect. 5.4.

At the moment, the prominent role assigned to the UN in this regard and the particular emphasis on non-spacefaring countries suggests the strong involvement of developing countries.

In terms of outcome, international cooperation, rather than competition, will amplify the impact and maximise the status benefits that China would seek from the accomplishment of a manned lunar landing. While it is debatable whether the country will appear as the world's techno-political leader by achieving this target 60 years after Apollo with a solo mission, it can be argued that a cooperative undertaking led by China could more powerfully affect global perceptions (in particular if cooperation with developing countries is pursued) and confer on the country the accolade of leading humanity, rather than individual countries, to new celestial bodies.

By inviting other nations to share the challenges and opportunities of future space exploration and by presenting this endeavour as the antithesis of a space race, China would underpin the image of a benign and benevolent great power, committed to human progress and determined to play a primary role in the future efforts of the international community to explore outer space. In short, such behaviour would allow it to display an important marker of leadership, a sign that, quite ironically, was also understood by the initiator of the space race, US President J.F. Kennedy, in his speech at the United Nations 3 weeks before his death: *Why should the United States and the Soviet Union, in preparing for such expeditions, become involved in immense duplications of research, construction, and expenditure? Surely we should explore whether the scientists and astronauts of our two countries—indeed of all the world—cannot work together in the conquest of space, sending some day in this decade to the moon not the representatives of a single nation, but the representatives of all of our countries.*[54]

In sum, a cooperative undertaking to the Moon in which China would take the lead could really act as a powerful turning point for the creation of a new *space consensus*—the Beijing space consensus—in the perception of future generations, making China *the* indispensable pillar for building the global partnership required to achieve the ambitious goals of future space exploration.

Paradoxically, it is this scenario, rather than an open and direct confrontation, that would pose the greatest challenge to the leading space powers, and particularly the USA, but would, at the same time, provide unprecedented opportunities.

[54] Addressed by President John F. Kennedy to the UN General Assembly. 20 September 1963. US Department of State. Web. http://www.state.gov/p/io/potusunga/207201.htm. Accessed 15 May 2014.

3.4 The Scientific and Technological Drivers

Although politically driven, China's ambitions to land its taikonauts on the Moon would clearly not be limited to the narrow pursuit of political benefits in both the domestic and international arena. Besides this rationale, there are other significant motivations that might eventually push China towards fulfilling lunar ambitions. These are *in primis* represented by the scientific and technological advances the implementation of this programme would bring and by all the tangible and intangible benefits such advances would deliver in turn.

When in 1978 Deng Xiaoping launched the campaign of the "Four Modernisations" in agriculture, industry, science and technology, and national defence,[55] he explicitly remarked: "The key to the Four Modernisations is the modernisation of science and technology. Without modern science and technology, it is impossible to build modern agriculture, modern industry or modern national defence. Without the rapid development of science and technology, there can be no rapid development of the economy".[56]

3.4.1 The Scientific Objectives

Since the launch of the *gaige kaifang* reforms in 1978, China has made great strides in scientific development, and the achievements realised after 35 years of reforms are impressive. Chinese elites, however, clearly perceive the country's level of and output in scientific research as still lagging far behind that of the most industrialised countries in many respects; they do not match its ancient and current stature.

In a sign of the priority and ambition attached to this matter, the country's highest academic institution in the natural sciences, the CAS, has stated: "although the total number of scientific research and publications has been increasing gradually in recent years, as a country with the largest population and the fastest economic growth, and noting that China had once added many marvellous pages to the history of human civilisation, the number of contributions from China to science are obviously incompatible with its position in the world and far less than what is required to sufficiently support the nation's development".[57]

The solution recommended was to boost scientific research in selected areas, and among the various disciplines, space science was identified as "the most promising

[55] The "Four Modernisation" policy was officially approved by the Central Committee of China's Communist Party in December 1978.

[56] Deng Xiaoping's speech at the opening ceremony of the national conference on science". 18 March 1978. In Selected Works of Deng Xiaoping. Volume 2 (1975–1982). Web. http://archive.org/stream/SelectedWorksOfDengXiaopingVol.1/Deng02_djvu.txt. Accessed 1 April 2014.

[57] *Cit.* Guo, Huadong, Ji Wu (eds) (2010). *Space Science and Technology in China: A Roadmap to 2050.* Chinese Academy of Sciences, Science Press (Springer), Beijing. p. 34.

area in which to make great discoveries"[58] and thus one of the research fields essential to fostering China's development and to achieving original scientific research results. The declared aspiration is that "space science research in China will be fully improved to the point such that China, with some remarkable scientific achievements, can make decisive contributions to human civilisation".[59]

In view of this explicit recognition by Chinese scientists that the manned spaceflight programme is "a major programme to put forward the comprehensive development of space science in China",[60] a series of valuable—although rather general—scientific objectives that could be pursued by a manned lunar exploration programme have been identified.

According to a widespread classification in use within the International Space Exploration Coordination Group (ISECG)—recently joined by China[61]—these comprise the categories of science *of*, science *from*, and science *on* the Moon.[62]

Science of the Moon includes geophysical, geochemical, and geological research, directed to having a better understanding of the origin and evolution of the Moon and, more generally, to deepening understanding of the solar system.[63] *Science from* the Moon can be seen as part of a broader/wider (long-term) rationale, which encompasses the exploitation of the Earth's natural satellite environment. It will be discussed in further detail in Sect. 3.7. *Science on* the Moon embraces numerous disciplines (e.g. biology and exobiology, medicine, material sciences, physics[64]) and includes investigations of the effects of the lunar environment on robotic instruments and equipment and—more importantly—on humans.[65]

Embarking on a manned lunar landing programme would be of great help in better investigating survival performance and human abilities in outer space. The lunar conditions, especially gravity, radiation, and magnetic variation, are relevant factors influencing human health, safety, and working ability.[66] Chinese scientists

---

[58] *Ibid.* p. 34.

[59] *Ibid.* p. 50.

[60] *Ibid.* p. 50.

[61] Officially, China joined the ISECG as a fully fledged member in occasion of the International Space Exploration Forum (9 January 2014).

[62] "The Global Exploration Strategy. The Framework for Coordination". International Space Exploration Coordination Group. April 2007. Web. http://www.globalspaceexploration.org/c/document_library/get_file?uuid=119c14c4-6f68-49dd-94fa-af08ecb0c4f6&groupId=10812

[63] "Global Exploration Roadmap". International Space Exploration Coordination Group. April 2007. Web. https://www.globalspaceexploration.org/c/document_library/get_file?uuid=119c14c4-6f68-49dd-94fa-af08ecb0c4f6&groupId=10812

[64] Ideally, also high-energy particle physics, which is facing some limitations in carrying out experiments on Earth, could be performed on the Moon. See Guo, Huadong, Ji Wu (eds) (2010). *Space Science and Technology in China: A Roadmap to 2050*. Chinese Academy of Sciences, Science Press (Springer), Beijing: p.16.

[65] *Ibid.* p. 18.

[66] The gravity of the Moon is only 1/6 of the Earth, the magnetic field is less than 1/1000 that of the Earth, while space radiation is significantly increased.

have already remarked how these changes in the environment, including the reduction in gravity and in the magnetic field and exposure to space radiation, will be crucial to gaining a deeper understanding of the unknown aspects of the life process. From a long-term perspective they will also afford lessons in sustaining a human presence beyond Earth and learning how to live and work on other celestial bodies. As explicitly stated by the CAS in its *Roadmap to 2050*: "the exploration of human survival under extra-terrestrial environmental influences will be our long term scientific pursuit, which will direct the research in different fields of space science and is always the essential scientific goal of manned spaceflight".[67]

3.4.2 The Technological Driver and China's Innovation Efforts

Alongside fostering China's efforts to pursue scientific achievements, a manned lunar landing programme would clearly act as a formidable driving force for the development and demonstration of advanced technologies. There is in fact a strong mutually beneficial interplay between the two objectives, usually summarised in the formula: "science leads technology and technology promotes science".

To be sure, the intention to advance space technology and related high technology would not be limited to supporting the scientific objectives of a manned lunar programme, but would more broadly be a reflection of China's deep-rooted desire to become a technologically innovative powerhouse.

Since the late 1990s the term indigenous innovation has been heralded by political leaders as an all-embracing elixir for China's structural problems and has been elevated to a strategic level equal to Deng Xiaoping's "reform and opening-up policy" during the Hu-Wen administration.[68] By August 1999, President Jiang Zemin was to remark: "In today's world, the core of each country's competitive strength is intellectual innovation, technological innovation and high-tech innovation".[69] As a result, by the mid-2000s indigenous innovation was implemented as a fully fledged policy priority.

The core of this "indigenous innovation" policy lies in the 2006 state-issued *Guidelines on National Medium- and Long-Term Program for Science and*

[67] Guo, Huadong, Ji Wu (eds) (2010). *Space Science and Technology in China: A Roadmap to 2050*. Chinese Academy of Sciences, Science Press (Springer), Beijing: p. 50.

[68] As Premier Wen Jiabao remarked: "We fundamentally have to rely on two main drivers, one, to persist in the promotion of opening and reforms, and two, rely on the progress of science and technology and the strengths of innovation". McGregor, James (2010). "China's Drive for Indigenous Innovation. A Web of Industrial Policies". Global Regulatory Cooperation Project—US Chamber of Commerce. APCO worldwide. Web. http://www.uschamber.com/sites/default/files/reports/100728chinareport_0.pdf

[69] Statement of President Jang Zemin; August 1999. Quoted from: *Ibid*. p. 10.

Technology Development.[70] Aimed at "laying the foundation for China to become a science and technology powerhouse by the middle of the twenty-first century", the guidelines identify a number of ambitious targets to support the development of an indigenous innovation system.

In particular, the document recommends reaching the target of having R&D expenditures equal 2.5 % of China's GDP by 2020, in order to reduce the country's reliance on foreign technology to 30 % or below.[71] In addition, the guidelines anticipate that by the same year the number of patents granted to Chinese nationals will be ranked among the top five in the world and China will have developed a number of breakthrough technologies thanks to the implementation of 16 megaprojects in 8 technological fields.[72] These are biotechnology, information technology, advanced materials, advanced manufacturing, advanced energy technology, marine technology, laser technology, and aerospace technology.

The manned spaceflight and lunar exploration programmes are both part of these 16 megaprojects. In the sense that these two endeavours represent precursor functional programmes for a manned lunar landing, it might be anticipated that the development of a national innovation system will be one of the central leitmotivs pushing China to embark upon such an ambitious endeavour. For Chinese scientists and engineers, it will be a powerful instrument through which indigenous technological prowess can be demonstrated and important breakthroughs achieved, eventually catapulting the country onto a technological leading position in the international arena.

Such technological development will also enable—so at least is the hope in Beijing—breaking the country's dependence on foreign technology and shifting away from the "made in China" paradigm to reach that of "created in China" or "innovated in China", while eventually acting as a leading force for China's economic growth, social development, and improvement of its national security.

---

[70] See The State Council of the People's Republic of China. The National Medium- and Long-Term Program for Science and Technology Development (2006–2020). An Outline. Beijing, China. 2006. Available at: http://sydney.edu.au/global-health/international-networks/National_Outline_for_Medium_and_Long_Term_ST_Development1.doc. See also: Segal, Adam. "China's Innovation Wall. Beijing Push for Homeground Technology". Foreign Affairs Snapshots. 28 September 2010. Web. http://www.foreignaffairs.com/articles/66753/adam-segal/chinas-innovation-wall.

[71] To compare reliance on foreign technology, in 2006 it was estimated to be 60 %, while 2006 gross expenditure on R&D was 1.3 % of China's GDP. See Government of the PRC Official Web Portal. http://www.gov.cn/english/2006-02/09/content_184426.htm.

[72] More precisely, the plan earmarks eight fields of technology in which 27 breakthrough technologies are to be pursued. It has also to be underlined that while the guidelines identified the goals and specific sector to focus on, it was the 11th Five-Year Plan (2006–2010) for High-Technology Industries that formally detailed the 16 megaprojects. Government of the PRC "China issues guidelines on sci-tech development program". Chinese Government's Official Web Portal. 9 February 2006. Web. http://www.gov.cn/english/2006-02/09/content_184426.htm. Accessed 23 December 2013.

The specific technologies to be developed in regard to a manned lunar exploration programme and further considerations on China's innovation efforts will be presented in Sects. 4.2.3 and 5.3, respectively.

3.5 Socio-economic Benefits

As mentioned above, space plays a central role in China's national development strategy, and the advances in science and technology (S&T) expected to spring from the implementation of a lunar endeavour would ultimately be directed to support this strategy: in particular, they would be designed to foster a process of sustained dynamic growth, generate a strong positive impact on China's overall economy, and deliver significant socio-economic benefits to its citizens.

3.5.1 Technological Innovation and Economic Development

Although China is one of the fastest-growing countries in the world, the growth model that has allowed its miraculous performance over the past 30 years now appears to be at a crossroads (see Sect. 5.1.1). As argued by numerous economists and scholars, this "extensive" model has not focused on productivity gains, but on factor accumulation, in other words, on the expansion of the quantity of inputs—especially workforce and capital—in order to increase the quantity of outputs.[73]

Chinese authorities, perfectly aware that reliance on extensive growth is gradually becoming detrimental because it exhausts resources and is subject to diminishing returns, have thus been pushing for an alternative growth model based on *inspiration* (technology innovation) rather than *perspiration* (manufacturing capacity).[74]

The importance of technological innovation to China's future growth pattern has been emphasised in recent years by leading Chinese macroeconomists, including Justin Yifu, former chief economist of the World Bank (WB). In an interesting contribution he stated that the potential for sustained and dynamic economic growth in China mostly depends on technological progress,[75] the other determinants on

[73] This interpretation is offered by economists such as George Friedman and institutions like the IMF. For other economists, it was, on the contrary, the enhancement of productivity levels that allowed Chinese exponential growth. For this interpretation, see Hu, Zuliu, and Mohsin S. Khan (1997). "Why China grows so fast? International Monetary Fund, Economic Issues 8. Washington D.C. Web. https://www.imf.org/EXTERNAL/PUBS/FT/ISSUES8/issue8.pdf

[74] Krugman, Paul (1994). "The Myth of the Asian Miracle". Foreign Affairs. Vol. 73 (6).

[75] From the perspective of the production functions, four macro-determinants determining economic growth can be identified: factors of production, the industrial structure, technology, and institutions.

which the country has so far relied being to a large extent subject to the speed of technological change and bound to become less relevant.[76]

In concrete terms the development of the numerous technologies, systems, hardware, and services required for the implementation of a lunar endeavour will in the first instance boost the growth of China's aerospace industries, both in terms of factors of production allocations (labour and investment) and of high value-added industrial upgrading. This will in turn strengthen the country's overall industrial capabilities, enhance national technical competence, and stimulate a knowledge-based economy. Eventually these achievements will also offer China the ability to further compete in the global economy, even when its competitive advantage—namely, low-cost manufacturing capacity—dwindles.

In terms of industrial development, benefits are not only expected to accrue to the companies directly involved in the implementation of the programme. The development of aerospace technology will support related industries, including information technology, as well as electronics, materials, and machine manufacturing. It will also indirectly strengthen other space-related industries. It is, for instance, anticipated that the high stability, reliability, and accuracy of the launch services required for a manned lunar landing will also improve confidence in the technological prowess of China's commercial satellite launch services. As several analysts have noted with regard to the Shenzhou programme, manned missions are important "evidence of the new level of sophistication that China can offer international clients".[77] The underlying message that can be conveyed through a manned lunar landing is clear and powerful: if Chinese technology is reliable enough to launch its taikonauts to the Moon, it should be trusted to succeed in launching a client country's satellite.

Similarly, other components and services developed for this programme could also spur the country's efforts to achieve additional market share in space technology.

High-technology employment opportunities and the development of highly skilled scientific and technical cadres can also be envisaged. According to estimates, there are currently more than 300,000 people working on the manned spaceflight programme alone; but it is clear that the expertise required for the implementation of a manned lunar exploration programme is likely to generate a huge number of specialists through China's universities, thus enhancing both the

[76] As Justin Yifu clearly states, "technological progress means higher productivity. So, even when the industrial structure and factors of production remain unchanged, with better technology the economy's output and growth will improve as well". Land and natural resources are basically fixed and the growth of labour is rather limited. Even if capital accumulates at great speed, the law of diminishing return means that unless there is technological progress, returns will decline.... And technological progress is also a prerequisite for institutional improvement. Yifu, Justin. "China's potential for sustained dynamic growth". In Leonard, Mark (ed) (2012). *China 3.0*. European Council on Foreign Relations, London.

[77] *Cit*. Dellios, Rosita (2005). "China's space program: A strategic and political analysis", Culture Mandala: The Bulletin of the Centre for East–west Cultural and Economic Studies Vol. 7 (1).

quality and quantity of the specialised workforce. A lunar endeavour could potentially also act as a powerful catalyst to attract scientists and engineers of the Chinese diaspora back to the country, along with S&T talent from other countries.

Indirect benefits are expected to arise with regard to regional and local development. Admittedly, this would not represent a new objective. During the 1960s US President Lyndon Johnson "consciously employed the lunar landing programme as the political vehicle through which the southern United States—a region generally economically behind other US regions—could be drawn into the national economy by investing funds and facilities in the region. The long term economic effects were enormous and continued even after the Apollo programme shut down".[78] Likewise, China could use its investments in a future lunar exploration programme as elements in its macroeconomic planning to boost the development of selected areas. Research has, for instance, revealed that the construction of the Xi'an Aerospace Cluster has confirmed this logic.[79] The new launch site for the human spaceflight programme to be opened on Hainan Island is also intended to boost local development, both in terms of industrial infrastructure and tourism-related services.[80]

3.5.2 *From the Moon to the Earth: Societal Benefits*

As an important growth driver, a manned lunar landing programme would not only enable economic expansion in terms of macroeconomic production functions, it would also yield other tangible and intangible paybacks for Chinese society. Relevant quality-of-life benefits can be expected in a number of areas (e.g. health, transportation, consumer goods, information technology) from the diffusion of technological innovation and breakthroughs.

What benefits will eventually appear as a result of a lunar endeavour are hard to predict in detail. However, if one considers the numerous and unforeseen spin-off

[78] *Cit.* Handberg, Roger, and Zhen Li (2007). *Chinese Space Policy. A study in domestic and international politics*. Routledge, New York: p. 133.

[79] See Siddivò, Marisa, and Alessandra De Chiara (2012). "High-tech industry clustering in less favoured areas: International comparison of two aerospace industrial districts in China and Italy". Journal of Science and Technology Policy in China Vol. 3 (2): 164–190. The research in particular demonstrates that the rationale for the construction of the Xi'an Cluster is definitely not an efficiency-driven process but is the outcome of policymakers' plans to redress interregional economic disparities.

[80] Worth mentioning is that a visitor centre and a space theme park—containing, among others, spaceflight simulators, a model of a lunar landscape and some Shenzhou cabins—will be built nearby, thus stimulating the tourism in the region. Vick, Charles. "Hainan/Wenchang". Global Security. 21 June 2010. Web. http://www.globalsecurity.org/space/world/china/sanya.htm. Accessed 18 April 2014.

products resulting from the exploration missions of the NASA,[81] ESA, and the Japan Aerospace Exploration Agency (JAXA),[82] China should find considerable additional incentives for pursuing its ambitious lunar ambitions. For instance, many technologies involved in the Apollo programme have been transferred to other fields, generating new products on which people now rely in their daily lives: CT scanning imaging technology and the laptop both originated from the Apollo programme. Similarly, it is realistic to expect that the technologies to be developed for a lunar exploration programme will eventually filter down from space through numerous industries and enter into the terrestrial marketplace. Their commercialisation will enhance the quality of life of the Chinese people, open business opportunities, and possibly even create entirely new industries and markets,[83] thus enlarging the sphere of China's economic activity.

An intangible but particularly valuable benefit offered by a highly visible endeavour such as the manned lunar exploration programme would be its intrinsic ability to act as a source of inspiration for the enrolment in technical education in China. The ascent curve of educational achievements recorded in the USA during the implementation of the Apollo programme (see Fig. 3.1) is persuasive evidence of this.

Like the Apollo programme, China's manned lunar exploration programme could potentially inspire Chinese students to pursue careers in science, technology, engineering, and mathematics (STEM), just as the pianist Lang Lang's international celebrity has driven millions of Chinese children to the piano. In addition, as underlined by the ISECG, having a visible space exploration programme will also send a message to students that they have the possibility of exciting long-term careers in science and technology.[84] In the long run, this potential could also make China the country with the best qualified workforce and perhaps even generate an oversupply of human resources in S&T.

Promotional activities to popularise space and stimulate interest in STEM have already been initiated as part of China's human spaceflight missions. On 20 June 2013, for instance, China's second female astronaut Wang Yaping delivered an interactive video lecture from the country's space laboratory Tiangong-1 to 330

[81] See "Spinoff database published by the National Aeronautics and Space Administration". National Aeronautics and Space Administration. Web. http://spinoff.nasa.gov/spinoff/spinsearch?BOOL=AND&ALLFIELDS=&CENTER=&BOOLM=AND&MANUFACT=&STATE=&CATEGORY=&ISSUE=&Spinsort=ISSUE Accessed 10 April 2014.

[82] See "Spinoff from Japan's Aerospace Technology. JAXA Industrial collaboration and coordination centre". Japan Aerospace Exploration Agency. Web. http://aerospacebiz.jaxa.jp/en/spinoff/. Accessed 10 April 2014.

[83] To date, more than 400 space technologies have been transferred on the ground and applied to a number of fields like education and medical care.

[84] "Benefits Stemming from Space Exploration". International Space Exploration Coordination Group. August 2013.

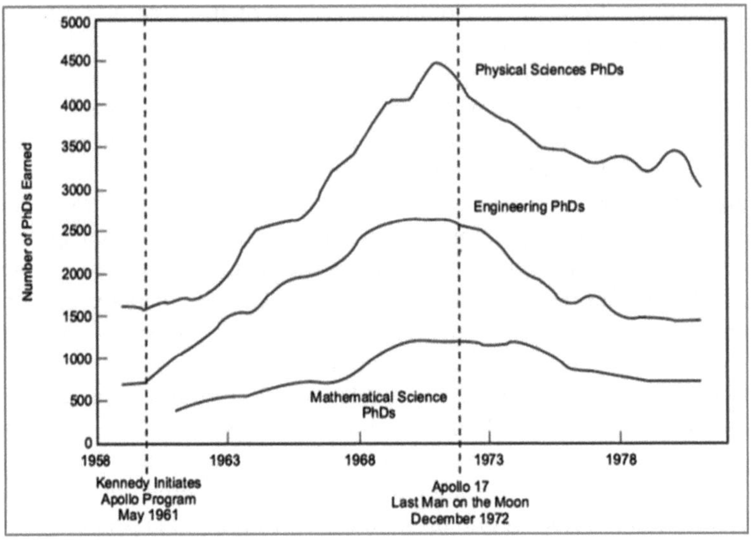

Fig. 3.1 Apollo programme's impact on educational achievements. *Source*: International Space Exploration Coordination Group

primary and secondary school students in Beijing.[85] The lecture was about motion in microgravity environments and the surface tension of liquids in space, as well as the concepts of weight, mass, and Newton's law. Here lesson was accompanied by various demonstrations.[86] An estimated audience of 60 million students and teachers around China watched the live broadcast on TV!

When commenting on China's first space lecture, Liu Cixin, one of the country's bestselling science fiction writers remarked[87]: "I wish that one day a lecture can be delivered from the Moon to demonstrate the rise of Earth",[88] clearly showing how a mission to the Moon would be deeply inspirational not only in terms of education and workforce development but also in broader sociocultural terms. In this regard it can be noted that China's manned spaceflight programme has already produced a dramatic positive impact on Chinese society, which is reflected, for instance, in a blossoming of Chinese space science fiction and the establishment of a considerable

[85] "Shenzhou X astronaut gives lecture". China Daily. 20 June 2013. Web. http://www.chinadaily.com.cn/china/2013shenzhoux/2013-06/20/content_16638698.htm. Accessed 12 May 2014.

[86] "Space lecture inspires dreams of the universe". Xinhua News. 21 June 2013. Web. http://news.xinhuanet.com/english/china/2013-06/21/c_124894304.htm. Accessed 12 May 2014.

[87] Liu Cixin has gained immense popularity, both in China and abroad, for "Three Bodies", a dystopian trilogy of space science fiction.

[88] Quoted from Zhi, Chen. "Space lecture inspires dreams of the universe". Xinhua news. 21 June 2013. Web. http://news.xinhuanet.com/english/china/2013-06/21/c_124894304.htm. Accessed 12 May 2014.

number of space-related civic groups.[89] These forms of entertainment are growing in popularity and relevance, so they are also bound to become a powerful outreach tool, which, in time, will allow the Chinese people to further identify with future space endeavours. A lunar endeavour is likely to have an even more profound influence on cultural and intellectual life in China—and around the world.

3.6 The Security Dimension: Dual-Use Technologies and Comprehensive National Power

Although China's motives for a lunar programme do not respond to the same strategic logic of the US–Soviet space race of the 1960s, this does not mean, however, that security-related considerations have totally disappeared.

As mentioned, the considerable advances in S&T offered by the implementation of a lunar endeavour would not only provide considerable industrial, economic, and social benefits; they might also yield indirect military benefits that will be manifest mainly in the development of dual-use technologies. Indeed, many authors have emphasised this dimension. According to Erich Seedhouse, for instance, the development of dual-use technologies represents the *real why* of China's manned spaceflight programme and ultimately of its ambition to send taikonauts Moonwards. As he says, "Beijing's most important justification and motivation for pursuing a manned space program is based firmly in the military arena, which is not surprising, since national security remains a potent justification for the large expenditures demanded by a space program".[90]

In view of the fact that most space technologies are dual use in nature and that China's space programme has very significant involvement the PLA, this type of interpretation has come to dominate Western political perceptions of China. The possible implementation of a manned lunar landing programme has therefore been largely interpreted as an indirect means of enhancing Chinese military capabilities, projecting them in the international arena and ultimately posing a threatening challenge to US dominance.

Even if the development of dual-use technologies and enhancement of military capabilities are a pertinent driver for a lunar endeavour, there are nonetheless important observations to be made in this respect.

First, the assumption that the high-level political support required for the implementation of the programme is motivated uniquely by military benefits appears far too black and white. In fact, nothing at this early stage points to a preponderance of military considerations, as the military benefits would be hard to

[89] For an extensive review on the effects of Chinese space science fiction on the aerospace industry and society, see Solomone, Stacey (2013). *China's Strategy in Space*. Springer, New York.

[90] *Cit.* Seedhouse, Erich (2010). *The New Space Race. China vs the United States*, Springer—Praxis Publishing, Chichester, UK: p. 12.

predict. Currently, any major such benefits would be expected to come from launcher technology, Deep Space Network (DSN) technology, and robotic technology. The mastery of heavy-rocket propulsion (particularly in terms of rocket stability, reliability, and accuracy) is an important enabler for enhancing intercontinental ballistic missile (IBMC) capability, and deterrence is for China as desirable in the twenty-first century as it was in the 1950s. DSN technology, which is used to track and communicate with the spacecraft and astronauts in a Moon mission, could instead be deployed for military intelligence purposes: in particular, high-speed data networking and data processing facilities are believed to offer valuable contributions for enhancing command, control, communications, computers, intelligence, surveillance, and reconnaissance (C4ISR) capabilities. As for the development of robotic technology, "the logic is simple: if you can operate a robot on the Moon, you can operate it in an enemy state or on the battlefield".[91] Future warfare, scholars have explained, "is expected to see significant usage of robotic technology in various forms and robots developed for a Moon mission could be easily modified for the purposes of military usages".[92] Other military benefits can be more generally envisaged in terms of overall infrastructure, management, and advancement of expertise.

In sum, military benefits from a manned lunar exploration can certainly be seen. However, these would not be unique to a manned, as opposed to an unmanned, programme.[93] The development of dual-use technologies could certainly be achieved by unmanned missions and considerably more cheaply. In addition, seeking such a demonstration does not appear strategically consistent, considering the capabilities of countries such as the USA and Russia. As the Chinese scholar Sun Dangen has remarked, the military projection of these technologies "would be like *throwing an egg against the rock* and will never be a strategic option for the descendants of the military strategist Sun Tzu".[94]

In any case, as Fiona Cunningham has explained—"separate space programmes exist for the PLA to exploit space technology for military purposes".[95] Thus, investing in a costly and complex programme like manned lunar exploration in pursuit of mere military benefits is the least plausible explanation.

Rather than acting as the main driver for a lunar endeavour, the development of dual-use technologies should instead be considered a "supporting driver" at best.[96] It is worth noting in general that as both a developing country and a developing

[91] *Cit.* Lele, Ajey (2010). "An Asian Moon race". Space Policy 26 (4): 227.

[92] *Ibid.*

[93] Cunningham, Fiona (2009). "The Stellar Status Symbol: True Motives for China's Manned Space Program". China Security Vol. 5 (3): 73–88.

[94] *Cit.* Dangen, Sun (2006). "Shenzhou and Dreams of Space". China Security Vol. 2 (1): p. 61.

[95] *Cit.* Cunningham, Fiona (2009). "The Stellar Status Symbol: True Motives for China's Manned Space Program". China Security Vol. 5 (3): 73–88.

[96] This interpretation is to a large extent confirmed by the fact that, from their beginning, efforts to establish the Shenzhou manned spaceflight programme have been driven by the scientific community, rather than the military. See Sect. 4.1.

power China must by necessity maintain a coordinated approach to increase its strength. The Chinese government and scholars emphasise the pursuit of what they define as *Zonghe guoli* or comprehensive national power (CNP).[97] The term refers to the total sum of powers or strengths of a country in economy, military affairs, science and technology, education and resource, and its influence.[98] While the pursuit of military security—which obviously also requires the development of military space capabilities—is clearly an essential component, it can by no means be considered the only one; the economic, social, and political security goals previously emphasised are indeed equally—or in several cases even more—substantial.

A further point is that China's pursuit of military space capabilities does not necessarily imply a militarily imposed objective of challenging US dominance via a new space race.

Such claims are, rather, shaped by US security calculations, in which China simply appears to have replaced the USSR as the main threat. On the Chinese side they cannot be substantiated, however. For China, security-related motivations, although crucial, are not directly designed to project China's rising *hard power* in the face of the USA: instead they are shaped by its past experience of humiliation and subjugation during the "century of shame" and are thus directed to securing the country against external threats, rather than to challenging US military hegemony.

As underlined by the PLA military experts Chang Xianqi and Sui Jungin, "China is modernising its national defence to satisfy its most basic needs to avoid being at the mercy of others".[99] It is thus in the light of this "never again" determination, and not as a quest for domination, that the enhancement of military space capabilities should be more properly inscribed.[100] Needless to say, when the objective is to avoid being at the mercy of others, the monopolisation of outer space by another country cannot be accepted; the implication being rather clear: if the USA more or less explicitly opposes China's pursuit of military security, then all that remains will be for China to respond.

Strategically, however, there are no good incentives for China to engage in a general space/arms race with the USA. In spite of the dramatic acceleration of China's space programme in recent years, there is still a remarkably wide gap between the two countries in terms of financial resources allocated, and parity appears some way off. Given the extensive asymmetry also in terms of

---

[97] Seedhouse, Erich (2010). *The New Space Race. China vs the United States*, Springer—Praxis Publishing, Chichester, UK: p. 39.

[98] Hu, Angang (2007). *Economic and Social Transformation in China: Challenges and Opportunities*. Routledge, New York: p. 34.

[99] Xianqi, Chiang, Jungin Sui (2006). "Active exploration and peaceful uses of outer space". China Security Vol. 2 (1): p. 22.

[100] It has even been noted that the lack of transparency in China's space programme—and more in general of power structures and its policymaking processes—is also due to the always present fear about divulging information to foreigners who will use it to exploit China, rather than the idea of maintaining a strategic advantage.

technological and military space capabilities, for Chinese policymakers there is no certainty of winning such a race, while the risk of overstretching resources and being lured into a Soviet-style bankruptcy is perceived as high and thus calls forth prudence. Indeed, as Rosita Dellios has noted, the USSR's catastrophic attempt to keep up with the USA in the development of a Strategic Defense Initiative—or Star Wars—still "weigh[s] on the mind of Chinese government planners"[101] and provides a relevant argument against the case for general direct competition with the Americans.

In addition, engaging in an open competition with the USA over space dominance would deprive China of the only strategic advantage it has for dampening Washington's vastly superior military space capability, namely, its asymmetric deterrence. Based on stealth and deception, this is ultimately made possible by its relatively low reliance on space assets.[102] It is clear that this advantage will be rapidly lost if an arms race in space eventually spreads out. Hence, it is clearly more convenient to delay the strategic contest for the time being.

To conclude, engaging in a space arms race with the USA does not appear a strategically or economically plausible choice for China, and serious doubts can be cast on the political willingness to pay the price for such an adventure.

3.7 Lunar Environment Exploitation

An additional rationale that may push China's policymakers to pursue their lunar ambitions is the possibility of exploiting the Moon's environment. In many regards, the Earth's natural satellite presents a unique, precious environment that could potentially be utilised in different ways.

The first possibility is the prospect of exploiting the Moon's natural resources, particularly minerals. This idea has already been envisaged by a large part of China's scientific community, which believes that the Moon could serve as a new supplier of energy and resources "to support sustainable development for humans and society".[103] Besides rare earth elements, titanium and uranium, the Moon is thought to be rich in helium-3, a light, nonradioactive isotope of helium that has been deposited in the upper layer of the Moon's regolith by the solar wind over

---

[101] *Cit.* Dellios, Rosita (2005). "China's space program: A strategic and political analysis", Culture Mandala: The Bulletin of the Centre for East–west Cultural and Economic Studies Vol. 7 (1).

[102] As underlined by Mark Hilborne, a state with few assets in space has less to fear from offensive actions than countries (like the USA) completely dependent on them. Hilborne, Mark (2015). "The impact of China's Rise in Space". In Al-Ekabi, Cenan, Blandina Banares, Peter Hulsroj, Arne Lahcen (eds). *ESPI Yearbook on Space Policy 2012/2013. Space in a Changing World.* Springer, Vienna.

[103] *Cit.* Zhao, Huanxin, and Lei Zhao. "China shoots for the Moon". Xinhua News. 2 December 2013. Web. http://news.xinhuanet.com/english/china/2013-12/02/c_132933884.htm. Accessed 10 May 2014.

billions of years. Ideally, helium-3 could be processed into fuel for commercial nuclear fusion and the production of clean energy, thereby compensating for finite oil and coal reserves on Earth.

For instance, the possibility of mining the lunar surface to obtain helium-3 was explicitly stated by Long Lehao, vice president of the China Academy of Launch Vehicle Technology (CALT) in a 2010 contribution,[104] and by a number of scientists at CAS.[105] In a recent interview released prior to the launch of the Chang'e 3 mission, Ouyang Ziyuan, a top CAS scientist, confirmed that a long-term rationale of China's lunar programme is to obtain helium-3, as it "is the perfect fusion energy source to replace oil and gas and solve human beings' energy demand for around 10,000 years".[106]

However, obtaining helium-3 from lunar regolith is an extremely difficult task, and it is questionable whether this ambition is financially and technologically viable for China. As noted by many analysts, under present circumstances it would require more energy to retrieve helium-3 and bring it back than it would yield, making it commercially unviable.[107] ESA international relations specialist, Karl Bergquist, has emphasised that the possibility of mining the Moon remains for China "many, many years away".[108]

This target is nevertheless of particular political significance for Chinese scientists. Energy is one of China's most critical "Achilles heels", and it is hence normal that Beijing policymakers are highly sensitive to solutions proposed by their scientific community in this regard. In this light, what the helium-3 proposal reveals is the efforts of China's scientific community to persuade policymakers to pursue a manned lunar landing programme without scientists always committing on too concrete goals. Despite the abundance of declarations, there is currently no trace of the helium-3 goal in the *Roadmap to 2050* issued by the CAS or in any other policy document. All the same, were China to succeed in exploiting the Moon's natural resources, it would be a paradigm shift of such a fundamental nature that the

[104] 龙乐豪. 关于中国载人登月工程若干问题的思考. (Long Lehao. "On Issue of China Manned Lunar Exploration"). Missile and Space Vehicles No. 6. 2010. Web. http://wenku.baidu.com/view/21d5423a5727a5e9856a619b.html

[105] This position however is not included in the roadmap to 2050 proposed by China nor in any other official documents.

[106] He also underlined that there are many ways humans can use the Moon, like the utilisation of a belt of solar panels to support the whole world. He summed up his vision for lunar exploration by saying "There are so many potential developments—it's beautiful—so we hope we can fully utilise the Moon to support sustainable development for humans and society. Quoted from: Shukman, David. "Why China is fixated on the Moon". BBC News. 29 November 2013. Web. http://www.bbc.co.uk/news/25141597. Accessed 11 May 2014.

[107] Lasker, John. "Race to the Moon for Nuclear Fuel". Wired.com 15 December 2006. Web. http://www.wired.com/science/space/news/2006/12/72276. Accessed 11 May 2014.

[108] Connor, Neil. "Mining the moon is pie in the sky for China, experts say". Agence France-Presse. 15 December 2013. Web. http://phys.org/news/2013-12-moon-pie-sky-china-experts.html. Accessed 12 May 2014.

implications are impossible to predict either at the national or the international level.

Another way to benefit from the unique environment of the Moon would be offered in the field of space science by the construction of telescopes for space astronomy. There would be many advantages in placing a telescope on the lunar surface. The Moon has no wind, a rarefied atmosphere, a "radio-quiet" environment, and nights lasting for approximately 14 days. It thus offers a stable platform for observing the universe.[109] For the purposes of radio astronomy in particular, the far side of the Moon is the cleanest place in the solar system, as the Moon itself blocks all interference from Earth.[110] This makes astronomers particularly interested in constructing a lunar-based low-frequency radio telescope to capture the signals emanating from the formation of the first stars, billions of years ago.[111]

In contrast to the exploitation of lunar mineral resources, CAS and other space stakeholders have explicitly recommended a clear commitment in this regard. The implementation of multi-waveband astronomical observations that include large lunar-based astronomical telescope arrays is envisaged in the long-term objectives of its *Space Science Roadmap to 2050* as a means that will allow China "to play a leading role in the international astronomical frontier fields and make historical contributions to the exploration of the universe".[112]

A third and quite futuristic objective that has been widely discussed by both Chinese and international scientists lies in the possibility of building a launching infrastructure on the lunar surface. Luan Enjie, a senior adviser to China's lunar exploration programme (CLEP), has said that one of the ultimate aims of the lunar programme is to use the Moon as a "springboard" for deep-space exploration.[113] Thanks to its low gravity (1/6 that of the Earth), the Moon could in fact become an ideal launch platform for future interplanetary flights. It should be emphasised that for Chinese scientists, this objective would be part of a wider goal, aimed at building a lunar base and hence colonising the Moon with a permanent human presence. The Moon would for China ultimately become the first place for humans to learn to live on another celestial body and a staging post to reach other planets in the solar system.

---

[109] See Carrasco, Jose' Manuel, Jordi Bernabeu, and Eugenia Colell. "The Moon as an astronomical platform". In Proceedings of the Fourth International Conference on Exploration and Utilisation of the Moon. European Space Agency. 2000: pp. 79–86.

[110] *Ibid.* p. 79.

[111] "Global Exploration Roadmap". International Space Exploration Coordination Group. April 2007: p. 17.

[112] Guo, Huadong, Ji Wu (eds) (2010). *Space Science and Technology in China: A Roadmap to 2050*. Chinese Academy of Sciences, Science Press (Springer), Beijing: pp 61–62. It is also recommended that between 2030 and 2050, China will establish a space physics observation platform.

[113] Zhao, Huanxin, and Lei Zhao. "China shoots for the Moon". Xinhua News. 2 December 2013. Web. http://news.xinhuanet.com/english/china/2013-12/02/c_132933884.htm. Accessed 10 April 2014.

3.8 Summary Overview

As can be seen from the reflections provided throughout this chapter, the rationales and objectives guiding China's leadership towards a possible manned lunar exploration programme prove to be complex and multifaceted. In order to summarise them, an overview is presented in Table 3.1.

Table 3.1 Summary of China's rationales and motivations for manned lunar exploration

Objectives	Subgoals
Domestic political benefits	• Boost nationalism and reinforce social cohesiveness • Retain and enhance the CCP's legitimacy • Mobilise and gain population commitment • Symbolise the "great rejuvenation" of the Chinese nation and inspire the population
International motivations	• Enhance Chinese prestige and international status • Get "a seat at the table" of the international powerhouses • Become a "dispenser" of international cooperation • Use the lunar endeavour to build political axes and reinforce, rather than isolate, the Chinese international position • Underpin the image of a leader by example • Take a lead in future international efforts in human space exploration
S&T development	• Generate scientific knowledge • Stimulate advances and achieve breakthroughs in space science (science of, from, on the Moon) • Demonstrate technological prowess • Achieve technological breakthroughs • Spur the creation of an indigenous innovation system
Socio-economic benefits	• Support a qualitative shift in China's growth model • Foster the development of the aerospace industry • Enhance the qualification level of the workforce and create high-tech employment opportunities • Boost economic development at regional/local levels • Stimulate the creation of spin-offs and technology transfers and enlarge the sphere of economic activity • Act as a source of sociocultural inspiration
Security-related calculations	• Development of dual-use technologies and military space capabilities (e.g. launcher, DSN, and robotic technologies) • Enhance the country's comprehensive national power
Lunar environment utilisation	• Exploit the Moon's mineral resources (helium-3?) • Build a lunar-based launch infrastructure for future interplanetary flights • Build telescopes for space astronomy and a lunar base

Chapter 4
China's Way to the Moon

The objective of this chapter is to assess China's long-term ambitions for a manned lunar landing. The analysis is comprised of two main sections. In the first, an extensive review of the precursor functional programmes for embarking upon a lunar endeavour—in particular of the manned spaceflight and the lunar exploration programmes—is provided. This will in turn be used as a basis for discussing the current state of play of Chinese lunar plans. More specifically, the second part will set out considerations of the skills and hardware development required for the implementation of the programme and an assessment of how the overall organisation of such a programme might be managed and structured. Some thoughts on the potential mission configuration will also be provided.

4.1 Precursor Functional Programmes

An analysis of the broader Chinese exploration programme is important because the missions and projects involved will be necessary preliminaries to reaching the ultimate target of a taikonaut landing on the Moon. Indeed, from a technology viewpoint in particular, both the human spaceflight programme and the lunar exploration programme constitute two precursor functional programmes through which China is acquiring most of the critical skills for embarking upon such an ambitious endeavour. Rendezvous and docking capabilities, the development of telemetry and control systems, mastery of human spaceflight and extravehicular activities, and the development of multiple launch sites and proper life support systems are all critical skills which China is acquiring thanks to these programmes. Thus, the narrative history of China's possible manned lunar programme has its origins in these endeavours.

In addition, given the opaque nature of China's space programme, a review of these projects will offer a contribution to understanding how future plans for going to the Moon will be organised and implemented. Finally, the analysis should be of

M. Aliberti, *When China Goes to the Moon...*, Studies in Space Policy 11,
DOI 10.1007/978-3-319-19473-8_4

value by identifying the most critical areas, in terms of hardware and skills development, for achieving the target of a manned lunar landing.

4.1.1 The Chinese Manned Spaceflight Programme

On 11 June 2013, a Chinese spacecraft blasted off from a launch site in the Gobi Desert for the longest and most ambitious space mission China had ever embarked on. Three taikonauts, Nie Haisheng (Commander and Pilot), Zhang Xiaoguang (Flight Engineer), and Wang Ya-Ping (Mission Specialist), would orbit around the Earth for 15 days and dock with a Chinese space laboratory, sent into orbit 2 years earlier, where they would conduct a series of medical and space technology experiments. For Chinese President Xi Jinping, who witnessed the launch, it was "a sacred and glorious mission", which was carrying the "space dream of the Chinese nation, and represent[ing] the lofty aspirations of the Chinese people to explore space".[1]

The successful mission was the cause of great celebration in China. It was an impressive result for a country that just a few decades before looked trapped in the spirals of underdevelopment and by 30 years of political turmoil during the Mao Zedong era. This achievement was, however, in fact the culmination of a long-standing and increasingly complex programme, whose origins—interestingly enough—date back to the beginning of the space age.

4.1.1.1 The Background: From Project 714 to Project 921

Human spaceflight had been widely discussed in China since the early 1960s,[2] and the first proposal to put astronauts into space was officially drafted at the 13th meeting of the Central Special Committee on 10 August 1965.[3] The Committee proposed naming the first human spaceflight *Shuguang-1* (literally Dawn-1). Soon after the successful launch of China's first satellite (Dong Fang Hong I—The East Is Red) on 24 April 1970, President Mao Zedong, Premier Zhou Enlai, and Defence Minister Lin Biao officially approved the draft proposal for the programme, and the human spaceflight effort was then referred to as "Project 714".[4] In the same year,

[1] "Xi wows bigger strides in space exploration. Space Daily. 25 June 2013. Web. http://www.spacedaily.com/reports/Xi_vows_bigger_stride_in_space_exploration_999.html. Accessed 10 December 2013.

[2] China, following a Soviet practice, started to make flights with biological cargoes and animals. China's first space dog, Xiao Bao, flew into the atmosphere on 15 July 1966.

[3] Zhongyang, Zheng (2007). "The origins and development of China's manned spaceflight programme". Space Policy Vol. 23 (3): 167–171.

[4] In China, space projects are usually named after the year and month or month and date of their approval (in this case, project 714 stands for July 14). Kulacki, Gregory, and Jeffrey Lewis. (2009). *A Place for One's Mat: China's Space Program, 1956-2003*. American Academy of Arts and Sciences, Cambridge, MA: p. 20.

Tsien Hsue-Shen, later to be known as the father of China's space programme, asked the Chinese Air Force to recruit the first group of astronauts, with the intention of training them for the first human mission in the newly established Beijing Institute of Space Medicine.[5] China was thus the third country in the world to select a squad of astronauts. The original plan foresaw that the first flight would take place at the end of 1973.

However, the destructive impact of the Cultural Revolution (1966–1971) would eventually also hit the space sector hard. As well as pushing the country close to bankruptcy, it created a climate of violence and distrust, which was also targeted at the leaders of the space programme. Among these was the main sponsor of project *Shuguang*, the Minister of Defence, Lin Biao, who died in a plane crash, after his alleged (and failed) coup d'état (which, curiously, was code-named project 714).

As a result of these events, the space programme was put on the back burner. In the spring of 1972, Mao Zedong declared that "Earthly need must come first", and Project 714 had to be terminated.[6] With the country shifting its focus to development objectives in the late 1970s and early 1980s, except for the development of the Long March rockets, investments in China's space activities were limited to the building of application satellites.[7]

Nevertheless, the final stages of the US–Soviet competition of the mid-1980s would eventually refuel the debate. Although the focus of the debate was the broader role of S&T in China's national development, one of its eventual outcomes was the resuscitation of the human spaceflight programme.[8] In March 1986, following the advice of four leading scientists, the government launched Project 863, a state-sponsored, strategic high-tech R&D initiative aimed at accelerating research and development in seven key areas of S&T: automation, biotechnology, energy, information technology, lasers, new materials, and space technology. The space section of the programme was focused on the technologies associated with an Earth-orbiting space station, its crew transportation system and launch vehicle. However, the programme was intended only to produce preliminary conceptual studies rather than engineering development of any particular system.[9]

In February 1987, an expert group on Plan 863-2 was established to define long-term goals for the space sector. "Plan 863-2 led to two sub-group studies: 863-204 was for a new manned spacecraft and launcher and 863-205 was for a manned space station".[10] For the rest of the 1980s, top Chinese space professionals and

---

[5] A total of 19 astronauts were selected, beginning a 2-year training programme in May 1971. Harvey, Brian (2013). *China in Space. The Great Leap Forward.* Springer, New York: p. 259.

[6] Quoted from: *Ibid.* p. 261.

[7] See Lan, Chen. "Lunar Exploration". Dragon in Space. Web. http://www.dragoninspace.com/planetary/lunar-exploration.aspx. Accessed 10 October 2013.

[8] *Cit.* Kulacki, Gregory, and Jeffrey Lewis. (2009). *A Place for One's Mat: China's Space Program, 1956-2003.* American Academy of Arts and Sciences, Cambridge, MA: p. 22.

[9] *Ibid.* p. 23.

[10] Harvey, Brian (2013). *China in Space. The Great Leap Forward.* Springer, New York: p. 263.

policymakers engaged in a series of internal debates over whether China should embark on a hugely expensive human spaceflight programme and what means of transport should be adopted.[11] Although the construction of a space station as the ultimate, long-term target was not a point of contention, a large part of the political leadership continued to question the need for a human spaceflight programme at all. While some of them argued that human spaceflight was not the most appropriate way to develop Chinese space's capabilities, others claimed that China lacked the necessary resources for embarking upon such an ambitious endeavour. As a result, the senior political leadership remained hesitant to make a full commitment for some time.[12]

It was only in the early 1990s that the new Chinese leadership, led by President Jiang Zemin, took the view that the time was right to revive the human spaceflight programme.[13] The programme was eventually approved by the Politburo of the CCP in September 1992, under the code name *Project 921*.[14] Because of its sensitivity, the programme was not confirmed or publicly announced until the end of the decade.

The original Project 921 foresaw an unmanned launch by 1998, a manned launch by 2002, a small space station by 2007, and a Mir-class station by 2010.[15] Too ambitious to be realised even by a fast-rising power like China, the roadmap ended up being significantly modified. The ultimate goal of building a permanently manned Earth-orbiting space station was thus eventually moved to 2020.

Despite delays and rescheduling, this plan remains the roadmap today, more than 20 years later. It has three stages. In the first stage, China would send humans to fly in low Earth orbit on board the *Shenzhou* spacecraft. The second phase would develop and test the techniques and technologies required for building a space station, including extravehicular activity (EVA) and orbital rendezvous and docking. Two temporarily man-tended single-module space laboratories were to be launched as technology demonstrators for the future space station. In the final phase, a 90-tonne space station would be constructed in low Earth orbit. The space station should be capable of supporting a crew of three astronauts living and working in orbit continuously. Crews would be ferried between the space station and Earth on board the Shenzhou spacecraft every 6 months, and the station would be refuelled and resupplied by unmanned cargo vehicles.

---

[11] Kulacki, Gregory (2012). "Why China is building a space station". Union of Concerned Scientists. Web. http://www.ucsusa.org/sites/default/files/legacy/assets/documents/nwgs/why-china-is-building-a-space-station-06-12-12.pdf.

[12] *Ibid.*

[13] Lan, Chen. "Project 921". Dragon in Space. 14 January 2013. Web. http://www.dragoninspace.com/ humanspaceflight/project921.aspx. Accessed 12 October 2013.

[14] The project name stands for 21 September of that year.

[15] Harvey, Brian (2013). *China in Space. The Great Leap Forward.* Springer, New York: p. 263.

4.1.1.2 Organisational Structure and Key Components

The programme has a two-line management structure. Along one line sit the technical decision-makers, with the chief designer and chief engineer at the top. Along the other line, the administrative or managerial line, sit officials of the various government and military organisations and aerospace contractors.

In order to coordinate the efforts of the different stakeholders within these managerial lines, in 1992, a special Human Spaceflight Project Office (HSPO) was established under the direct leadership of the General Armaments Department (GAD). The office, which reports back directly to the State Council and the CCP Central Committee, is made up of a board of directors.[16] Its members include a deputy director of the GAD and high-level officials of the Chinese Academy of Sciences (CAS), the China National Space Administration (CNSA), the China Aerospace Science and Technology Corporation (CASC), the China Aerospace Science and Industry Corporation (CASIC), and the China Electronics Technology Group Corporation (CETC).[17] The membership of the HSPO is indicative of the intricate web of stakeholders involved in the management of the programme, as explained in Chap. 2. The joint meeting of this board of directors is responsible for setting the general direction of the human spaceflight programme and for taking decision on major issues during the implementation phase.

Under the board of directors, there are eight technical committees, each responsible for the management of one of the eight key systems of Project 921. Each committee is managed via the two-line system, with a high-ranking official representing administrative management and a senior designer and a senior engineer for technical decision-making. To coordinate the works of the different committees and strengthen the overall management of the programme, the China Manned Space Agency (CMSA) was established.[18] CMSA is also responsible for representing the government, carrying out international cooperation, news releasing, and media work.

The development of eight main key systems of Project 921 is assigned to different contractors. The main tasks, the respective responsibilities, and the contractors are summarised in Table 4.1.

---

[16] Lan, Chen. "Project 921". Dragon in Space. 14 January 2013. Web. http://www.dragoninspace.com/ humanspaceflight/project921.aspx. Accessed 12 October 2013.

[17] See the official website of the China Manned Space Engineering: http://en.cmse.gov.cn/list.php?catid=34.

[18] CMSA is made up of five divisions: Science and Technology Planning Division, Overall System Design, Infrastructure Construction Division, International Cooperation Divisions, and Information and Publicity Division. *Ibid.*

Table 4.1 Shenzhou programme main systems[a]

Components	Responsibility	Contractors
Astronauts (Project 921-1)	Astronaut selection and training; astronaut medical monitoring and support; development of spacesuits; spacecraft life support and environment control systems	PLA 507th Institute
Space applications (Project 921-2)	Onboard scientific experiment packages and development of payloads	Chinese Academy of Sciences
Manned spacecraft (Project 921-3)	Development of the Shenzhou space capsule	China Academy of Space Technology (CAST)
		Shanghai Academy of Spaceflight Technology (SAST)
Space Laboratory (Project 921-4)	Development of the Tiangong space laboratory	China Academy of Space Technology (CAST)
		Shanghai Academy of Spaceflight Technology (SAST)
Launch vehicle (Project 921-5)	Development of the LM-2 F launch vehicle	China Academy of Launch Vehicle Technology (CALT)
Launch site (Project 921-6)	Construction and operations of the manned mission launch site	Jiuquan Satellite Launch Center
		Wenchang Satellite Launch Center
TT&C system (Project 921-7)	Operations of the spacecraft tracking and communications network	Xi'an Satellite Control Center
Recovery site (Project 921-8)	Operations of the spacecraft recovery system	PLA General Armaments Department

[a]Lan, Chen. "Project 921". Dragon in Space. 14 January 2013. Web http://www.dragoninspace.com/humanspaceflight/project921.aspx. Accessed 12 October 2013

4.1.1.3 Phase I: The Shenzhou Missions (1999–2010)

In the implementation of the first phase of the programme, Russian cooperation proved to be indispensable for both the development of the spacecraft and training of the future taikonauts. In 1994, Russia started to sell some of its advanced aviation and space technology to the Chinese government. The following year, an agreement was signed between the two countries for the transfer of Russian Soyuz spacecraft technology to China.[19] Included in the agreement was training, provision of Soyuz capsules, life support systems, docking systems, and space suits. In 1996 two Chinese astronauts, Wu Jie and Li Qinglong, began training at the Yuri Gagarin Cosmonaut Training Center in Russia. After training, the two returned to China and proceeded to train other Chinese astronauts at sites near Beijing and Jiuquan.[20]

The hardware and information sold by the Russians led to significant modifications of the original spacecraft, which in 1994 was named Shenzhou ("divine vessel").[21] As a result, the design and many of the key technologies of Shenzhou were heavily modelled on the Soyuz. However, as extensively documented, although it is true that Shenzhou follows the general configuration of Soyuz, it is not a mere copy; indeed, Chinese officials have missed no opportunity to point out that Shenzhou was "made in China".[22] The main differences between the two spacecraft concern the weight, the width, the length, and the diameter. In addition, the docking system and the overall internal volume present substantial differences from Soyuz. Table 4.2 shows Shenzhou's main characteristics in comparison with those of Soyuz.

Like its Russian counterpart, Shenzhou comprises three modules: an orbital module at the front, a re-entry capsule in the middle, and a service/propulsion module in the back. The orbital module contains crew-serviced equipment and on-orbit habitation and is equipped with its own propulsion and control systems, so as to allow autonomous flight in orbit. The cabin is designed to provide the astronauts with air, at a temperature of 17–25 °C and humidity of 30–70 %.[23] The re-entry module contains the spacecraft's instrument panel, storage space, and seats for three, possibly four, taikonauts.[24] Finally, the service module provides the electrical power, control, and propulsion for the spacecraft in orbit. It comprises four re-entry rockets with variable thrust, 28 manoeuvring engines with variable thrust, two solar panels, and radiators to discharge heat.[25]

Shenzhou was first developed in two prototype versions and then in three different variants, according to the tasks the different missions had to carry out. These are summarised in Table 4.3.

---

[19] "China and the Second Space Age". FUTRON Corporation. 15 October 2003: p. 7. Web. http://www.futron.com/upload/wysiwyg/Resources/Whitepapers/China_n_%20Second_Space_Age_1003.pdf. Accessed 20 December 2013.

[20] *Ibid.* p. 7.

[21] *Ibid.* p. 6.

[22] Harvey, Brian (2013). *China in Space. The Great Leap Forward.* Springer, New York: p. 269.

[23] "Shenzhou spacecraft Information". Spaceflight 101. Web. http://www.spaceflight101.com/shenzhou-spacecraft-information.html. Accessed 15 October 2013.

[24] See Seedhouse, Erich (2010). *The New Space Race. China vs the United States*, Springer – Praxis Publishing, Chichester, UK: p. 177.

[25] For a more detailed analysis of the Shenzhou spacecraft, see "Shenzhou spacecraft Information". Spaceflight 101. Web. http://www.spaceflight101.com/shenzhou-spacecraft-information.html. Accessed 15 October 2013. See also Harvey, Brian (2013). *China in Space. The Great Leap Forward.* Springer, New York: pp. 267–271.

Table 4.2 Shenzhou and Soyuz: a comparison[a]

Characteristics	Shenzhou	Soyuz
Complete spacecraft		
Weight	7.8 tonnes	7.21 tonnes
Length	9.15 m	6.98 m
Diameter	2.8 m	2.6 m
Propulsion module		
Weight	3 tonnes	2.95 tonnes
Propellant	1.1 tonnes	900 kg
Length	2.94 m	2.3 m
Diameter	2.8 m	2.2 m
Base	2.8 m	2.72 m
Solar panels	Two of 24 m^2	Two
Descent module		
Weight	3.2 tonnes	3 tonnes
Length	2.5 m	1.9 m
Diameter	2.5 m	2.17 m
Orbital module		
Weight	2 tonnes	1.3 tonnes
Length	2.8 m	2.2 m
Diameter	2.8 m	2.25 m
Solar panels	Two of 12 m^2	None

[a]*Source*: Harvey, Brian (2013). China in Space. The Great Leap Forward. Springer, New York: p. 270

Table 4.3 Shenzhou variants

Shenzhou variant	Missions	Launches
Flight test prototype	Shenzhou 1	1
Unmanned prototype	Shenzhou 2, 3, 4	3
Solo flight variant	Shenzhou 5, 6	2
EVA variant	Shenzhou 7	1
Docking variant	Shenzhou 8, 9, 10	3

The first Shenzhou mission was successfully launched on 20 November 1999. The other three unmanned prototype missions then followed (one in 2001 and two in 2002) before the launch of China's first manned spaceflight mission.

Launched by a Long March (LM)-2F rocket from the Jiuquan Satellite Launch Center on 15 October 2003, the Shenzhou-5 mission made China the third country—after Russia and the United States—to independently launch a human into space and represented the key achievement of phase 1 of project 921. During its mission, PLA Colonel Yang Liwei orbited the Earth 14 times and landed safely after 21 h of flight near the re-entry site in Inner Mongolia. Although neither the launch nor the re-entry was broadcast live on television, because of the government's fear of embarrassment in case of failure, the mission was immediately the

Table 4.4 List of Shenzhou missions

Mission	Date	Launcher	Objectives	Crew	Duration	Recovery
Shenzhou 1	19/11/1999	LM-2F	Experimental prototype flight test	Unmanned	21 h	21/11/1999
Shenzhou 2	10/01/2001	LM-2F	Prototype flight test	Unmanned	7 days	16/01/2001
Shenzhou 3	25/03/2002	LM-2F	Unmanned flight test	Unmanned	7 days	01/04/2002
Shenzhou 4	30/12/2002	LM-2F	Unmanned flight test	Unmanned	7 days	05/01/2003
Shenzhou 5	**15/10/2003**	**LM-2F**	**First manned flight, one-man crew**	**Yang Liwei**	**1 day**	**16/10/2003**
Shenzhou 6	12/10/2005	LM-2F	Two-man crew, multiday flight	Fei Junlong Nie Haisheng	6 days	17/10/2005
Shenzhou 7	27/09/2008	LM-2F	Three-man crew, first EVA	Zhai Zhigang Liu Boming Jing Haipeng	3 days	28/09/2008
Shenzhou 8	*01/11/2011*	*LM-2F*	*Unmanned docking with Tiangong-1*	*Unmanned*	*19 days*	*17/11/2011*
Shenzhou 9	*16/06/2012*	*LM-2F*	*First crewed visit to Tiangong*	*Jing Haipeng* *Liu Wang* *Liu Yang*	*13 days*	*29/06/2012*
Shenzhou 10	*11/06/2013*	*LM-2F*	*Second crewed visit to Tiangong*	*Nie Haisheng* *Zhang Xiaoguang* *Wang Yaping*	*15 days*	*26/06/2013*

subject of numerous demonstrations of patriotism and was hailed as a triumph of China's science and technology.[26] With this first manned flight, China did indeed achieve important breakthroughs in 13 key technologies, including re-entry lift control of the manned spacecraft, emergency rescue, soft landing, module separation, and heat prevention.[27]

Following this historic flight, Shenzhou has accomplished five more missions, each one aiming to advance Chinese spaceflight capabilities and achieve a new milestone in space. A summary overview of these missions is provided in Table 4.4.

---

[26] See Johnson-Freese, Joan (2005). "Space Wei Qi. The Launch of Shenzhou V". Naval War College Review Vol. 57 (2).

[27] Seedhouse, Erich (2010). *The New Space Race. China vs the United States*, Springer – Praxis Publishing, Chichester, UK: p. 149.

One of the most striking features of the sequence of launch dates and objectives of the different Shenzhou missions is the steady but slow pace of development. As Brian Harvey put it:

> although Western observers might have expected China to have developed its manned space programme at a pace seen in the US and Soviet Union in the late 1950s and early 1960s, this was not the case. The development of the manned program was characterized by considerable caution, with four full unmanned missions (1999, 2001, two in 2002) before a single astronaut was put on board for a short mission (2003). Even then, the pace of the program was slow, with two years passing before the next multi-day mission of two astronauts (2005), another three before the spacewalk mission (2008), and a further four before the first flight to the space station (2012). The slow, "conservative" pace was rewarded with comparatively incident-free missions.[28]

Gregory Kulacki has remarked that some Chinese space professionals think the pace demonstrates an excess of caution in the Chinese political leadership, which has a very low tolerance of failure.[29] The approach has nonetheless been "purposeful and economic, each manned mission representing a substantial step forward, with very little repetition of earlier achievements". [30]

4.1.1.4 Phase II: The Tiangong Programme (2011–2017)

The second phase of Project 921 envisions the creation and launch of space laboratories, in order to develop and demonstrate "the technologies and techniques required for building a permanently-manned multi-modular space station".[31] These techniques include space rendezvous and docking, EVA, and long-term space living.

Tiangong-1, "heavenly palace", is the first of these experimental test-beds. It is a single-module Earth-orbiting spacecraft intended for the development and practice of orbital rendezvous and docking techniques—and for this reason is commonly referred to as a "target vehicle". Its primary role was technology demonstration rather than operational use. However, it was also intended to be temporarily occupied by visiting taikonauts as an experimental orbital station for experiencing orbital life and conducting scientific experiments.

[28] Harvey, Brian (2013). *China in Space. The Great Leap Forward.* Springer, New York: p. 307.

[29] Kulacki, Gregory (2012). "Why China is building a space station". Union of Concerned Scientists: p. 8. Web. http://www.ucsusa.org/sites/default/files/legacy/assets/documents/nwgs/why-china-is-building-a-space-station-06-12-12.pdf.

[30] Harvey, Brian (2013). *China in Space. The Great Leap Forward.* Springer, New York: p. 307.

[31] It has to be remarked that the second phase of Project 921 is just an intermediary step towards the ultimate goal of building a permanently crewed space station. See Chen, Lan. "Project 921". Dragon in Space. 14 January 2013. Web. http://www.dragoninspace.com/ humanspaceflight/project921.aspx. Accessed 12 October 2013.

The spacecraft, which is roughly 10-m long and weighs 8500 kg,[32] is made up of two cylinder-shaped sections: a service section and a living section, the latter slightly wider than the former. The two compartments are connected via a 1.1-m-long transition cylinder. The service section, derived from the Shenzhou service module, is 3.3 m in length and 2.5 m in diameter. It accommodates the orbital manoeuvring engine and different subsystems (electrical, environmental control, communications, and propulsion) and is provided with two four-panel solar wings (3 m × 10 m) attached to its exterior.[33] Attached to this section, there is a dish antenna for communication with the ground station.[34] The living section, which is 5 m in length and 3.35 m in diameter, has 15 cubic metres of habitable volume, equipped with sleeping stations, exercise gear for visiting crews, as well as the control panel for flight control and station keeping.[35] The docking system has a ringlike capture structure, collocated at the front end of the living section, allowing it to dock with the Shenzhou spacecraft. It is very similar to the system (APAS-75) developed by the Soviet Union for its Soyuz spacecraft during the 1970s.[36]

Completed in August 2010, the final version of Tiangong-1 arrived at the Jiuquan Satellite Launch Center in late June 2011, scheduled for launch in late August. However, a setback caused by a launch failure of another version of the Long March rocket meant the launch was postponed for a month.[37] Tiangong-1 was finally launched on 29 September 2011.

A noteworthy aspect of the mission is the TT&C system used to maintain continuous contact with the space laboratory. As well as being tracked by the different ground stations in China, Tiangong-1 was also followed by a series of tracking ships named *Yuan Wang* (literally "long view") and by eight overseas stations: Swakopmund (Namibia), Malindi (Kenya), Karachi (Pakistan), Santiago (Chile), Alcantara (Spain), Aussaguel (France), and the Kerguelen Islands and Dongara (Australia). These have been made available to China by cooperation agreements with third-party countries and organisations like ESA.

---

[32] The Space Foundation (2013). The Space Report 2013 | The Authoritative Guide to Global Space Activity. The Space Foundation, Colorado Springs.

[33] Lan, Chen. "*Tiangong*-1". Dragon in Space. Web http://www.dragoninspace.com/humanspaceflight/tiangong1.aspx. Accessed 20 October 2013.

[34] Communication with the ground station is not direct, but occurs via the *Tian Lian* data relay satellite.

[35] The Space Foundation (2013). The Space Report 2013 | The Authoritative Guide to Global Space Activity. The Space Foundation, Colorado Springs.

[36] Harvey, Brian (2013). *China in Space. The Great Leap Forward.* Springer, New York: p. 14. The visiting crew could enter and exit the spacecraft via a 0.8-m-diameter tunnel inside the docking port. During a rendezvous docking, Tiangong-1 would act as the passive spacecraft, travelling in a "backside first" position, with its docking port pointing backwards, so that the chasing Shenzhou spacecraft could make a V-bar approach from behind. Chen, Lan. "Tiangong-1". Dragon in Space. Web http://www.dragoninspace.com/humanspaceflight/tiangong1.aspx. Accessed 20 October 2013.

[37] Harvey, Brian (2013). *China in Space. The Great Leap Forward.* Springer, New York: p. 14.

While in orbit and waiting for its first rendezvous and docking, Tiangong-1 carried out scientific experiments and activities. Instruments on board the module included:

- A polarised gamma ray telescope to test solar activity, cosmic structure, origin, and evolution
- An imaging spectrometer to take pictures of the Earth, track pollutants, and measure gases in the atmosphere
- An external exposure platform to test optical electronics and conduct material tests
- A high-precision atomic clock to test theories of gravity
- A three-dimensional microwave altimeter to measure the height of water in the oceans
- A boiler to test microgravity fluid physics, material formation, and mechanics[38]

Having successfully achieved the first part of the programme, the next step was to carry out an orbital docking of a spacecraft with Tiangong. Since this would be the first such mission for China, it was decided not to risk a crew, but to send an unmanned spacecraft instead.

The Shenzhou 8 mission was conceived for this purpose. Launched on 31 October 2011, Shenzhou 8 docked with the Tiangong on 2 November, at a time meticulously calculated so that the docking operation could take place while the two spacecraft flew over Chinese territory: in this way, ground-based tracking stations could provide more intensive tracking and telemetry support.[39] The two spacecraft remained locked together for 12 days. A second docking exercise occurred on 14 November, when Shenzhou undocked from Tiangong for 30 min, before being reattached. Finally, on 16 November, Shenzhou withdrew from Tiangong for the second time and began operations for re-entry to Earth (17 November 2011).

As noted by Chen Lan, "although unmanned, the spacecraft was exactly identical to the one that would have been used in the following manned mission".[40] In addition, two dummy astronauts were carried on the Shenzhou spacecraft, to check the proficiency of the life support system. It is also worth mentioning that, in its third seat, Shenzhou 8 carried an experimental payload produced jointly with DLR and ESA. SIMBOX (Science in Microgravity BOX) was a 25-kg experimental box intended to make immunology tests and to test the production of food, oxygen, and clean water in anticipation of long-duration spaceflights.[41]

---

[38] See Yim, Shing-Yik *et al* (2012). "Current Development of Manned Lunar Landings". Proceedings of the International Astronautical Congress, Naples, Italy, October 1-5, 2012. Paper: IAC-12.A5.1.10.

[39] Lan, Chen. "Shenzhou-8". Dragon In Space. Web. http://www.dragoninspace.com/shenzhou/shenzhou8.aspx. Accessed 15 October 2015.

[40] *Ibid.*

[41] For an extensive review of China–Germany cooperation on SIMBOX, see Braun, Markus (2013). "SIMBOX on Shenzhou 8: German-Chinese Cooperation in Biomedical Space Research". In: Lan, Chen, and Jacqueline Myrrhe (eds). *Go Taikonauts. All about China's space programme.* Issue 9. June 2013.

Finally, installed on a microchip was the so-called "Dream of Thousands": contributions by more than 40,000 people who, in an Internet competition launched by *China Space News*, had been invited to write their dreams of the future. While this might seem a tiny detail, the initiative is a further confirmation of how the Chinese authorities constantly work to sell the space programme domestically.[42] In terms of achievements, the mission provided useful engineering experience for the subsequent manned rendezvous and docking missions and, of course, for the construction of the forthcoming space station.

The first crewed visit to Tiangong-1 was on 18 June 2012, when the Shenzhou 9 spacecraft, carrying a three-person crew, docked with the module. On that day, the Tiangong station was declared operational, and Chinese President Hu Jintao personally went to mission control to formally congratulate the three taikonauts in a telecast.[43] There was much cause for celebration, as the mission represented an important new milestone for the country's space programme: it made China the third country after the USSR and the United States to inhabit a human-made module in orbit around the earth!

Besides the scientific and medical experiments, one of the main objectives of the mission was to test manual docking operations (un-docking and re-docking): on 23 June 2012—in conjunction with national celebrations of the Dragon Boat Festival—Shenzhou was undocked from Tiangong under the manual control of Liu Wang. It retreated to a distance of 400 m and then re-docked. The exercise was repeated a second time before the return to the Earth (29 June 2012).

A second crewed spacecraft, Shenzhou 10, visited Tiangong-1 the following year, in June 2013. During their 15-day mission—the longest mission to date in China's human spaceflight programme—the three taikonauts once again tested both the automatic and manual docking operations[44] and demonstrated that Tiangong could support astronauts for more extended periods.[45] Shenzhou 10 also flew around the station to increase experience with rendezvous and proximity operations, enabling safer manoeuvres around it on future missions.[46] In addition, several medical and space technology experiments were carried out, and Wang Yaping, China's second woman in space, gave lectures to 60 million school students while in orbit.

---

[42] The first missions to Tiangong-1 have been in addition documented by a film, "Flying", issued by the August First Film Studio—the cultural wing of the PLA—and distributed throughout China.

[43] Both the launch and the work of the astronauts in the space laboratory were covered everyday on Chinese television.

[44] "Chinese Astronauts manually dock spacecraft". Space Daily. 23 June 2013. Web. http://www.spacedaily.com/reports/Chinese_astronauts_manually_dock_spacecraft_999.html. Accessed 10 November 2013.

[45] "Twilight for Tiangong" Space Daily. 25 June 2013. Web. http://www.spacedaily.com/reports/Twilight_for_Tiangong_999.html. Accessed 10 November 2013.

[46] The Space Foundation (2013). The Space Report 2013 | The Authoritative Guide to Global Space Activity. The Space Foundation, Colorado Springs.

Table 4.5 Tiangong missions

Mission	Date	Objectives
Tiangong-1	29/09/2011	Launch of the target vehicle
Shenzhou 8	31/10/2011	Rendezvous and unmanned docking with Tiangong-1
Shenzhou 9	16/06/2012	Automatic and manual docking operations
Shenzhou 10	03/06/2013	Second crewed visit to Tiangong-1

The media reported that, during a video call with the three taikonauts, President Xi Jinping declared: "the space dream is part of the dream to make China stronger. With the development of space programs, the Chinese people will take bigger strides to explore further into the space".[47]

According to the Chinese Manned Space Agency (CMSA), Shenzhou 10 was the first "operational" flight mission in the human spaceflight programme, pursuing four main objectives: (a) to ferry the crew and materials between Earth and the Tiangong-1 space laboratory and test the performance of the Shenzhou human capsule and its docking system; (b) to further test the crew's ability to fly, live, and work in the Shenzhou-Tiangong spacecraft complex; (c) to demonstrate the adaptability and efficiency of the crew in the space environment and broadcast a classroom lesson to Chinese students; and (d) to further test coordination between different systems in the human spaceflight programme.[48]

With the successful completion of the Shenzhou 10 mission, China was expected to retire Tiangong-1 at the end of 2013 by de-orbiting it in the Pacific Ocean.[49] As of 1 April 2014, the spacecraft remains in orbit, although no additional mission is foreseen. The Tiangong-1 missions and their respective objectives are summarised in Table 4.5.

A second space laboratory, Tiangong-2, is due to launch in 2015, followed by the Shenzhou-11 mission in May 2016.[50] Originally built as a backup to Tiangong-1, it will be similar to its predecessor but with important innovations. First of all, it will support a three-person crew for up to 20 days, somewhat longer than Tiangong-1.[51] Second, it will have an "improved design, which will feature an orbital fuelling system to enable the space laboratory to be refuelled of air, water and propellant by a cargo vehicle. Finally, it will probably be used to test also the robotic arm system

---

[47] "Xi vows bigger strides in space exploration". Space Daily. 25 June 2013. Web. http://www.spacedaily.com/reports/Xi_vows_bigger_stride_in_space_exploration_999.html. Accessed 10 November 2013.

[48] "Tiangong-I /Shenzhou-X Manned Spaceflight Mission". China Manned Space Engineering. Web. http://en.cmse.gov.cn/list.php?catid=207. Accessed 12 November 2013.

[49] "Twilight for Tiangong" Space Daily. 25 June 2013. Web. http://www.spacedaily.com/reports/Twilight_for_Tiangong_999.html. Accessed 10 November 2013.

[50] "China plans to launch Tiangong-2 space lab around 2015". Space Daily. 27 June 2013 http://www.spacedaily.com/reports/China_plans_to_launch_Tiangong_2_space_lab_around_2015_999.html. Accessed 10 November 2013.

[51] The Space Foundation (2013). The Space Report 2013 | The Authoritative Guide to Global Space Activity. The Space Foundation, Colorado Springs.

which will be fitted on the future space station".[52] The spacecraft will not only serve as a technological demonstrator for the future space station but will also carry scientific payloads, like POLAR (a Chinese-Swiss collaboration to detect black holes and conduct tests of quantum gravity theories) and BANXING 2 (a microsatellite carried onboard Tiangong-2 and then deployed in an orbit that will make it fly in formation with the mother ship).[53]

A third and last space laboratory (Tiangong-3) was originally planned for launch later in the decade, around 2017. [54] However, according to the latest plan released soon after the Shenzhou-10 mission, this vehicle has been cancelled.

4.1.1.5 Phase III: Towards a Permanent Space Station

The two Tiangong Space laboratories, as well as the previous human spaceflight missions onboard the Shenzhou capsule, are not stand-alone projects but integral parts of a 30-year plan to build a modular and permanently crewed space station by the early 2020s. It is anticipated that the station will operate in a 400–450-km low Earth orbit (LEO) for at least 10 years.

Originally, the station was conceived to be similar in size to Russia's former Mir space station,[55] but in 2011, the design was modified by the China Manned Spaceflight Engineering Office to make it more like the ISS in scale.[56] The station will consist of several fixed modules: the core module, the docking hub with a six-port node, the experiment modules, and a truss structure on which the solar panels will be attached. Shenzhou capsules and the cargo vehicles can be considered removable modules.

The first part of the station, an experimental core module, is scheduled to launch from a new launch site on Hainan Island around 2018/2020. It will provide "the main living quarter for the on-board crew and also serve as the main flight control

[52] Based on the design of the *Tiangong* space lab, the cargo ship is 3.35 m in diameter, has a total mass of less than 13,000 kg, and can carry up to 6000 kg of payload. A single docking port allows the ship to be docked with the space station to deliver both 'wet' and dry cargo. The automated cargo spacecraft, which has yet been named, will be used to transport three types of cargo to the space station: air, water, and propellant which are required for the maintenance of the station itself, food and other materials for the astronauts onboard the station, and equipment for scientific researches and experiments. The ship may also be capable of assisting the space station for orbit maintenance using its own propulsion system. Lan, Chen. "Cargo Vehicle". Dragon In Space. Web. http://www.dragoninspace.com/humanspaceflight/cargo-vehicle.aspx Accessed 1 5 October 2015.

[53] Harvey, Brian (2013). *China in Space. The Great Leap Forward*. Springer, New York: p. 359.

[54] *Ibid*. p. 23.

[55] The Space Foundation (2013). The Space Report 2013 | The Authoritative Guide to Global Space Activity. The Space Foundation, Colorado Springs.

[56] Harvey, Brian (2013). *China in Space. The Great Leap Forward*. Springer, New York: p. 24.

and management centre for the entire station".[57] An onboard airlock will allow astronauts doing EVAs to exit and re-enter the space station. Two robotic arms mounted on the module will move equipment and supplies around the station and capture other modules and visiting spacecraft going to berth with the core module.[58] The primary robotic arm, designed and built by CAST, will have a payload capacity of 25 tonnes. The smaller secondary arm will be built by Harbin Institute of Technology.

This core module will be attached to a base hub provided with five additional docking ports (two on the X-axis, two on the Y-axis, and one on the Z-axis) through which various additional modules and space vehicles can be docked. The two lateral docking ports (Y-axis) on the docking hub will be occupied by two experiment modules designed to accommodate scientific instruments and equipment. The two modules will be similar in size and mass, but with slightly different functions. Experiment module 1 will consist of a resource section, the pressurised section, and an airlock. Besides scientific experiment facilities, it will also be equipped with a secondary flight control and management system in case of a malfunction in the main control system on the core module.[59] Experiment module 2 will consist of a resource section, a pressurised section, and a non-pressurised section, in which a large astronomical telescope will be installed.[60]

A regenerative environment control and life support system will be used, and a new type of EVA space suit will be developed for use on the station. The Shenzhou human capsule will be docked with the space station on the X-axis docking port during the crew visit. The second X-axis docking port at the other end of the core module can be used to receive new supplies from cargo vehicles or visits by the spacecraft of international partners.

Once the space station is permanently manned, in addition to the visiting Shenzhou spacecraft, a second Shenzhou spacecraft will need to be docked with the station permanently to serve as a "lifeboat". This will most likely use the free Z-axis docking port.[61]

In November 2012, it was reported that during the operational phase of the space station programme, a second core module may be launched to dock on the X-axis port of the base hub, expanding the station from a "T-shape" configuration to a

---

[57] Lan, Chen, and Jacqueline Myrrhe (eds). *Go Taikonauts. All about China's space programme.* Issue 9. July 2013: p 10.

[58] *Cit.* "Space Station". SinoDefence. Web. http://sinodefence.com/space-station/. Accessed 15 October 2013.

[59] Lan, Chen, and Jacqueline Myrrhe (eds). *Go Taikonauts. All about China's space programme.* Issue 9. July 2013: p 13.

[60] *Ibid.* p. 13.

[61] For a detailed review of the future CSS and the building-up process, see Lan, Chen (2013). "Tiangong-1 Revisited. Learning to Operate a Space Station, with a Grand Vision in Mind". In: Lan, Chen, and Jacqueline Myrrhe (eds). *Go Taikonauts. All about China's space programme.* Issue 9. July 2013: p 10–12.

Fig. 4.1 Representation of China's future space station

"cross-shape" configuration.[62] An unofficial rendering of the future Chinese space station is provided in Fig. 4.1.

There are already several indications of the scientific experiments to be conducted. There are now more than 200 projects on the candidate list, and CAS has planned for six "science platforms" to be installed on the station:

- A Space Exposure Experimental Platform, for experiments in radiation biology, materials science, new components and materials, astronomy, space physics, and environment
- A Variable Gravity Experimental Platform, providing opportunities from 0 to 2 G for experiments in biology, complex fluids, materials science, and medicines
- A High Temperature and Combustion Science Experimental Platform
- A High Microgravity Level Experimental Platform, for experiments in laser cooling atomic clocks, the verification of gravity, the equivalence principle, crystals, fluid science, laser, and optical diagnostics
- A Life and Ecology Experimental Platform, a greenhouse for cell and tissue cultivation to cultivate plants, raise animals, and test the disposal of waste gases and water
- A Protein Engineering Experimental Platform, for experiments with protein macromolecules, liquid and gas diffusion, and protein structures and functions[63]

[62] "Chinese Space Station". Dragon in Space. Web http://www.dragoninspace.com/humanspaceflight/space-station.aspx. Accessed 20 October 2013.

[63] *Cit.* Guo, Huadong, Ji Wu (eds) (2010). *Space Science and Technology in China: A Roadmap to 2050*. Chinese Academy of Sciences, Science Press (Springer), Beijing: pp. 87–88.

A number of experiments will be assigned for the future taikonauts themselves, among them:

- Psychology of crew and individual performance in an isolated, confined, and hostile environment
- First aid, space sickness, immunity, and telemedicine
- Physical resistance to weightlessness, addressing bone loss, atrophy, and cardiovascular deconditioning
- Resistance to radiation hazards, cancers, gene mutations, and pharmacological protectors
- Controlled ecological life support systems: food production, the balance of oxygen and nitrogen, the recycling and regeneration of water
- Fire safety—prevention, detection, control, and suppression[64]

In terms of scientific research and experiments, the announced intention to make the planned space station available to "global scientists" is also worthy of note.[65] This was confirmed at the 55th session of UN COPUOS in June 2013 and at the 64th IAC held in Beijing in September 2013, where Chinese space professionals reiterated their willingness to open the station to international partners for cooperation, including joint experiments, joint manned missions, the docking of foreign visiting vehicles, and additional pressurised modules built and launched by future partners. As mentioned above, potential cooperative undertakings on board the CSS are intended to primarily come under the UNOOSA-led Human Space Technology Initiative (HSTI).[66]

On 27 September 2013, in a special session at the 64th IAC Congress, a spokesperson for the CMSA stated that the CSS would be completed by 2022—2 years later than previously reported.[67] Furthermore, on 31 October 2013, the CMSA released the official logo of the China Manned Space Programme—a logo referencing both the character "中" (*zhong*), which indicates the Central Kingdom, and the basic T-shape of the space station. The names of the station, its modules, and the cargo vehicles were also made public[68]:

[64] *Cit. Ibid.* p. 89.

[65] "Chinese space station to benefit the world". China Daily. 17 June 2012. Web. http://usa.chinadaily.com.cn/china/2012-06/17/content_15507227.htm. Accessed 7 October 2013.

[66] Lan, Chen, Bill Carey, and Theo Pirrard (2013). "Welcome to Beijing for Space. Report from the 64th International Astronautical Congress". In: Lan, Chen, and Jacqueline Myrrhe (eds). *Go Taikonauts. All about China's space programme*. Issue 10. December 2013: p. 16.

[67] *Ibid.* p. 16. For further information on the international cooperation on the CSS, see Sect. 3.3.

[68] "China Manned Space Program Logo and Names of Space Station and Cargo Ship Officially Released". China Manned Space Engineering. 31 October 2013. Web. http://en.cmse.gov.cn/show.php?contentid=1354. Accessed 20 November 2013.

- China's manned space station as a whole will continue to be named "Tiangong".
- The core module will be named "*Tianhe*" (Harmony/Peace in Heaven).
- Experiment Module-1 will be named "*Wentian*" (Greet the Heaven).
- Experiment Module-2 will be named "*Xuntian*" (Cruise the Heaven).
- The cargo spaceship will be named "*Tianzhou*" (Heavenly Vessel).

4.1.2 The Lunar Exploration Programme

Together with the manned spaceflight programme, the Chinese Lunar Exploration Programme (CLEP) is the second functional project paving the way to a future manned lunar landing.

4.1.2.1 The Background

Ambitious plans with regard to exploration of the Moon started to circulate during the 1970s, but it was only in the early 1990s that the idea of robotic lunar exploration began to be seriously considered. This was mainly thanks to international developments in lunar space exploration. By 1990, thanks to the successful launch of its *Hiten* lunar probe, Japan had succeeded in breaking the monopoly of the superpowers on missions to the Moon. In addition, in 1994, NASA launched its *Clementine* lunar exploration mission, and a number of countries that China considered peers—like India—were also considering initiating similar programmes. Many Chinese policymakers and scientists felt compelled to establish the country's own lunar exploration programme, fearing that China would again fall behind other nations and lose its voice at international level.[69]

In 1992, a group of prominent scientists proposed sending a metal emblem to the Moon's surface by 1997 in order to reach the double target of celebrating the handover of Hong Kong to China and of catching-up with the parallel US and Japanese lunar efforts. In 1993, a feasibility study conducted for the newly established CNSA indicated that the Long March 3A launch vehicle could reach the Moon. Notwithstanding the decision of central government policymakers to give priority to the more ambitious manned programme and delay the Moon

[69] CAS top leading scientists, including Ouyang Ziyuang, explicitly lamented that "in the face of global competition, if China were indifferent, the country would fall behind internationally and lose its voice". *Cit.* Besha, Patrick (2010). "Policy making in China's space program: A history and analysis of the *Chang'e* lunar orbiter project". Space Policy 26 (4): 216.

probe,[70] the Chinese scientific community—and CAS in particular—continued to press for government commitment.

In 1995, CAS issued a report entitled "The Necessity and Feasibility of China's Lunar Exploration Programme", which introduced the concept of a lunar orbiter based on the *Dong Fang Hong* 3 satellite bus,[71] while 2 years later, in April 1997, three members of the CAS submitted to the State Council a new document entitled "Recommendations for the Development of China's Lunar Exploration Programme".[72]

As a result of this "policy entrepreneurship" by CAS scientists, in 1998, a group of experts of the now-defunct Commission of Science, Technology, and Industries for National Defense (COSTIND) was appointed to initiate a preliminary study on a lunar exploration programme.[73] The following report, "Overall Design and Key Technology Elements of a Lunar Exploratory Robot", set three main objectives for a future Chinese lunar probe: (1) improve the knowledge of the lunar surface; (2) monitor the solar wind, radiation, and meteors from the lunar surface; (3) analyse lunar rocks with an onboard laboratory to detect the presence of the element helium-3.[74]

In the autumn of 2000, the scientific objectives for the mission finally received formal approval from the State Council and the State Planning Commission,[75] while, at the same time, China's main space-related institutions started to hold a series of conferences on the topic.[76]

With the publication of a White Paper in November 2000, lunar exploration finally became the subject of a national policy statement. Among its medium-term priorities, the document clearly identified the exploration of deep space, centring on the Moon. Lunar exploration was defined as "the first step in China's deep space exploration effort".[77] The programme was officially approved on 28 February 2003 under the name Project 211.[78] It received final approval in January 2004 from Premier Wen Jiabao.

---

[70] The decision to postpone the project was for the government motivated by the lack of strong scientific motivations. *Ibid.*

[71] Besides presenting the concept of a lunar orbiter based on the DFH 3 satellite bus, the report was emphasising the possibility to exploit the Moon as a source of Helium-3. *Ibid.*

[72] Lan, Chen. "Lunar Exploration". Dragon in Space. Web. http://www.dragoninspace.com/planetary/lunar-exploration.aspx. Accessed 25 October 2013.

[73] *Ibid.*

[74] Harvey, Brian (2013). *China in Space. The Great Leap Forward.* Springer, New York: p. 311.

[75] Besha, Patrick (2010). "Policy making in China's space program: A history and analysis of the Chang'e lunar orbiter project". Space Policy Vol. 26 (4): 218.

[76] In May 2000, in his speech titled "China's Space Exploration in the 21st Century" during the first International Space Week, Luan Enjie, Director of the China National Space Administration (CNSA), revealed for the first time that the country was planning to explore the Moon.

[77] *Cit.* Government of the People's Republic of China. "China's Space Activities in 2011". White Papers of the Government of China. Beijing, China. 29 December 2011. Web. http://www.china.org.cn/government/whitepaper/node_7145648.htm.

[78] The name stands for "the first space project approved in the 21st century". Generally the name of programmes refers to the date of their inception.

When looking at the process that led to the eventual approval of CLEP, the most striking feature is the substantial role and influence exercised by CAS scientists in pushing their policy proposal upon decision-makers. The approval of CLEP was in fact ultimately brought about by the tireless advocacy of these scientists with inside access to high levels of government.

4.1.2.2 CLEP Organisational Structure

In order to lead the project and coordinate the work of the different civilian and military stakeholders involved in its implementation, on 19 February 2004, a Leading Small Group and its related office was set up by the CCP's Central Committee and the State Council.[79] This group, which is similar to the Human Spaceflight Project Office in terms of responsibility and functions, is at the very top of the chain of command. The only related office superior to it is the Lunar Probe Project Office set up within the CCP's Central Committee.

The LSG is headed by SASTIND Director and Central Committee member Zhang Yunchuan and includes leaders of the Ministry of Science and Technology, Ministry of Finance, General Armaments Department of the PLA, Chinese Academy of Sciences, and China Aerospace Science and Technology Corporation (CASC).[80] The LSG distributed key leadership positions for overseeing the execution of the project as follows: a General Commander (Luan Enjie), a Chief Designer (Sun Jiadong), and a Chief Scientist (Ouyang Ziyuan). As openly recognised by Chinese scientists themselves, this "Chang'e Iron Triangle" (*Chang'e tie sanjiao*) forms the core programmatic leadership of the project.[81] Deputy Chief Commanders and Deputy Chief Designers from CNSA, CASC, COSTIND, and CAS were subsequently appointed.[82]

In addition, in 2005 a Lunar Exploration and Engineering Centre for organising the mission was established—under the authority of this *Chang'e Iron Triangle*—while a Lunar and Planetary Science Research Centre was set up within the CAS to guide the scientific efforts.[83]

The LSG Office which, in common with the Human Spaceflight Project Office (HSPO), follows a "two-line command structure" is responsible for managing the various subsystems of the programme. CLEP comprises five main subsystems: the lunar exploration spacecraft; the launch vehicle; the telemetry, tracking, and control

---

[79] See Sect. 2.2.1.

[80] Besha, Patrick (2010). "Policy making in China's space program: A history and analysis of the Chang'e lunar orbiter project". Space Policy Vol. 26 (4): 219.

[81] Ouyang, Ziyuan *et al* (2007). "The origins and development of China's manned spaceflight programme". Space Policy 23 (3): 167–171. See also Zheng, Yongchun *et al* (2008). "China's Lunar Exploration Program: Present and future". Planetary and Space Science 56 (7): 881–886.

[82] Besha, Patrick (2010). "Policy making in China's space program: A history and analysis of the Chang'e lunar orbiter project". Space Policy Vol. 26 (4): 219.

[83] Harvey, Brian (2013). *China in Space. The Great Leap Forward*. Springer, New York: p. 313.

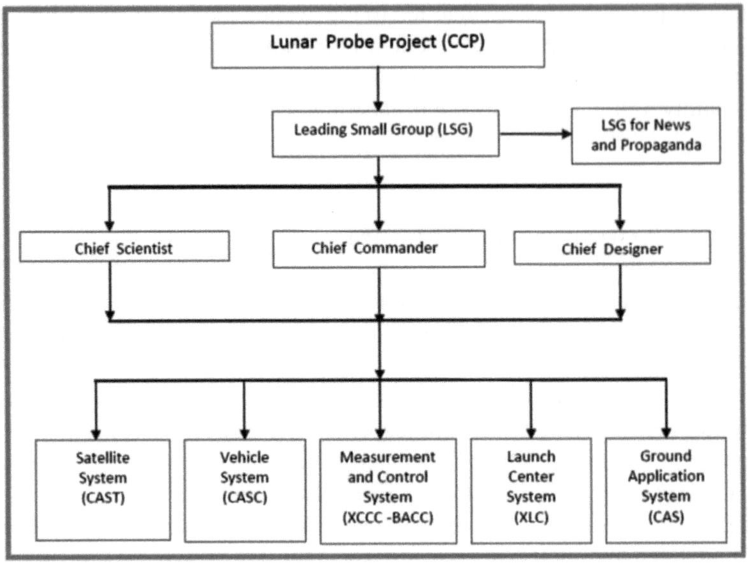

Fig. 4.2 CLEP organisational structure

(TT&C) system; the launch site; and the ground application system. For the development of each of these systems, there is a distinct prime contractor:

- The CAST for the lunar orbiter
- The CALT and SAST for the launcher
- The Xi'an Satellite Monitoring Center and Beijing Aerospace CCC for the TT&C system
- The Xichang Satellite Launch Center for the launch infrastructure
- The CAS for the application system

Finally, in June 2007—i.e. a few months before the launch of the first probe—COSTIND set up an additional Leading Small Group on news propaganda to boost and coordinate the media message, further confirming the societal value of the lunar project.[84] As already mentioned, it was to enhance the social impact of the mission that the lunar spacecraft was named after Chang'e, the Moon goddess of ancient Chinese mythology.

A representation of the CLEP's organisational structure is presented in Fig. 4.2.

4.1.2.3 Mission Implementation

The general plan to conduct long-term lunar exploration, as approved in January 2004 by Premier Wen Jiabao, contemplates a three-step strategy: (1) the successful

[84] Besha, Patrick (2010). "Policy making in China's space program: A history and analysis of the Chang'e lunar orbiter project". Space Policy Vol. 26 (4): 220.

launch and orbit of a probe to map the Moon's surface, (2) the landing of an unmanned lunar rover on the Moon, and (3) a lunar sample return mission to carry lunar material back to Earth.[85]

Orbiting Stage The first stage, which covered the period 2003–2007, was aimed at developing a probe with the main tasks of testing lunar orbiting technology and the deep-space telemetry and tracking network. The mission was launched on 24 October 2007. Two weeks later, on 5 November, the probe reached its orbit around the Moon where it stayed for 16 months.[86]

Beyond the engineering objectives of developing and testing lunar orbiting technology and the Deep Space Network technology (two essential preconditions for the subsequent rover and sample return phases), the mission had the following scientific tasks: (a) to obtain a three-dimensional stereo image of the lunar surface; (b) to determine distribution of a number of useful mineral elements and to estimate their abundance; (c) to survey the thickness of lunar soil and to evaluate the presence of helium-3, among other elements; and (d) to explore the environment (particles and radiation) between the Moon and Earth.[87]

To achieve the mission's goals, Chang'e's payload included five types of scientific instruments: a stereo camera and spectrometer imager, a laser altimeter, a microwave radiometer, the gamma and X-ray spectrometers, and a space environment monitoring system. In order to collect, process, store, and transmit the scientific data from various payloads, a special payload data management system was included.[88]

The mission officially ended in October 2008, and the spacecraft was ultimately destroyed on 1 March 2009 with a planned crash into the lunar surface. At both domestic and international level, China's image was strongly boosted by this success, as the mission combined a high level of technical expertise with substantial scientific outcomes (the mission succeeded in making a map of the lunar surface, analysing its chemistry and thickness, and characterising the lunar environment).[89]

Chinese Premier Wen Jiabao personally showed the first Moon images captured by Chang'e in a presentation ceremony at the Beijing Aerospace Control Center on

---

[85] See "The Chang'e-1 Project. China's Lunar Exploration Program". China National Space Administration. 8 March 2007. Web. http://www.cnsa.gov.cn/n615709/n772514/n772543/93747.html. Accessed 20 November 2013.

[86] See Huang, Jiangchuan, *et al* (2012). "Research and Development of Chang'e-2 Satellite," Proceedings of the International Astronautical Congress, Naples, Italy, October 1-5, 2012. Paper: IAC-12-A3.2A.5.

[87] Huixian, Sun *et al.* (2005). "Scientific objectives and payloads of Chang'E-1 lunar satellite". J. Earth Syst. Sci. 114 (6): 789–794.

[88] *Ibid.*

[89] For more detailed information on the Chang'e scientific and technological achievements, see "Chang'e-2". Earth Observation Portal. Web. https://directory.eoportal.org/web/eoportal/satellite-missions/c-missions/chang-e-2. Accessed 10 January 2014.

26 November 2007,[90] while newspapers around the country heralded the success of "China's 2000-year-old dream".[91] A complete high-resolution map of the Moon was published in December 2009, making China the third country to publish its own Moon maps, with the bonus of three-dimensionality.

Landing Stage The second and current stage, from 2008 to 2014, initially foresaw the development of a probe for a soft landing and a rover for lunar surface inspection. In 2009, however, it was decided that a first probe based on the vehicles of the preceding phase should be developed to test additional technologies that would better pave the way for the subsequent rover and sample return missions. As a forerunner to the Chang'e-3 mission, the overall objective of the Chang'e-2 was thus to demonstrate key technologies as far as possible, to reduce the risk during Chang'e-3's landing on the Moon.[92]

There were six key techniques to be tested:

1. Direct injection into an Earth–Moon transfer orbit by a launcher
2. Brake technology
3. Technology of 100 km × 15 km orbit manoeuvre and orbit measurement
4. High-definition imaging for the preselected landing area for Chang'e-3 (the Bay of Rainbows)
5. X-band TT&C measurement system
6. Other new technology, such as the high-speed Lunar–Earth data transmission and the brand-new landing camera[93]

Built as a backup to Chang'e-1, this second lunar probe was broadly similar to its predecessor, yet had important differences. These are summarised in Table 4.6.

Chang'e-2 was launched on 1 October 2010, coinciding with the celebration of China's National Day. Thanks to the more powerful rocket used (LM 3C), it reached the Moon four days later, without circling the Earth first like its predecessor.[94] On 5 October, it entered a 12-h lunar orbit, where it started to collect data. Twice—on 27 October 2010 and on 23 May 2011—Chang'e-2 carried out an orbital manoeuvre to enter a 100 × 15 km elliptic orbit, so that the spacecraft could get closer to the lunar surface and obtain higher-resolution images of the

[90] "Chinese Premier unveils first picture of the Moon". People's Daily. 26 November 2007. Web. http://english.people.com.cn/90001/90783/6309606.html. Accessed 20 November 2013.

[91] Patrick Besha, "Policy making in China's space program: A history and analysis of the Chang'e lunar orbiter project", Space Policy 26. August 2010: p 220.

[92] *Cit.* Huang, Jiangchuan, *et al* (2012). "Research and Development of Chang'e-2 Satellite". Proceedings of the International Astronautical Congress, Naples, Italy, October 1-5, 2012. Paper: IAC-12-A3.2A.5.

[93] *Ibid.*

[94] Lan, Chen. "Chang'e-2 Mission". Dragon in Space. Web. http://www.dragoninspace.com/planetary/change2.aspx. Accessed 10 November 2013.

Table 4.6 Chang'e-1 and Chang'e-2 comparison

Item	Chang'e-1	Chang'e-2
Launch vehicle	LM-3A	LM-3C
Launch mass	2350 kg	2480 kg
Launch trajectory	S-GTO	Earth–Lunar Transfer Orbit
Earth to Moon cruise time	12 days	5 days
Working orbit	200 km × 200 km	100 km × 100 km/15 km
Spatial resolution	120 m	7 m/1.3 m
TT&C	S-band	S-band /X-band
Extended mission	Crash into lunar surface	L.2/deep-space exploration

planned landing spot for the Chang'e-3 rover.[95] In addition, these manoeuvres enabled it to "test the coordination capacity between navigation, control and propulsion systems when the spacecraft operated on the far side of the Moon where it could not be monitored".[96]

On 8 June 2011, Chang'e-2 left lunar orbit for the L2 Earth–Sun Lagrangian point, to conduct scientific observations and test deep-space tracking and control capabilities for future possible explorations of Mars and the poles of the Sun. The L2 was reached after 77 days of flight, in late August, and the probe settled into a "parking orbit" circling around L2.[97]

After spending 235 days at L2, Chang'e-2 departed on 15 April 2012 heading further into deep space, so as to provide an important technical basis for the successful implementation of China's future deep-space exploration, especially trajectory design, deep-space tracking, and telecommunications.[98]

The second phase of CLEP culminated with the Chang'e-3 mission. In November 2009, the Chang'e-3 concept design was approved by SASTIND and CAS. The programme entered the prototype development stage that was eventually completed in March 2012. By that time, the simulated hovering and soft landing of

[95] Harvey, Brian (2013). *China in Space. The Great Leap Forward.* Springer, New York: pp. 317–323.

[96] Huang, Jiangchuan, *et al* (2012). "Research and Development of Chang'e-2 Satellite". Proceedings of the International Astronautical Congress, Naples, Italy, October 1-5, 2012. Paper: IAC-12-A3.2A.5.

[97] Lan, Chen. "Chang'e-2 Mission". Dragon in Space. Web. http://www.dragoninspace.com/planetary/change2.aspx. Accessed 10 November 2013.

[98] *Cit.* Huang, Jiangchuan, *et al* (2012). "Research and Development of Chang'e-2 Satellite". Proceedings of the International Astronautical Congress, Naples, Italy, October 1-5, 2012. Paper: IAC-12-A3.2A.5.

the lunar landing vehicle and field test of the lunar rover had proved positive. In mid-May 2013, Chang'e-3 started its final major testing before being shipped to the launch site in August.[99]

Chinese planners had laid out a set of three engineering and four scientific objectives for the Chang'e-3 mission. The former were to (a) master the key technologies of lunar soft landing and the lunar rover; (b) establish basic capabilities of robotic lunar landing, surface survey, and deep-space TT&C; and (c) establish an overall engineering system for lunar exploration.

As for the scientific objectives, these were to (a) survey the topography and geological structure of the Moon; (b) analyse the content and distribution of its mineral and chemical elements; (c) survey the space environment between the Moon, the Earth, and the Sun; and (d) carry out optical astronomy observations from the Moon.[100]

Launched on 2 December 2013 atop a LM 3-B rocket, the Chang'e-3 spacecraft—which consists of two modules, the Service Module and the Lunar Landing Vehicle—was first parked into a 100×100 km lunar orbit. After separating from the Service Module, the Lunar Landing Vehicle descended to a 100×15 km, 45° inclined elliptical orbit. After reaching the 15-km perigee, the vehicle ignited its thrusters to reduce its velocity and slowly descended to 4 km above the Moon surface, at which point its engine shut down for a free fall onto the lunar surface.[101]

Immediately after its successful landing on 14 December 2013, the spacecraft deployed the six-wheeled lunar rover named Yutu (Jade Rabbit), which started to explore the surrounding areas. Yutu had a design life of 90 lunar days and could explore an area of 3 km^2, with a maximum travelling distance of 10 km. It had a total mass of 1200 kg and carried a 20-kg payload.[102] The vehicle was capable of navigating autonomously, avoiding obstacles, and selecting the most optimised routes and locations for exploration activities. Onboard equipment included a ground penetrating radar for detecting the structure and depth of the lunar soil and an optical telescope for lunar-based observations. The different onboard cameras could capture images of the lunar surface. A robotic arm allowed *Yutu* to collect lunar soil samples for analysis. The vehicle was able to transmit images and data back to the Earth in real time.[103]

---

[99] For a detailed description of Chang'e-3 development, see Lan, Chen. "Small Step for *Yutu*, Giant Leap for China. Chnag'e-3 historic Lunar Landing". In: Lan, Chen, and Jacqueline Myrrhe (eds). *Go Taikonauts. All About Chinese Space Programme*. Issue 11. February 2014: 8–15.

[100] *Ibid.* pp. 11–12.

[101] See "Chang'e-3 Mission". Dragon in Space. Web. http://www.dragoninspace.com/planetary/change3.aspx. Accessed 15 October 2013.

[102] *Ibid.*

[103] For a more detailed description of Chang'e-3 major scientific payloads, see Lan, Chen. "Small Step for *Yutu*, Giant Leap for China. Chnag'e-3 historic Lunar Landing". In: Lan, Chen, and Jacqueline Myrrhe (eds). *Go Taikonauts. All About Chinese Space Programme*. Issue 11. February 2014: 12.

As of 1 December 2014, the mission is still formally underway. Although the majority of its engineering objectives have been successfully proven, the set of scientific tasks has not yet been fully accomplished; as on 25 January 2014, Yutu entered hibernation mode, as a result of a "mechanical abnormality".[104] Despite wake-up attempts in March, the mechanical malfunctions have reportedly not been resolved. It is worth noting, however, that a Chang'e-4 vehicle, featuring an identical design as Chang'e-3, was also built to serve as a backup to Chang'e-3. No formal decision on the destination of this spacecraft—currently placed in a clean room—has been to date announced. Presumably, it could be used to repeat the Chang'e-3 mission in 2015 or used to test technologies and procedural techniques in view of the upcoming sample return stage.

Sample Return Stage In the third and final phase, which presumably will cover the period 2015–2020, a small capsule, Chang'e-5, will soft-land on the lunar surface, collect samples using newly developed sampling and drilling machines and robotics, and return the samples to Earth. The mission is expected to launch around 2017 and will be preceded by a "pathfinder" mission, performed by the test capsule Chang'e-5 T1.

SASTIND initially planned to launch this test spacecraft in 2015. Surprisingly, it was launched ahead of schedule, on 23 October 2014. For this mission, a capsule was carried aboard a Chang'e-2 bus to lunar orbit, where it circled the Moon before returning to Earth and conducting an atmospheric re-entry at a speed much higher than that of an Earth-orbiting spacecraft. The capsule was successfully recovered on 31 October 2014, marking a new historic achievement for the Chinese space programme. With this mission, China became the third country, after the USSR and the United Sates, to accomplish a round-trip mission to the Moon.[105] As reported by Xinhua News, the main task of this "pathfinder" mission was to test technology likely to be used in Chang'e-5 and allow mission specialists to understand the impacts of high-speed atmospheric re-entry on the forthcoming lunar sample capsule. [106]

The Chang'e-5 mission will use a lunar orbit rendezvous (LOR) method similar to that used by the Apollo programme. The spacecraft will be launched from the Hainan Satellite Launch Site atop an LM5 heavy-lift launch vehicle. It will perform a direct flight to a lunar orbit, where a smaller Lunar Landing Module (LLM)

---

[104] "China Exclusive: Control circuit malfunction troubles China's Yutu". Xinhua News. 3 March 2014. Web. http://news.xinhuanet.com/english/china/2014-03/01/c_133152096.htm. Accessed 10 March 2014.

[105] David, Leonard. "China's 1st Round-Trip Moon Shot Sets Stage for Bigger Lunar Feats". Space.com. Web. http://www.space.com/27661-china-moon-mission-sample-return.html. Accessed 1 December 2014.

[106] "New lunar mission to test Chang'e-5 technology". Xinhua News. 22 October 2014. Web. http://news.xinhuanet.com/english/sci/2014-10/22/c_133734632.htm. Accessed 1 December 2014.

Table 4.7 Future Chang'e missions

Spacecraft	Function	Scheduled launch
Chang'e-4	Backup of Chang'e-3	2015 (?)
Chang'e-5 T1	Pathfinder mission	2015 (launched October 2014)
Chang'e-5	Sample return	2017
Chang'e-6	Backup/sample return	2017–2019

carrying the Sample Return Capsule (SRC) will be released and soft-land on the Moon, while the mother ship will remain in orbit. [107]

As analyst Chen Lan has reported, "the LLM will collect the lunar soil samples and place them in the sealed SRC. The SRC will then take off from the lunar surface, using the LLM as a launch platform. The SRC will then rendezvous and dock with the mother ship in the lunar orbit, before returning to Earth. Once reaching the Earth orbit, the SRC will be separated from the mother ship and carry out an unpowered atmospheric re-entry. Once in the atmosphere, the capsule will deploy a parachute for a soft landing in the recovery zone in Inner Mongolia".[108]

Two models of the spacecraft will be built, with Chang'e-6 being a backup to Chang'e-5 in case the first mission fails. If Chang'e-5 is a success, many of the technologies for a possible manned lunar landing mission post-2020 will have been proven.

To conclude, the timeline of the scheduled future missions is shown in Table 4.7.

4.2 Towards a Taikonaut Landing

4.2.1 Rumours or Concrete Plans?

Speculation over a Chinese plan to land taikonauts on the Moon began to circulate in the early 2000s. It was fed by Chinese scientists and academicians themselves who, galvanised by their country's burgeoning growth and steady achievements in space, started to envisage not only the possibility of a manned expedition to the Moon but also in the longer term the creation of a lunar base. Ouyang Ziyuan, a top CAS scientist and member of the Leading Small Group for the lunar exploration programme, published a collection of essays in 2000 entitled "Academicians Envisioning the 21st Century". In it he wrote of his desire to send taikonauts Moon-wards and then establish a Moon base, which in time would evolve into a

[107] Lan, Chen. "Chang'e-5". Dragon in Space. Web. http://www.dragoninspace.com/planetary/change5-rv-test.aspx. Accessed 15 November 2013.

[108] *Ibid.*

full-scale lunar city.[109] He argued that a lunar base for China would be a great leap forward in national security, scientific progress, and national unity. In addition, he emphasised that the Moon could not only provide raw materials, but its low gravity could also make it a viable launching pad for galactic travel.[110] The book was nominally intended to popularise science and encourage Chinese children to study scientific disciplines. However, the suspicion that it might not just have been the utopian dream of academicians was at least partially confirmed by the presentation, at the 2000 Hannover World Exposition, of a model depicting Chinese taikonauts, with a rover, planting the PRC's flag on the lunar surface.[111] In the same year, conflicting leaks started to appear in the media with increasing frequency. In October 2000, *Associated Press* and Xinhua News reported that China was planning to create a permanent lunar base with the aim of mining the lunar soil for helium-3 and fuelling nuclear fusion plants on Earth. But, a few days later, *SpaceDaily* reported that a top scientist in the Chinese space programme had denied that any firm plan for a lunar landing had been agreed, although he also affirmed: "As long as the country has a plan and provides funding protection, Moon landing is not an insurmountable obstacle for China". [112]

Despite the rumours, in the 2000 edition of the White Paper, there was no trace of this ambitious project. Even among the long-term development targets (for the next 20 years or more), the document did not reveal any plans for the development of technologies directly related to the human exploration of the Moon. Chinese targets at that time were to "establish China's own manned spaceflight system and carry out manned spaceflight scientific research and technological experiments and studies of outer space".[113] Where space exploration was concerned, only robotic lunar exploration was mentioned.

In 2004, the lunar exploration programme was approved, comprising the three-step strategy of orbiting, landing, and sample returning a probe, described above. Although never publicly acknowledged, the human landing on the Moon surface was already conceived as the potential final (fourth) phase of this programme. Beyond the symbolic decision to name the project Chang'e, it was doubtlessly intentional that the official logo of the programme represented the Chinese character for Moon (月-*yue*') with two human footsteps at its centre. In this light, it could

---

[109] A description of the book's content can be found in Long, Wei. "Chinese Scientist Envisages Moon City In Early 21st Century". Space Daily. 23 October 2000. Web. http://www.spacedaily.com/news/china-00zzk.html. Accessed 10 September 2013.

[110] *Ibid.*

[111] See Harvey, Brian (2004). *China's Space Program. From Conception to Manned Spaceflight.* Springer, New York: p. 314.

[112] Long, Wei. "Chinese Scientist Envisages Moon City In Early 21st Century". Space Daily. 23 October 2000. Web. http://www.spacedaily.com/news/china-00zzk.html. Accessed 10 September 2013.

[113] Government of the People's Republic of China. "China's Space Activities in 2011". White Papers of the Government of China. Beijing, China. 29 December 2011. Web. http://www.china.org.cn/government/whitepaper/node_7145648.htm.

logically be argued that CLEP was from the beginning intended to be a preparatory stage, paving the way for a manned mission to the Moon. In the meantime, the parallel development of the Shenzhou programme was testing and mastering the human spaceflight capabilities required for the endeavour.

Interestingly, the public announcement of CLEP was made soon after the launch of the Vision for Space Exploration (VSE) by the Bush administration in January 2004, reinforcing the idea that China would compete with the United States to return to the Moon. In spite of the United States announced intention to open this new venture in space to international partners,[114] China's exclusion from participation in the ISS made it clear that the invitation would not be extended to Beijing, thus pushing Chinese policymakers to move ahead with the implementation of their national programme.

With the steady advancement of the Chang'e programme and the successful launch of the first lunar probe (2007), rumours about the existence of plans for a manned lunar landing quickly resurfaced, this time reinforced by the public statements of prominent space officials. At the 16th Humans in Space Symposium of the International Academy of Astronautics (IAA), held in Beijing in 2007, Jiang Liwei said: "There are many beautiful tales about the Moon in Chinese literature. It carries a special significance for us... Building a lunar base can not only expand our knowledge about the Moon, but marks a crucial step to realize a flight to Mars or farther planets".[115]

Although, in a subsequent press conference in October 2007, CNSA Chief Sun Laiyan once again denied the existence of any plans to send humans to the Moon, he also declared "So far, our moon mission only includes unmanned probing projects. But I believe one day China will for sure send its own astronauts to land on the Moon. I hope I can see it happen".[116]

---

[114] See Seedhouse, Erich (2010). *The New Space Race. China vs the United States*, Springer – Praxis Publishing, Chichester, UK. In his Vision for Space Exploration President Bush—interestingly in line with J. F Kennedy's speech of 1963—remarkably affirmed: "we will invite other nations to share the challenges and opportunities of this new era of discovery. The vision I outline today is a journey, not a race, and I call on other nations to join us on this journey, in a spirit of cooperation and friendship". See National Aeronautics and Space Administration. The Vision for Space Exploration. Washington DC, United States. February 2004. Web. http://www.nasa.gov/pdf/55583main_vision_space_exploration.pdf.

[115] "Astronauts share their experience". People's Daily. 22 May 2007. Web. http://english.peopledaily.com.cn/200705/22/eng20070522_376754.html. Accessed 11 September 2013.

[116] In another press conference held in October 2007, Luan Enjie, chief commander of CLEP, also told reporters that China had no plan or timetable for a manned moon landing for now. He underlined: "a manned moon landing is a project with great difficulties, high risks and huge investments. A wish-list approach is not the way to go about it. Many factors have to be taken into account to carry out such a project, such as economic budgets, technological level, and whether it is a must for current scientific studies. So, it's too early to talk about manned landings on the moon for the time being." Quoted from "China has no timetable for a manned lunar landing". Xinhua News. 26 November 2007. Web. http://news.xinhuanet.com/english/2007-11/26/content_7149107.htm. Accessed 11 September 2013.

Several documents have also been published by Chinese scientists. In a paper published in *Planetary Science* in early 2008, for instance, Ouyang Ziyuan and other prominent scientists suggested that sending humans Moon-wards was one of the main long-term perspectives of China's space programme. In the eyes of the world, the fact that China's scientific community was envisaging the prospect of a manned lunar landing suggested at least the existence of an internal debate on the programme, although the result was to confuse the scientists' campaign with real government intent.

The biggest traction for this mounting international speculation was provided by then NASA Administrator Michael Griffin who, after visiting China in 2006, assiduously warned US policymakers about China's ability (and intention) to beat America back to the Moon.[117] On the Chinese side, there was still no official policy statement. The international space community would wait two more years to see the first policy document related to a manned lunar landing.

4.2.2 *Official Documents*

4.2.2.1 CAS Roadmap to 2050

The very first official document envisioning a manned lunar landing dates back to 2009. In that year, the Chinese Academy of Sciences, issued an important report entitled "*Science and Technology in China—A Roadmap to 2050*". This roadmap had been initiated 2 years earlier. It covers over a dozen key areas, including ecological agriculture, health, security, biology, genetics, and information technology. A specific, separate sub-report addresses long-term planning for developing S&T in the space field and identifies a series of strategic goals in three areas: space science, space applications, and space technology. Three time horizons—immediate, medium term, and long term—are given for their achievement, to be reached respectively by 2020, 2030, and 2050. Figure 4.3 illustrates the proposed roadmap for science, applications, and technology to 2050.

In the section on space science, a manned lunar landing is identified as a medium-term target to be reached by 2030. The section also envisages the establishment of a lunar base as the logical step for the following 10 years. A list of specific technologies that will need to be developed in order to provide the essential technical support for the successful implementation of a manned lunar exploration programme is clearly identified. Thus, within the general strategic goal of space

[117] He explained that a manned lunar mission by China could be achieved without developing a Saturn V-class launch vehicle like NASA's planned Ares V. In concrete, he believed that once China developed the LM-5 launch vehicle by 2014, it could quickly conduct a manned lunar circumnavigation mission followed shortly thereafter by a manned lunar orbit mission. These missions would then set the stage for a mission to land *taikonauts* on the surface of the Moon between 2016 and 2020.

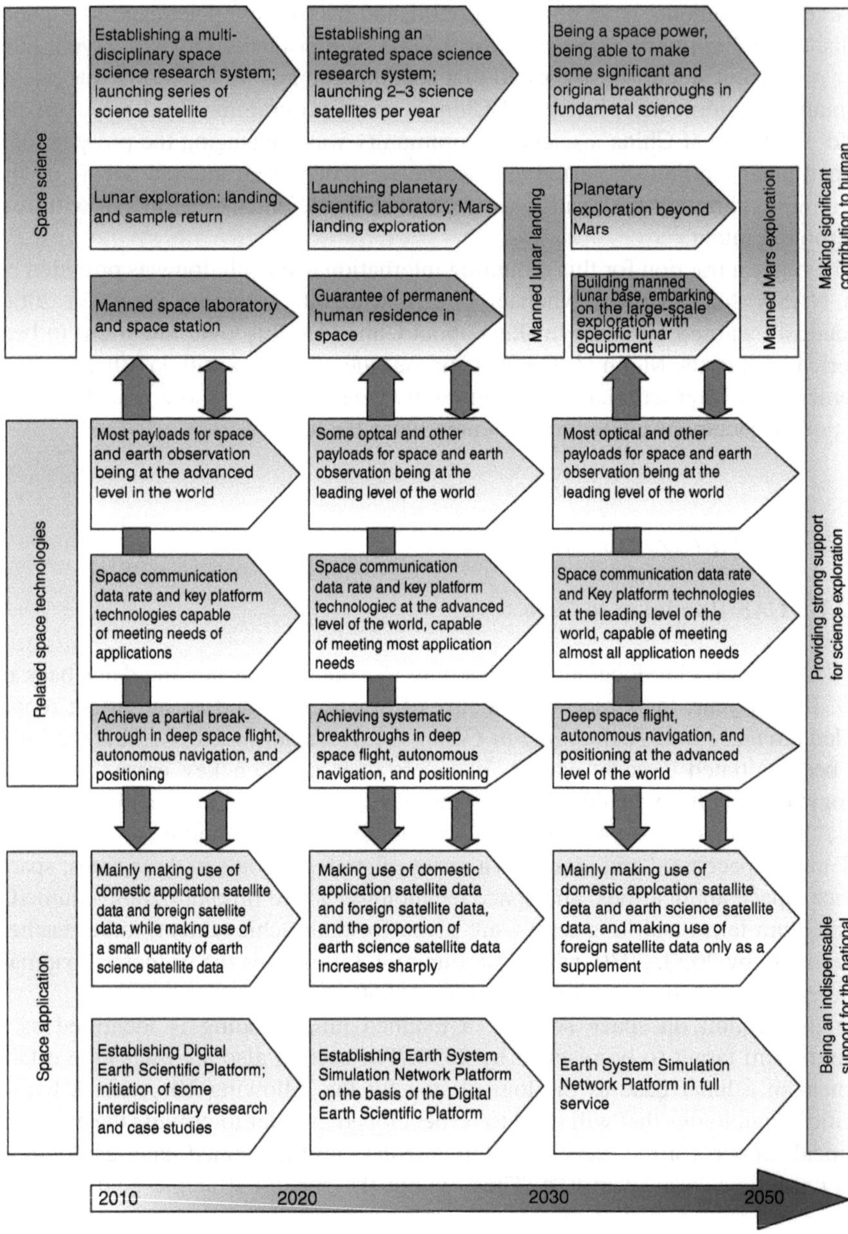

Fig. 4.3 CAS roadmap to 2050 (*Source*: CAS)

technology, the report identifies two strategic subgoals: namely, deep spaceflight technology and technologies for human residence in space (respectively, strategic subgoal 3.6 and 3.7 of the roadmap).[118] Deep spaceflight technologies that need to be developed include (a) autonomous navigation technology for spacecraft, (b) interplanetary propulsion technology, (c) energy-saving technology for interplanetary flight, and (d) power technologies for deep-space exploration. Concerning human residence in space, CAS identifies (a) technologies to guarantee a crew's long-term space exploration and (b) fire safety technology for manned spacecraft.

In spite of this clear statement of interest in a manned lunar landing by CAS, it is worth noting that the report does not represent the State Council's position: it is indeed just the roadmap proposed by an "autonomous" academic institution. However, as discussed in the previous chapter in the case of the Chang'e project, CAS scientists often exercise a substantial influence over the policymaking process in the national space programme. The unique position and status that CAS enjoys enable its scientists to gain attention at fairly high levels within government and give them the ability to push their ideas within the governmental agenda.

Interestingly, an analysis of the phrasing used in the document reveals highly political content that is clearly intended to influence Beijing policymakers. Thus, the document states: "in 2050, China may become the richest country in the world in terms of total GDP. China is thus supposed to take more responsibility in providing new knowledge by developing science and technology, making significant contributions to human civilisation and at the same time realising the great rejuvenation of the Chinese nation". If the government were to implement the proposed roadmap "the country will occupy a leading position in the world, such as many important discoveries and great breakthroughs will be made by Chinese, and China will also contribute a great deal to making Asia the epicentre for science and technology, thus the phrase 'made in China' will be replaced by 'created in China'."[119]

These arguments must have had a great impact on the mind-set of Beijing policymakers. Little surprise then that CAS advocacy of a human landing programme would soon be followed by a 2011 White Paper.

4.2.2.2 The 2011 White Paper and the 12th Five-Year Plan

At the end of 2011, China published its third space White Paper, "China's Space Activities in 2011".[120] As already mentioned, compared to previous versions, this

---

[118] Guo, Huadong, Ji Wu (eds) (2010). *Space Science and Technology in China: A Roadmap to 2050*. Chinese Academy of Sciences, Science Press (Springer), Beijing: pp. 63–68.

[119] *Ibid.* p. 7.

[120] Government of the People's Republic of China. China's Space Activities in 2011". White Papers of the Government of China. Beijing, China. 29 December 2011. Web. http://www.china.org.cn/government/whitepaper/node_7145648.htm.

one is clearer, as it provides straightforward information on short-term space priorities as well as some indications of the likely long-term ones.[121] To be sure, it still presents very bland statements. However, the possibility that a manned lunar mission could take place in the future is explicitly mentioned. Among the major tasks to be realised within the following 5 years, the document states that "China will conduct studies on the preliminary plan for a human lunar landing".[122]

This represents the first formal, official statement of interest in a manned lunar mission; as noted by Dean Cheng, "this is huge, since the white paper reflects governmental buy-in from the entire system".[123] Although the phrasing remains rather ambiguous, the bare fact of mentioning it may imply that—as happened for the manned spaceflight programme—the studies for a preliminary plan are already under way.

In addition, many of the principal commitments contained in the White Paper (e.g. completion of the Long March 5 by 2014, pre-research on a heavy launch vehicle (the Long March 9), the improvement of the TT&C network, construction of the new Hainan space port) are functional goals meeting the requirements of a manned lunar exploration programme. As asserted in the third section (Major Tasks for the Next Five Years) "in the next five years, China will strengthen the basic capacities of the space industry, accelerate research on leading-edge technology, and continue to implement important space scientific and technological projects, including human spaceflight, lunar exploration, [. . .] new-generation launch vehicles, and other priority projects in key fields". Concerning human spaceflight, the document states that "China will push forward human spaceflight projects and make new technological breakthroughs, creating a foundation for future human spaceflight".[124]

4.2.2.3 The Future

There are subtle indications that the final decision on whether to proceed with a manned lunar programme will be not made within the next Five-Year Plan (2016–2020), which will probably focus on CLEP's return missions (Chang'e-5

[121] David, Leonard. "China's New Space Exploration Vision Shoots for the Moon". Space.com. 20 January 2012. Web.http://www.space.com/14309-china-space-future-missions-moon.html. Accessed 23 September 2013.

[122] Government of the People's Republic of China. China's Space Activities in 2011". White Papers of the Government of China. Beijing, China. 29 December 2011. Web. http://www.china.org.cn/government/whitepaper/node_7145648.htm.

[123] "Scientist Targets 2024 for China's First Moon Walk". Space.com. 20 June 2006. Web. http://www.space.com/2515-scientist-targets-2024-china-moon-walk.html. Accessed 23 September 2013.

[124] Government of the People's Republic of China. China's Space Activities in 2011". White Papers of the Government of China. Beijing, China. 29 December 2011. Web. http://www.china.org.cn/government/whitepaper/node_7145648.htm.

Table 4.8 Critical skills for manned lunar exploration

	China	United States	Russia	Europe	Japan	India
Autonomous human spaceflight	P	D	D	L	L	L
Rendezvous	D	D	D	D	D	L
Docking	D	D	D	D	D	L
Extravehicular activity	D	D	D	L	L	L
Navigation to Moon orbit	P	D	D	D	D	D
Telemetry/control at the Moon	P	D	D	D	D	D
Human-rated heavy-lift launcher	P	D	D	L	L	L
Launch sites	D	D	D	D	D	D
Lunar lander	P	D	D	P	D	D
Re-entry capsule	P	D	D	L	L	L

D, fully developed or demonstrated; P, partially developed; L, lacking

and Chang'e-6) and preparation for the third phase of the manned spaceflight programme (launch of Tiangong-3 and completion/launch of the core module of the CSS). As suggested by US analyst Charles Vick,[125] the decision will more probably be put off to the 2021–2025 FYP, or even the 2026–2030 FYP, depending on the overall state of play in the technology development required for the implementation of the endeavour. Thus, no official name has been given to the programme yet. Chinese space professionals have referred to it as China's manned lunar landing programme (中国载人登月工程—*Zhôngguó zài rén dong yuè gônngchéng*),[126] which is thus the designation used within this book.

4.2.3 Skills and Hardware Development

Through the previously discussed precursor functional programmes, China has acquired or is acquiring most of the critical skills and hardware technologies to reach the ultimate target of a manned Moon landing. These capabilities are summarised in Table 4.8 and contrasted with the capabilities of other spacefaring nations.

As the table shows, relevant skills have already been mastered. However, there is still a number of critical hardware capabilities under development or only partially developed. These mainly include the launch system, the TT&C network for the Moon, and the lunar lander and re-entry capsule.

[125] Vick, Charles. "China's Crewed Lunar Landing & Lunar Base Aspirations". Global Security. 8 September 2009. Web. http://www.globalsecurity.org/space/world/china/piloted-lunar-landing.htm. Accessed 8 November 2013.

[126] 龙乐豪. 关于中国载人登月工程若干问题的思考. (Long Lehao. "On Issue of China Manned Lunar Exploration"). Missile and Space Vehicles No. 6. 2010. Web. http://wenku.baidu.com/view/21d5423a5727a5e9856a619b.html. Web 20 November 2013.

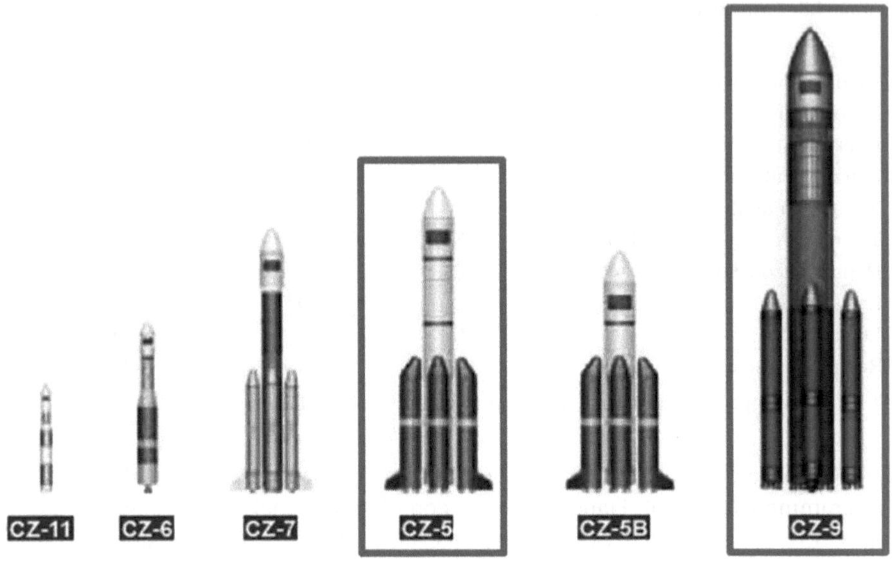

Fig. 4.4 The new Long March fleet

4.2.3.1 Launch Vehicle

The most discussed technology required is, perhaps, the future launch vehicle. A manned lunar landing requires a much improved launch capability. As early as 2004, China officially announced the development of a new heavy-lift launcher, the Long March 5, although the idea to develop it dates back to the early 1990s. As already mentioned, the LM5 is part of new modular fleet of launchers, which also comprises a light launcher (LM6, payload capacity of 10 tonnes to LEO), a medium-lift launcher (LM7, payload of 10–20 tonnes), and a super-heavy-lift launcher (LM9) (Fig. 4.4).

In its initial configuration, the Long March 5 (LM or CZ5), which will be similar in performance to the European Ariane 5, has been presented as a "55-m-tall rocket using liquid-hydrogen and liquid-oxygen main engines, flanked by four large strap-ons, weighing up to 800 tonnes, with a lift-off thrust of up to 1000 tonnes and able to place 23 tonnes in low Earth orbit or send 11 tonnes to geostationary orbit".[127]

Over the years, the rocket's design has undergone significant modifications, and eventually six different configurations have been developed, according to the different missions LM5 will be targeted to accomplish. The most powerful variant is the CZ5-B: it has a mass of 784 tonnes, two main stage engines and four large strap-ons with two engines each, and a payload capacity of 25 tonnes to LEO.[128] It is currently scheduled to become operational in 2015.

[127] *Cit. Ibid.* p. 362.

[128] "Changzheng 5". Dragon is Space. Web. http://www.dragoninspace.com/rocketry/cz5.aspx. Accessed 25 November 2013.

The development of LM5 has primarily been intended as the means of launching the large modules of the future Chinese space station, but, considering its liftoff thrust and payload capacity, it could hypothetically be used in the preparatory stages of a Moon landing[129] or to reach the Moon using an EOR mode, which implies assembling the manned lunar spacecraft in low earth orbit throughout multiple launches (see below).

However, it was clear from the beginning that the engines planned for this rocket would be too small and without the sufficient thrust for a "direct flight mode" or for a 1-launch LOR mode to the Moon. According to Li Peng, a top rocketry expert in China, to meet the requirements of a direct flight into the lunar injection orbit, at least a fourfold improvement in thrust of China's most powerful engine (the YF-100) is needed, up to 700 tonnes for the first stage, with a doubling of thrust on the upper stage.[130]

In spite of the lengthy delays in the development of the new engines of the LM5, in 2010, CALT Vice-President Liang Xiaohong confirmed that China was considering the development of a new launcher with a thrust at liftoff of 3000 tonnes.[131] This announcement revealed not only that China was seriously considering the idea of bringing its taikonauts to the Moon, but also that it was considering a single-launch flight mode as one way of reaching the target. The new launcher, which would fall into the same class as the American Saturn V and the Soviet N.1, eventually acquired the name of Long March 9 (LM9 or CZ-9). Although it is still at the design study stage, it is expected that, once approved for final development, one of the designs could "emerge for flight in 2020–2025 with the capability to launch Chinese astronauts to the surface of the Moon". [132]

In the preliminary design studies, the LM9 was to have a general liftoff thrust in the order of 3000 metric tonnes versus the Long March 5 liftoff thrust of 1000 tonnes. However, the early configuration of LM9 has undergone considerable change, since the envisaged liftoff thrust of 3000 tonnes would only have offered a "minimalist lunar expedition".[133] As recently as 2012, a more powerful configuration has been the subject of technical studies, with two options under

[129] According to Craig Covault, the Long March 5 appears positioned in the development flow to function like the US Saturn 1B rockets did in relation to the Saturn V in Apollo. That role was to use a powerful, but smaller launch vehicle to launch key elements of the program like the Apollo Command/Service Modules and Lunar Modules for test in Earth orbit. Covault, Craig. "First Look: China's Big New Rockets". America Space. 6 November 2012. Web. http://www.americaspace.com/?p=22881. Accessed 25 November 2013.

[130] Dingding, Xin. "Jumbo rocket design poses challenges". China Daily. 4 March 2013. Web. http://www.chinadaily.com.cn/china/2013npc/2013-03/04/content_16271514_2.htm. Accessed 23 November 2013.

[131] Perrett, Bradley (2010). "Longer Marches". Aviation Week & Space Technology Vol. 172 (11): 22–23.

[132] Covault, Craig. "First Look: China's Big New Rockets". America Space. 6 November 2012. Web. http://www.americaspace.com/?p=22881. Accessed 25 November 2013.

[133] Harvey, Brian (2013). *China in Space. The Great Leap Forward*. Springer, New York: p 362.

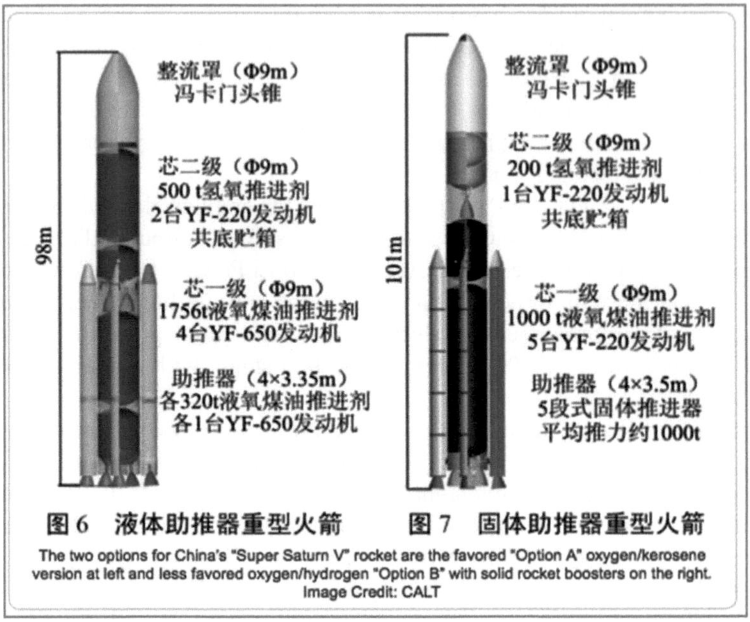

Fig. 4.5 The Long March 9 (options A and B)

consideration. The first option envisages an oxygen-/kerosene-powered rocket; the second oxygen/hydrogen engines with solid rocket boosters. Both are intended to generate more than 5000 metric tons of thrust (5200 for concept A, 5000 for concept B) and to have a payload capacity of 130 metric tonnes (Fig. 4.5).[134]

Several analysts have pointed out that "option A appears to be the preferred of the two options because its first stage uses liquid propellant strap on boosters, compared with 'Option B' that combines an oxygen/hydrogen core with solid rocket boosters, an area where China lacks experience".[135]

As part of the option A project, China has already started to develop "a large new oxygen/kerosene rocket engine called the YF-650 that stems directly from the Long March 5 in advanced production".[136] According to Vick, the YF-100, oxygen/kerosene engine with 120 metric tons of thrust for the new Long March 5, provides the technical basis for creating the 330-metric tons thrust YF-330 single thrust

[134] Vick, Charles P. "China's Long March 9 Manned Lunar Booster". Global Security. 2 September 2010 (updated 21 March 2013). Web. http://www.globalsecurity.org/space/world/china/cz-x.htm. Accessed 25 November 2013.

[135] Covault, Craig. "First Look: China's Big New Rockets". America Space. 6 November 2012. Web. http://www.americaspace.com/?p=22881. Accessed 25 November 2013.

[136] *Ibid.*

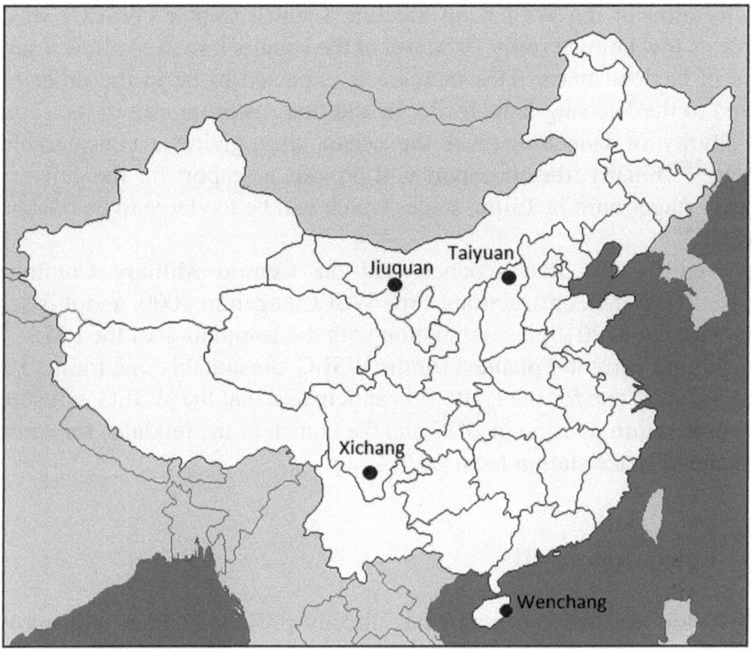

Fig. 4.6 China's launch site facilities

chamber engine. It in turn is being combined with a "second identical thrust chamber engine to create the YF-650 engine with 650 metric tons thrust".[137]

LM9 is currently in the final design development phase. It is expected that the development of the two engines will be completed in 2020 and that LM9 will become ready for launch between 2025 and 2030.[138]

4.2.3.2 Launch Site

A major infrastructural element of the future manned lunar exploration programme is obviously the launch site facilities. China already has three large launch sites (Jiuquan, Xichang, and Taiyuan),[139] but to meet the requirements of the new heavy launchers, it has started the construction of a new spaceport on the coast of China's southernmost island of Hainan. Figure 4.6 displays the location of China's current and forthcoming launch site facilities.

[137] Vick, Charles P. "China's Long March 9 Manned Lunar Booster". Global Security. 2 September 2010 (updated 21 March 2013). Web. http://www.globalsecurity.org/space/world/china/cz-x.htm. Accessed 25 November 2013.

[138] *Ibid.*

[139] Jiuquan is located in the Gobi desert, Xichang in Sichuan, and Taiyuan in the Shaanxi province, southwest of Beijing. Most of the launches have been performed by the Xichang launch site.

The location of the Wenchang Satellite Launch Center (WSLC) was chosen firstly for its low latitude (only 19° north of the equator), so as to allow a substantial increase in payload mass. This increase is expected to be in the order of 7.4 % compared to the Xichang launch site. In addition, its proximity to the shore offers the possibility of launching over the ocean, thus giving a considerable safety advantage.[140] Finally, the spaceport will provide a seaport for the delivery of the new rocket stages built in Taijin, stages which will be too large to be transported on rail tracks.[141]

Approved by the State Council and the Central Military Commission in September 2007, the construction of the WSLC began in 2009, and it is scheduled to be operational in 2015, in conjunction with the completion of the LM5.[142] Three launch pads are currently planned for the WSLC, presumably one for the LM5, one for the LM7, and one for the LM9. It is anticipated that the WSLC will support the lunar sample return mission in 2017 and the launch of the modules for construction of the manned space station from 2020.[143]

4.2.3.3 Lunar Spacecraft

The Shenzhou manned spacecraft has already provided China with most of the required spacecraft-related hardware to embark upon a manned lunar mission. Recall that it was originally based on Soyuz, which was specifically designed in the 1960s for flight to and return from the Moon. Like Soyuz, Shenzhou comprises three modules: an orbital module at the front, a re-entry capsule in the middle, and a service/propulsion module at the back. This modular configuration already makes Shenzhou suitable for supporting the rendezvous and docking operations required by a manned lunar landing mission. In addition, this division offers considerable weight (and fuel) advantages, as it minimises not only the amount of material needed to land on (and liftoff from) the Moon but also the material needed for the return to Earth, in particular related to heat shields, retro-propulsion systems, parachutes, and other equipment required for landing. This means that the different modules can be tailored independently. Further specifications of the Shenzhou modules, as relevant for a manned lunar landing, are presented in Table 4.9.

Shenzhou modules could easily be adapted to develop a future lunar variant, together with a scaled-up (modified) version of the lunar lander already developed for the Chang'e-3 mission, and that will be further advanced with the Chang'e-5

---

[140] Harvey, Brian (2013). *China in Space. The Great Leap Forward.* Springer, New York: p. 81.

[141] Lan, Chen. "Hainan Satellite Launch Site". Dragon in Space. 28 January 2013. Web. http://www.dragoninspace.com/facility/hainan.aspx. Accessed 30 November 2013.

[142] Vick, Charles. "Hainan/Wenchang". Global Security. 21 June 2010. Web. http://www.globalsecurity.org/space/world/china/sanya.htm. Accessed 25 November 2013.

[143] Lan, Chen. "Hainan Satellite Launch Site". Dragon in Space. 28 January 2013. Web. http://www.dragoninspace.com/facility/hainan.aspx. Accessed 30 November 2013.

Table 4.9 Shenzhou specifications

	Orbital module	Re-entry module	Service module
Mass	1500 kg	3240 kg	3000 kg
Length	2.80 m	2.50 m	2.94 m
Basic diameter	2.25 m	2.52 m	2.50 m
Design life	200 days	20 days	20 days
Habitable volume	8.00 m^3	6.00 m^3	–
Engine	4	–	4 × 2500 N

sample return mission. The scaleup will mainly require the expansion of the small lander of 1200 kg into a module of roughly 10 tons.

According to the US analyst Mark Wade, the development of a Shenzhou-derived lunar spacecraft, using an Earth Orbital Rendezvous mode (EOR, see below), would presumably expand the total mass of the spacecraft to 39 metric tons and consist of:

- A lunar landing stage of 28-metric tons gross [as opposed to the 10 tons proposed by others (4.5-metric tons empty)], including landing gear, which would land the spacecraft on the lunar surface and form the launch platform for the return spacecraft
- A Shenzhou-derived return spacecraft/ascent stage of 11-metric tons gross (5.5-metric tons empty), which would comprise the 1-metric ton orbital module (adopted for use as a cockpit for the crew during the landing manoeuvre), the 3-metric ton Shenzhou re-entry module (for two, possibly three, crew members), and a modified service module (7 metric tons including 5.5 metric tons of propellants)[144]

It can be expected that developing the spacecraft's hardware will pose few technological challenges, although a number of spacecraft-related techniques still need to be tested. The Chang'e lunar exploration programme is, however, providing valuable support in this regard. Numerous key techniques, such as injection into lunar orbit; orbital manoeuvres; coordination capacity between navigation, control, and propulsion systems; and the automatic soft landing, have already been mastered and demonstrated by the three Chang'e missions implemented so far. Other relevant techniques (e.g. automatic rendezvous in Low Lunar Orbit, ascent stage from the lunar surface, re-entry-related techniques) will probably be tested and refined during the upcoming Chang'e-5 and Chang'e-6 missions, respectively, planned for 2017 and 2019.

[144] Wade, Mark. "Chinese Lunar Base". Encyclopedia Astronautica. Web. http://www.astronautix.com/craft/chirbase.htm#more. Accessed 26 November 2013.

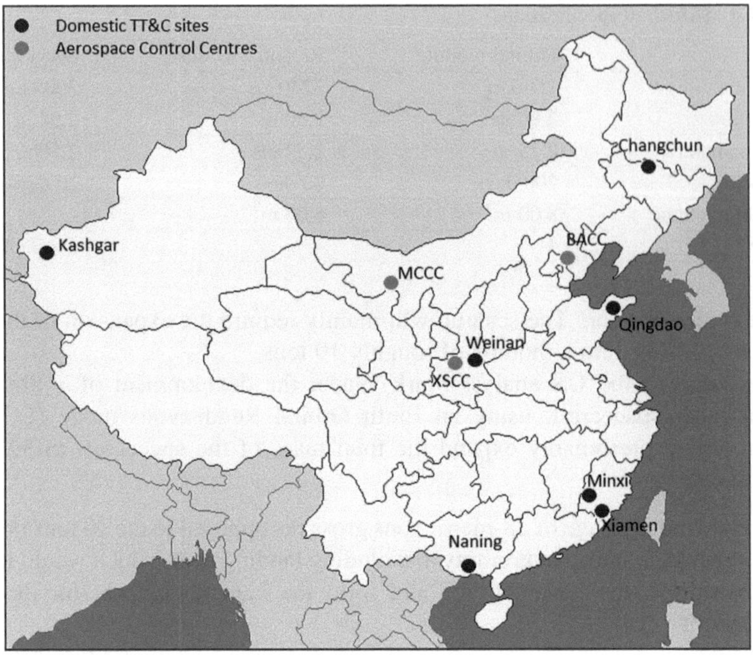

Fig. 4.7 China's domestic tracking infrastructure

4.2.3.4 TT&C and Communications Network

Additional technological skills that need to be developed for the successful implementation of a manned lunar landing programme are related to the TT&C network and communications system.

In this regard, China has already set up a unified and fairly good land-, sea-, and space-based TTC network for manned spaceflight, which is up to international standards and has been networked internationally. Consistent efforts have been made to improve the range of China's telemetry capabilities for the *Chang'e* lunar missions from 80,000 km to 400,000 km. This is the distance to the Moon in apogee. Further efforts will still need to be made in order to ensure TT&C and communication continuity.

At present, China's tracking infrastructure consists of "three aerospace command and control centres, a number of fixed and mobile TT&C stations, several TT&C stations abroad, five '*Yuan Wang*' tracking ships and three '*Tianlian*' relay satellites".[145] China's domestic tracking infrastructure is represented in Fig. 4.7.

[145] Vick, Charles. "Hainan/Wenchang". Global Security. 21 June 2010. Web. http://www.globalsecurity.org/space/world/china/sanya.htm. Accessed 25 November 2013.

Table 4.10 China's domestic station capabilities

	LEO	SSO	GEO	Manned spaceflight	Lunar exploration
Weinan	×	×	×	×	
Changchun	×	×			
Kashgar	×	×		×	
Minxi	×	×	×		
Xiamen	×	×	×	×	
Naning	×	×			
Qingdao	×			×	×
Mobile station 1	×			×	
Mobile station 2	×			×	

The three command and control centres, namely, Beijing Aerospace Control Center (BACC), Dongfeng Mission Command and Control Centre (MCCC), and Xi'an Satellite Control Center (XSCC), support manned spaceflight with complementary functions and tasks. The Dongfeng MCCC, located within the Jiuquan Satellite Launch Center, is mainly responsible for monitoring and managing the spacecraft during the launch and ascending stage of the flight. Once the spacecraft enters orbit, BACC takes over control of the flying spacecraft and remains in charge throughout the mission. The Xi'an Center (XSCC) acts as the communications hub, command, control, and main data processing centre, as well as administrative headquarters of China's TT&C network. [146]

As of 2013, China can rely on seven fixed ground stations and two mobile stations within its territory. These stations have different transmission capabilities, which are summarised in Table 4.10. At present, only four of the seven ground stations are able to support human spaceflight missions, while only the Qingdao tracking station's transmission capabilities have sufficient range to support a lunar exploration mission.

In order to overcome the initial lack of a worldwide network of ground-based tracking stations and to assist manned space programme and deep-space missions, China relies on a series of tracking and communications ships, named *Yuan Wang* (literally "long view"). The fleet, built between 1978 and 2008, currently comprises five tracking ships,[147] equipped with c-band and S-band antennas, arrays, satellite dishes, computer and control rooms, and support communications in the ultra-long and ultrahigh frequency bands.[148] Their coverage, however, is quite limited and they are expensive to operate. In addition, the meteorological conditions of the southern hemisphere's oceans potentially inhibit ship-based tracking of manned spaceflight missions from April to October.

[146] Lan, Chen. "Xi'an Satellite Control Centre". Dragon in Space. 14 January 2013. Web. http://www.dragoninspace.com/facility/xian.aspx. Accessed 27 November 2013.

[147] A total of six *Yuan Wang* has been built, but the two of them have been decommissioned and replaced by the new Yuan Wang 5 and 6.

[148] Harvey, Brian (2013). *China in Space. The Great Leap Forward.* Springer, New York: p. 65.

Given these numerous disadvantages, for the past decade, China has made consistent efforts to build overseas stations and sign mutual access agreements with third countries and organisations. The first Chinese overseas station was built on South Tarawa Island (Kiribati) in 1997, but following the recognition of the Republic of Taiwan by the Kiribati government, the PRC decided to dismantle it. In 2000 an agreement was signed with the Namibian government and a station was built in Swakopmund in 2001.[149] Two years later, China added a station in Karachi (Pakistan) and Malindi (Kenya). More recently, the stations in Dongara (Western Australia) and Santiago (Chile) were also made available to China and used to track China's space laboratory Tiangong-1.[150]

The *Tianlian* relay satellites (literally, "sky link") are an additional element in China's tracking infrastructure. They are specifically intended to support the human spaceflight programme—and in particular the operations of Tiangong and the future space station. They provide low-, medium-, and high-speed data communications between the spacecraft and ground stations in Ka- and S-band. The first Tianlian-1 satellite was launched in April 2008 to support the flight of Shenzhou-7, while Tianlian-1-02 was launched in July 2011 (before the launch of the Shenzhou 8 mission) and Tianlian-1-03 in July 2012.[151] Being positioned in GEO, the three satellites can provide nearly 100 % global coverage, but a second-generation series (Tianlian-2) is nonetheless being developed by CAST in order to provide higher capacity and longer service life.

Overall, the land-, sea-, and space-based tracking infrastructure China can now rely on is remarkable, as it allows the country to satisfy the increasing needs of its space programme. However, in spite of this emerging global network, China still needs to enhance its long-range TT&C capabilities and upgrade the existing infrastructure in order to meet the requirements of a manned lunar landing mission. With the current capabilities, the tracking infrastructure would not be able to ensure communication continuity between the spacecraft and the ground system for the entire duration of the mission. As recently reported by a chief engineer at the BACC, "there are still eight to ten hours a day during which China cannot track its deep space detectors. It is thus imperative to build a deep space monitoring station abroad in order to make up for blind measurements and realize round-the-clock monitoring for future deep space missions". [152]

---

[149] Vick, Charles. "Hainan/Wenchang". Global Security. 21 June 2010. Web. http://www.globalsecurity.org/space/world/china/sanya.htm. Accessed 25 November 2013.

[150] For tracking its Tiangong-1, China is relying on eight overseas stations: Swakopmund (Namibia), Malindi (Kenya), Karachi (Pakistan, 2003), Santiago (Chile), Alcantara (Spain), Aussaguel (France), and Kerguelen Islands and Dongara (Australia).

[151] "China launches Tianlian-1-03 to expand Space-Based Communication System". Spaceflight 101. 25 July 2012. Web. http://www.spaceflight101.com/tianlian-1-3-launch.html. Accessed 27 November 2013.

[152] Quoted from "Deep space monitoring station abroad imperative". Space Daily. 17 December 2013. Web. http://www.spacedaily.com/reports/Deep_space_monitoring_station_abroad_imperative_999.html. Accessed 28 November 2013.

As the 2011 White Paper has noted, one of the main tasks China will have to face in the next 5 years is to "improve its space TT&C network, build [new] deep-space TT&C stations, develop advanced TT&C technologies, and enhance its TT&C capabilities in all respects to satisfy the demands for remote TT&C".[153] International cooperation in this field will presumably also be pursued, as happened with the recent Chang'e-3 mission, in which ESA—thanks to its ESTRACK network—played a crucial role.[154] For China, the possible use of ESA's deep-space stations in Spain and Perth is currently under consideration.

4.2.4 Programme Systems, Organisational Structure, and Mission Configuration

4.2.4.1 Programme Systems

Looking at the constituents of the manned spaceflight programme and lunar exploration programme can provide valuable indications as to how the hardware development for a manned lunar landing mission will be organised and managed.

Although no official plan has been published and the lunar endeavour is still in its concept study stage, it can be expected that the programme will consist of seven main systems: launch vehicle, taikonauts, lunar lander and re-entry capsule, space applications, TT&C network, launch site, and recovery site. In a similar way to the *Shenzhou* and CLEP programmes, these systems will be developed by different stakeholders, as suggested in Table 4.11.

4.2.4.2 Programme Organisation

As for the overall organisational structure of the programme, it is likely that first a manned Lunar Exploration Program Office will be set up within the CCP Central Committee, as happened in 2004 for the Chang'e programme. Immediately under the authority of this office, the Human Spaceflight Project Office (HSPO)—or alternatively an ad hoc Leading Small Group (*Lingdao Xiaozu*) with specific responsibility and functions—will be established to form the core programmatic leadership of the programme.

As for the HSPO, the LSG and its related office will stay at the very top of the chain of command and will report back directly to the manned Lunar Exploration Program Office, the State Council, and the CCP's Central Committee. It would

---

153 Government of the People's Republic of China. China's Space Activities in 2011". White Papers of the Government of China. Beijing, China. 29 December 2011. Web. http://www.china.org.cn/government/whitepaper/node_7145648.htm.

154 "Helping China to the Moon". European Space Agency. 29 November 2013. Web. http://www.esa.int/Our_Activities/Operations/Helping_China_to_the_Moon. Accessed 30 November 2013.

Table 4.11 Programme systems

Components	Responsibility	Contractors
Launch vehicle (Moon Project-1)	Development of the CZ5 and CZ-9 launch vehicles	China Academy of Launch Vehicle Technology (CALT)
Taikonauts (Moon Project-2)	Taikonaut selection and training; taikonaut medical monitoring and support; development of spacesuits; spacecraft's life support and environment control system	507th Institute of the PLA/Astronaut Centre of China
Lunar lander/re-entry (Moon Project-3)	Development of the *Shenzhou* lunar lander and re-entry capsule	China Academy of Space Technology (CAST) Shanghai Academy of Spaceflight Technology (SAST)
Space applications (Moon Project-4)	Onboard scientific experiment packages	CAS
Launch site (Project-5)	Construction and operations of the manned mission launch sites	Hainan Spaceport
TT&C system (Project-6)	The operations of the spacecraft tracking and communications network	Xi'an, Dongfeng, and Beijing Satellite Control Centres
Recovery site (Project-7)	The operations of the spacecraft recovery system	PLA General Armaments Department

normally be made up of a board of directors (with a likely key leadership position assigned to a General Armaments Department official) and would comprise a two-line management structure. Along one line would be the technical decision-makers, with the chief designer and chief engineer at the top, and a series of technical experts at lower levels. Along the administrative or managerial line would sit officials of the various government and military organisations and aerospace contractors involved.[155] A joint meeting mechanism, responsible for decision-making on relevant issues during the implementation of the project, would be established between the two lines. Following the decisions taken at the joint meetings, the China Manned Space Agency (CMSA) would manage the project systematically through the entire process and would coordinate the work of different departments.

Under the authority of CMSA and the LSG, there would be seven technical committees, each responsible for managing one of the seven key subsystems of the

[155] The positions of chief commander and deputy chief commander will be likely taken by officials of the GAD, MIIT, CAS, and CASC, while the positions of chief designer and deputy chief designer will be likely taken by technical experts.

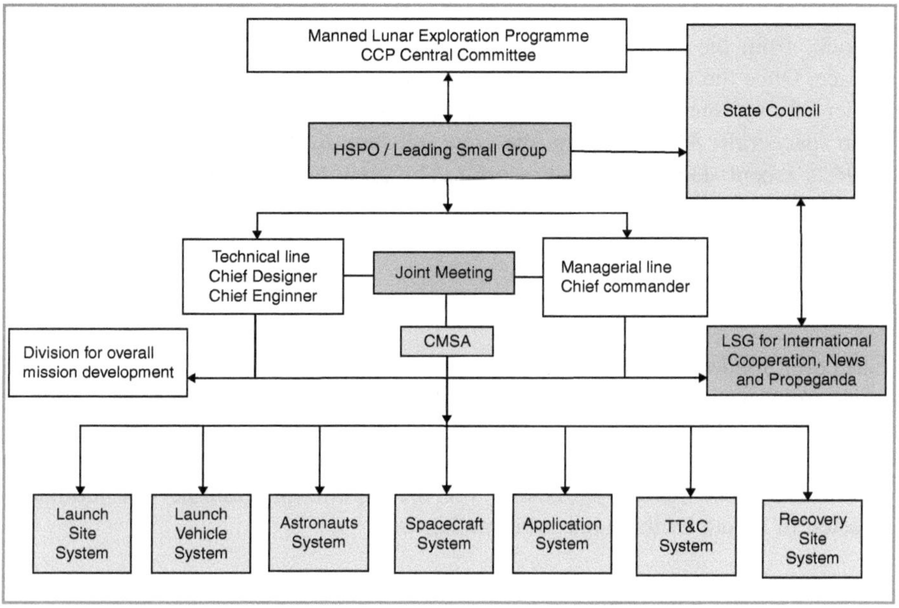

Fig. 4.8 Possible organisation for a manned lunar exploration (author's visualisation)

programme. Each committee would also be managed in the same two-line system, with a director for administrative management and a chief designer and a chief engineer for technical decision-making.

Presumably, an additional ad hoc LSG for news and propaganda and for international cooperation will be established, reporting directly to the State Council and to the Central Committee of the CCP. It will coordinate the news releases and media work as well as the implementation of potential cooperative undertakings (Fig. 4.8).

4.2.4.3 Mission Configuration

It is likely that China has already initiated the study of the most appropriate configuration for carrying out its manned lunar landing. Several "modes" could be relevant. These are briefly explained below, taking worldwide experience and programme proposals into account.

Lunar Orbit Rendezvous

Lunar orbit rendezvous (LOR) was the methodology utilised by the Apollo programme and planned for the Soviet N1-L3 programme. In a LOR configuration, the lunar lander and the main spacecraft are delivered in one or several launches into Low Lunar Orbit (LLO).[156] While in orbit around the Moon, the lunar lander undocks from the main spacecraft and then independently descends to the lunar surface. Once the mission on the Moon is completed, the ascent stage of the lunar lander lifts off again into lunar orbit and performs an orbital rendezvous with the main spacecraft. After docking, the crew transfers to the re-entry capsule and the lander's ascent stage is ditched in orbit. The crew finally returns home using the re-entry capsule.[157]

Earth Orbit Rendezvous

Earth orbit rendezvous (EOR) makes use of multiple launches (proposals generally swing from two to four) to put the different components of the lunar spacecraft in an Earth Parking Orbit, then assembles them through the use of rendezvous and docking, and finally pursues a direct flight to the Moon. The EOR methodology was initially considered by NASA for the Apollo programme but ultimately rejected in favour of the lunar orbit rendezvous.

Earth Orbit Rendezvous–Lunar Orbit Rendezvous

This methodology combines the EOR and LOR modes, launching the different modules with several launches (from three to four) to rendezvous, dock and assemble in both Earth and Moon parking orbits. In a 3-launch configuration, a heavy-lift launcher would first transport the lunar lander into an LLO (or alternatively LEO). A second and third launch would then carry a trans-lunar injection (TLI) stage and the spacecraft into LEO, where they would rendezvous and dock, before heading to the LLO. Once the LLO was reached, the spacecraft would rendezvous with the lunar lander. The lunar lander would then undock from the spacecraft and independently descend to the lunar surface. Similar procedures to the LOR mode would be performed for the ascent stage and return to Earth. The EOR–LOR configuration has recently been proposed by NASA for its "Constellation" programme, as well as by the Russian RKK Energia, the EADS Astrium, and

[156] While in a 1-launch mode the different modules would fly together into lunar orbit, in a multiple launch mode, they would primarily need to perform rendezvous and docking in a lunar parking orbit for complete assembly. See Li, Wenqin *et al* (2013). "Manned Lunar Launching Mode and the Requirement for Heavy Launch Vehicle". Proceedings of the 64th International Astronautical Congress, Beijing, China, September 23-27, 2013. Paper: IAC-13-D2.P.25.

[157] Shing-Yik, Yim *et al*. "Current Development of Manned Lunar Landings". Proceedings of the 63rd International Astronautical Congress, Naples, Italy, October 1-5, 2012. Paper: IAC-12. A5.1.10.

the Japanese Mitsubishi Heavy Industries, as the optimal solution for future manned lunar landings.[158]

Direct Flight

A direct flight mode envisages no Earth–Moon orbit "staging node" and no rendezvous but only a single rocket launch. In a direct ascent configuration, a huge rocket would launch the spacecraft directly to the lunar surface, where it would land tail first and then launch off the Moon back to Earth. The United States and former Soviet Union initially proposed direct flight mode for their manned lunar landing programmes, based on rockets having a payload capacity of more than 200 tons.[159] In both cases, the idea was soon rejected in favour of an LOR mode.

4.3 Concluding Remarks

There is still no indication as to which configuration China might eventually intend to utilise for landing its taikonauts on the Moon. Although a direct flight would probably seem the most natural way to travel to the Moon and back, in fact, it would involve extremely difficult technological feats that still make it the least appealing option.[160]

Presumably, consideration has also been given to an EOR mode, since that would not require a Saturn V-class launcher, but a couple of the less powerful LM5s. The EOR mode would have to be performed with multiple launches to ensure that the weight of the payload would be within the capabilities of the LM5. In the 2010 paper, "On Issues of China Manned Lunar Exploration", Long Lehao—former vice-director of the Chinese lunar exploration programme—identified the EOR configuration as the most suitable option (among the four) for a Chinese lunar landing mission.[161]

---

[158] Li, Wenqin *et al* (2013). "Manned Lunar Launching Mode and the Requirement for Heavy Launch Vehicle". Proceedings of the 64th International Astronautical Congress, Beijing, China, September 23-27, 2013. Paper: IAC-13-D2.P.25.

[159] Shing-Yik, Yim *et al.* "Current Development of Manned Lunar Landings". Proceedings of the 63rd International Astronautical Congress, Naples, Italy, October 1-5, 2012. Paper: IAC-12. A5.1.10.

[160] For a detailed description of NASA considerations regarding the flight configuration of the Apollo programme, see Brooks, Courtney G., James M. Grimwood, and Loyd Swenson (1979). "Chariots for Apollo: A History of Manned Lunar Spacecraft". National Aeronautics and Space Administration. Web. http://history.nasa.gov/SP-4205/ch3-2.html. See also Hansen, James R (1992). "The Rendezvous That Was Almost Missed: Lunar Orbit Rendezvous and the Apollo Program". NASA Langley Research Center Office. Web. http://www.nasa.gov/centers/langley/news/factsheets/Rendezvous.html. Accessed 4 December 2008.

[161] They identified four different approaches to the Moon, most of which require multiple launches of Long March 5 boosters. Two of the methods would require three launches. See Day, Dwayne

More recently, an EOR–LOR combination has been recommended instead by a group of scholars from Tsinghua University in a paper issued at the 2012 International Astronautical Congress (IAC). NASA has also envisaged the use of an EOR–LOR mode as the most likely configuration for a Chinese manned lunar landing programme. In its notional concept of operations, NASA has estimated that such a mission will require four LM5 launches. The first two launches would carry a TLI and the lunar lander into LEO, where they would dock and head to a lunar parking orbit. The third and fourth launches would subsequently deliver another TLI and the Shenzhou spacecraft into LEO. The two would also rendezvous and then head to the lunar parking orbit. There, the Shenzhou spacecraft would rendezvous and dock with the lunar lander, which would eventually be utilised by two taikonauts for the descent to the Moon. [162]

However, the recently announced development of the LM9 now subtly suggests that China may also intend to pursue a manned lunar landing through an LOR mode. Thanks to the significant improvement in the liftoff thrust of the forthcoming LM9 (from 3000 to 5200 metric tons of thrust), potentially the LOR could be performed by a single launch, similar to the Apollo missions. According to several experts of the China Academy of Launch Vehicle Technology (CALT), in a paper issued at the 2013 IAC, a single-launch LOR configuration is preferable, as it will avoid "the risks of multiple rendezvous and docking operations and has higher reliability".[163] In addition, a major advantage of LOR compared to direct flight is that only the small, lightweight lunar lander, and not the entire spacecraft, would have to land on the Moon.

Although no indication has been given yet, the LOR methodology is presumably becoming the preferred/most likely configuration. One reason for this view is that a robotic LOR is planned for the Chang'e-5 and Chang'e-6 lunar sample return missions. These two missions could serve as pathfinders for testing the techniques required by a future manned LOR mission. Potentially, a manned lunar circumnavigation, similar to that performed in 1968 by Apollo 8, could also follow and test lunar orbit manoeuvres and re-entry-related techniques.[164]

A. "The new path to space: India and China enter the game". The Space Review. 13 October 2008. Web. http://www.thespacereview.com/article/1231/1. Accessed 5 December 2013. See also Pace, Scott. "China's Human Spaceflight Program: Achievements and Prospects". Presentation at the Heritage Foundation. Washington DC. 8 October 2008. Web. http://www.thespacereview.com/archive/1231a.pdf. Accessed 5 December 2013.

[162] Pace, Scott. "China's Human Spaceflight Program: Achievements and Prospects". Presentation at the Heritage Foundation. Washington DC. 8 October 2008. Web. http://www.thespacereview.com/archive/1231a.pdf. Accessed 5 December 2013.

[163] Wenqin, Li *et al* (2013). "Manned Lunar Launching Mode and the Requirement for Heavy Launch Vehicle". Proceedings of the 64th International Astronautical Congress, Beijing, China, September 23-27, 2013. Paper: IAC-13-D2.P.25.

[164] Vick, Charles. "*Shenzhou Circumlunar Missions*. Global Security". 4 September 2009. Web. http://www.globalsecurity.org/space/world/china/piloted-lunar.htm. Accessed 25 November 201.

Chapter 5
China and the Moon: Endogenous Conditioning Factors

Concrete plans and strong motivations for reaching the Moon are not, on their own, sufficient for China to send its taikonauts there. The high complexity of the manned lunar exploration programme involves a number of conditioning factors and prerequisites that must be fulfilled in order to succeed in this endeavour.

By considering "China going to the Moon" as a dependent variable, it is necessary to identify the series of independent variables that could ultimately affect Chinese capacity to carry out its manned lunar exploration programme.

This book has identified four macro-variables influencing China's space ambitions: socio-economic, political, technological, and international variables. This chapter will focus on what can be regarded as the endogenous conditioning factors, while Chap. 6 will assess the international ones. Far from predicting China's future in each of these domains, the following sections aim to assess why, how, and to what extent the considered variables could affect China's lunar exploration programme. These factors will be summarised in the last section, which will also try to answer the question of whether China can or cannot go to the Moon on its own and assess why it might not be willing to embark upon a solo mission.

5.1 Socio-economic Variables

One of the most evident and influential variables affecting Chinese plans to go to the Moon is economic in nature and deals with the need for China to continue its economic growth throughout the next decade. In fact, an indispensable precondition for the success of the manned lunar exploration programme is steady and increasing investment during the implementation phase of the programme. It goes without saying that this expenditure will be more easily supported if the curve of China's economic growth shows an upward trend.

Historical comparisons are often misleading; however, it is worth noting that, at the peak of the Apollo programme, the United States was investing almost 0.8 % of

M. Aliberti, *When China Goes to the Moon...*, Studies in Space Policy 11,
DOI 10.1007/978-3-319-19473-8_5

its GDP.[1] Such huge expenditure was ultimately made possible and politically assisted by the fact that during the 1960s the US economy was expanding rapidly.

Although China is one of the fastest growing countries in the world and now has a GDP that in real terms is equivalent in size to that of the United States in the late 1960s,[2] how sustainable and enduring this growth will be should be investigated and an assessment made of whether the country can succeed in executing such a highly expensive programme. To address these issues, a concise overview of China's growth path will first be presented, which will in turn be used to analyse the country's future potential and challenges. Subsequently, the way these challenges will affect the country's plans to go to the Moon will be assessed.

5.1.1 Overview of China's Growth Model

For more than three decades, China has been the world's fastest growing major economy. The scale, scope, and speed of its economic growth are unprecedented. Since 1978, GDP has grown by an average of about 10 % a year (with several peaks of more than 14 % a year), thus allowing the country to experience rapid economic and social development, to progressively overtake the world's economic powerhouses, and to play an increasingly influential role in the global economy. With a population of 1.35 billion and a GDP of $9240 trillion in 2013,[3] China recently became the second largest economy in the world. It also succeeded in overtaking the United States as the world's largest manufacturer and exporter, in having world primacy in foreign direct investment (FDI) inflows and amount of foreign reserves, and in lifting more than 600 million people out of poverty.[4]

If one considers that at the beginning of the 1970s the country was near bankruptcy, social instability was widespread and the political system was close to collapse, these results look even more impressive (Fig. 5.1).

Many and sometimes divergent explanations have been offered for this miraculous economic performance. It should first be recalled that, since the "*gaige*

[1] According to the official records (NASA website: http://history.nasa.gov/Apollomon/Apollo.html), NASA budget increased from $500 million in 1960 to a high point of $5.2 billion in 1965. In that year, NASA funding level represented 5.3 % of the federal budget and roughly the 0.8 % of the US GDP. In total, between 1959 and 1973, NASA spent $23.6 billion on human spaceflight, exclusive of infrastructure and support, of which nearly $20 billion was for Apollo (equivalent to $110 billion in 2010 terms).

[2] This comparison has been made though a computational analysis provided by Wolfram Alpha website. See http://www.wolframalpha.com/input/?i=United+States+real+GDP+1969. China has a GDP of $8221 trillion in nominal terms and of $4522 trillion in real terms. United States in 1969 had a GDP of $1041 trillion in nominal terms and of $4711 trillion in real terms.

[3] "Country at a glance: China". The World Bank (2013). Web. http://www.worldbank.org/en/country/china. Accessed 17 November 2014.

[4] See Appendix A for additional economic indicators on China.

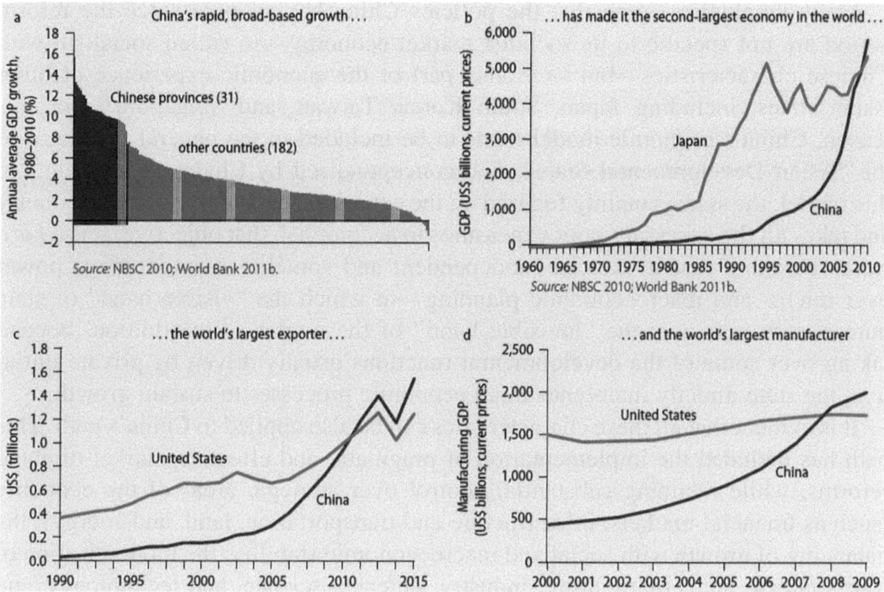

Fig. 5.1 China's impressive economic performance

kaifang" reforms launched by Deng Xiaoping, China has been moving from a closed and centrally planned socialist state towards an open and market-oriented economy, increasingly integrated into the global economy. It would, however, be misleading to interpret this shift as a conversion towards the neoliberal policies promoted by the so-called Washington Consensus institutions—the World Bank, the International Monetary Fund (IMF), and now also the World Trade Organization (WTO). The policy features of the Washington Consensus—including "indiscriminate opening to trade and capital flows; increasingly independent central banks; tighter and pro-cyclical fiscal and monetary policies"[5]—did not at all fit the country's developmental needs.[6] China refused therefore to embrace them and the state continued to play a fundamental role in creating and managing the conditions for growth.[7]

[5] Cit. Wiesbrot, Mark. "2016: When China overtakes the US". The Guardian. 27 April 2011. Web. http://www.theguardian.com/commentisfree/cifamerica/2011/apr/27/china-imf-economy-2016. Accessed 28 August 2013.

[6] Mazzei, Franco, Vittorio Volpi (2010). *La rivincita della Mano Visibile. Il Modello economico asiatico e l'Occidente*. Egea, Milano: p. 140.

[7] As also underlined by some analysts, it is true that China's rapid growth in the last three was assisted by a steady integration with the global economy and a rapid expansion of trade and foreign investment. But all these factors were in fact carefully managed by the state, to make sure that they fitted in with the government's development goals. Wiesbrot, Mark. "2016: When China overtakes the US". The Guardian. 27 April 2011. Web. http://www.theguardian.com/commentisfree/cifamerica/2011/apr/27/china-imf-economy-2016. Accessed 28 August 2013.

Many sinologists agree that the policies China has adopted since the reform period are not specific to its socialist market economy—so-called socialism with Chinese characteristics—but were also part of the economic experience of other Asian states, including Japan, South Korea, Taiwan, and Singapore.[8] For this reason, China's economic model tends to be included in the general paradigm of the "Asian Developmental State", first conceptualised by Chalmers Johnson.[9] In this model, the state is mainly focused on the economic development of the country and takes all the necessary policy measures to accomplish that objective. It is a *hard state*—meaning that it exercises independent and sometimes autonomous power over micro- and macroeconomic planning—in which the "visible hand" of state bureaucracy acts over the "invisible hand" of the market.[10] In addition, beyond taking over some of the developmental functions usually driven by private initiative, the state directly intervenes in all economic processes to sustain growth.

It is evident that all these characteristics can be also applied to China's path. This path has included the implementation of pragmatic and effective market-oriented reforms, while retaining substantial control over strategic areas of the economy (such as financial markets, infrastructure and transportation, land, and energy); the balancing of growth with social and macroeconomic stability; the modernisation of four strategic areas (agriculture, industry, defence, science, and technology); and the adoption of an export-oriented industrialisation (EOI) strategy. Since the 1980s, China's central and regional governments have worked on a three-step development strategy, which has involved regional development initiatives, key national projects for supporting infrastructure development on a massive scale, and the creation of strategic industries, assets, and technological capabilities.[11] While guiding the development of commercial enterprises and market incentives, China has also prevented foreign actors from competing with certain domestic industries.

To conclude on this point, the maintenance of the critical balance between liberalisation and continued government control, alongside the government's gradual reforms and a controlled but rapid integration into the global economy, can

[8] A valuable contribution in this debate has been offered by Bruno Amable, who provided a typology of capitalism based on the theory of institutional complementarity. Among the different models, he identified an Asian model of capitalism. See Amable, Bruno (2003). *The Diversity of Modern Capitalism*. Oxford University Press, Oxford. (See in particular Chap. 3).

[9] Johnson, Chalmers (1982). *MITI and the Japanese Miracle*. Stanford University Press, Stanford. Johnson applied this model to explain the Japanese economic boom. However, the model has been subsequently extended to interpret the growth of the so-called Asian Tigers (Hong Kong, Singapore, Taiwan, and South Korea), of the second generation's NIEs (Thailand, Malaysia, Indonesia, and Philippines), and finally of China.

[10] Mazzei, Franco, Vittorio Volpi (2010). *La rivincita della Mano Visibile. Il Modello economico asiatico e l'Occidente*. Egea, Milano.

[11] Rugman, Alan M., and Simon Collinson (2009). *International Business. 5th Edition*. Prentice Hall, Harlow: p. 620.

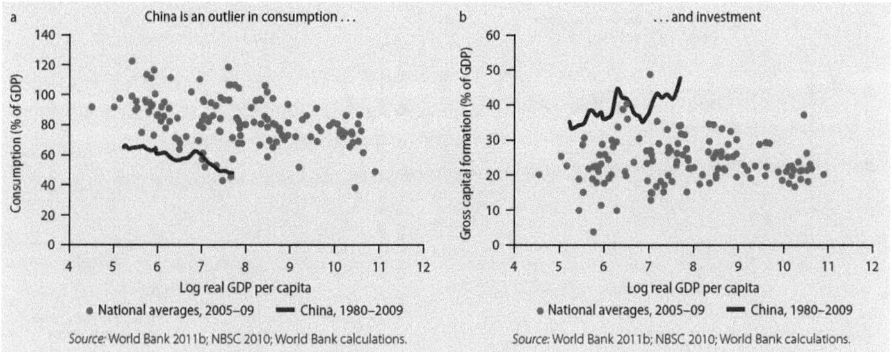

Fig. 5.2 China's consumption and investment rate (WB)

Table 5.1 Investment sources

Year	State budget (%)	Internal loan (%)	FDI (%)	Self-raising (%)
1981	28.1	12.7	3.8	55.4
2008	4.3	14.5	2.9	78.3
2012	4.6	12.6	1.1	81.7

ultimately be considered the major factors responsible for the economic success currently experienced by China.[12]

From a strictly economic point of view, China's impressive performance has been explained by its reliance on a high level of investment: for two decades, the gross fixed investment/GDP ratio has stood on average at the impressive rate of 49 % of Chinese GDP (see Fig. 5.2).[13]

With the liberalisation process centred on opening up the country in terms of inward and outward trade and FDI—through the creation, among others, of Special Economic Zones (SEZs)—China has proved highly attractive for multinational and transnational enterprises (MNEs), which have invested heavily in the country. However, contrary to most analyses—stating that China's performance is fully dependent on the impressive amount of foreign capital—FDI represents just a small portion of the total amount of investment. By simply looking at the composition of investments made public by the Chinese National Bureau of Statistics

[12] *Ibid.* p. 620.

[13] The World Bank and the Development Research Center of the State Council, P. R. China (2013). *China 2030: Building a Modern, Harmonious, and Creative Society*. The World Bank, Washington DC.

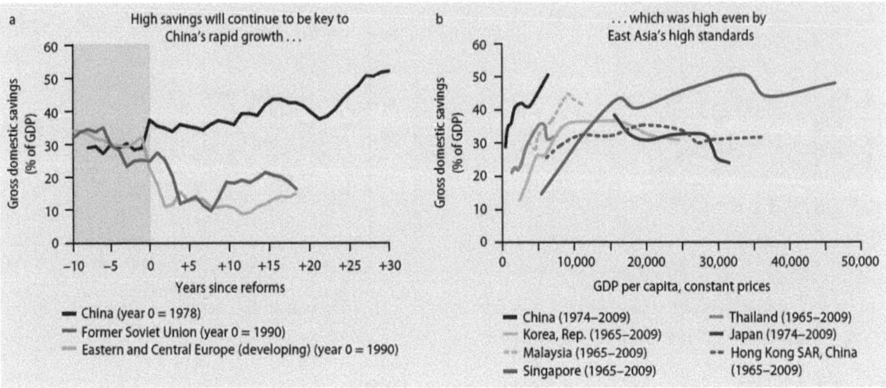

Fig. 5.3 Savings in China

(see Table 5.1),[14] it becomes evident that China's growth has not been dependent on FDI.[15]

Thus, although enormous in absolute terms and relevant in assisting Chinese growth, FDI cannot be considered the key factor for growth, which is on the contrary principally sustained by "self-raising" investments and internal loans. In fact, FDI accounts for only 1.1 % of total investment; the largest portion is represented by self-raising investments (81.7 %). Such investments are mainly realised, thanks to the large profits made by the SOEs. Given that SOEs are industries where entry and competition are limited, the SOE sector is highly profitable, and the large profits are systematically used to finance investment, since the opportunity cost of these funds is extremely low. Thus, a crucial role in laying the foundation and in assisting China's growth has been played by SOEs and ultimately by the government.

Alongside the high level of investment, China has been reliant on credit from domestic sources, also a sector that is still largely managed by the central government. Domestic investments and credit are ultimately sustained by a high corporate and private savings rate, which is settling at around 50 % of Chinese GDP (see Fig. 5.3).[16]

The reason for the high Chinese—and more generally high Asian—savings rates has been explained by many authors in terms of cultural factors, but from an

[14] Being aware of the many critics to Chinese official statistics, given the possibility of data manipulation, it has to be however stressed that also in this eventuality, data might only be slightly different.

[15] Chinese National Bureau of Statistics. "China Statistical Yearbook 2012" China Statistics Press. 2012. http://www.stats.gov.cn/tjsj/ndsj/2012/indexeh.htm.

[16] The World Bank and the Development Research Center of the State Council, P. R. China (2013). *China 2030: Building a Modern, Harmonious, and Creative Society*. The World Bank, Washington DC.

economic perspective, there are relevant motivations as well.[17] SOEs can save since they do not have to pay dividends to shareholders. In 2011, for instance, central government SOEs earned profits of RMB 913 billion but paid only RMB 77 billion in dividends.[18] Instead, most of the profits were saved to finance investment funds.[19] For their part, SMEs are to some extent obliged to save, given the difficulties in accessing credit. High household savings rates can ultimately be explained by a need to counter the lack of social protection (and difficulties in accessing services), given that the "*danwei*" (the work unit), which had assured housing, child and medical care, pensions, etc., collapsed during the reform period.

Although there is still no consensus among scholars, it can be said that in general terms China has adopted an *extensive growth model*, since it has not focused on productivity gains but on factor accumulation, in short, the expansion of the quantity of inputs—especially workforce—in order to increase the quantity of outputs.[20] The Chinese authorities are perfectly aware that reliance on extensive growth can be detrimental in the long term because it exhausts resources and is subject to diminishing returns; however, for China it has so far represented a cost-effective solution, given the seemingly inexhaustible supply of a low-cost workforce.

The overproduction generated by this extensive growth has largely been absorbed not by internal demand, which is growing yet still low, but by exports, thus making China an export-led economy. From a commercial point of view, China's competitive advantage in international trade rests on at least two pillars: low labour costs and an undervalued yuan (Renminbi, RMB).[21]

Cheap manufacturing has in practice allowed Chinese products to be easily exported worldwide and has helped China become a powerhouse in global manufacturing. According to research conducted by the US Department of Labor,[22] the average total remuneration cost in the manufacturing sector was

---

[17] Wei, Shang-Jin. "Why Do the Chinese Saves so Much?". Forbes. 2 February 2010. Web. http://www.forbes.com/2010/02/02/china-saving-marriage-markets-economy-trade.html. Accessed 5 September 2013.

[18] To give an idea of how extensive the SOE sector, it should be remarked that central SOEs represent just a small portion of the total number of SOEs. In 2007, there were 120,000 SOEs, of which only 22,000 were central. Knight, John and Sai Ding (2009). "Why does China Invest so much?" University of Oxford Discussion Paper Series No. 441.

[19] International Monetary Fund (2013). "People's Republic of China". IMF Country Report No.13/211: p. 31.

[20] This interpretation is offered by economists such George Friedman and institutions like the IMF. For other economists, it was on the contrary the enhancement of the productivity levels that allowed Chinese exponential growth. For this interpretation, see Hu, Zuliu and Mohsin S. Khan (1997). "Why China grows so fast?" International Monetary Fund, Washington DC.

[21] Mazzei, Franco, Vittorio Volpi (2010). *La rivincita della Mano Visibile. Il Modello economico asiatico e l'Occidente*. Egea, Milano: pp. 133–178.

[22] Banister, Judith (2013). "China's manufacturing employment and hourly labor compensation, 2002–2009". International Labour Comparison, US Bureau of Labor Statistics. Web. http://www.bls.gov/fls/china_method.htm. Accessed 2 September 2013.

$1.74 per hour in 2009. Beyond direct wages, this included social and medical insurance, employee benefits, pension payments, and housing funds. Compensation costs differ considerably between rural workers and urban workers: $1.15 per hour for the former, compared to $2.85 per hour for the latter. It should also be noted, however, that the majority of manufacturing workers are employed in rural areas (64.4 % vs. 34.6 % of urban workers).[23]

Despite a significant increase in the past decade—average pay per hour worked has tripled since 2002, passing from $0.57 in 2002 to $0.95 in 2006 and $1.59 in 2008—average pay per hour is only 5 % of that of the United States, whose average total hourly wage in 2009 was $34.19 per hour. Chinese hourly pay is also far below that of many its Asian neighbours: 5.7 % of pay in Japan ($30.3), only 11.5 % of the Republic of Korea hourly wage ($15.06), and 9.8 % of Singapore's ($17.54). The only pay rate approximately on par is that of the Philippines ($1.70).[24]

Alongside low-cost labour, the currency's low valuation is a second pillar in China's fruitful recipe for export-oriented production. Initially pegged to the US dollar at 2.46 yuan per dollar, during the 1980s the RMB was increasingly devalued by the Chinese authorities to stimulate the competitiveness of Chinese products. In 1994, the RMB reached the trough of 8.28 yuan per dollar, a peg that was then basically maintained until 2005, when, under strong pressure from Washington, it was lifted so as to allow the RMB to be gradually revalued.[25] At present, international institutions such as the IMF consider that, at 6.22 against the dollar, the RMB remains moderately undervalued against a broad basket of currencies.[26] The Chinese authorities, however, still refuse to institute a natural floating exchange rate, insisting that a steady economy, accompanied by a stable financial system and strong regulation, is the essential condition for full RMB convertibility. This implies that full convertibility is still far away.[27]

Similarly to other Asian countries, China's monetary policy is to maintain the RMB exchange rate low, so as to facilitate the creation of increasing surpluses in the balance of payments. These surpluses are only partially offset by the increasing bulimia of raw materials. Not by chance, the commercial balance presents a certain degree of asymmetry: it is characterised by significant surpluses with the high-income economies (United States and Europe) and a deficit with developing

[23] *Ibid.*

[24] The percentages are calculated on the basis of the data made available by the US Bureau of Statistics.

[25] Mazzei, Franco, Vittorio Volpi (2010). *La rivincita della Mano Visibile. Il Modello economico asiatico e l'Occidente*. Egea, Milano: pp. 153–154.

[26] Rabinovitch, Simon. "IMF says renminbi 'moderately undervalued'" Financial Times. 25 July 2012. Web. http://www.ft.com/intl/cms/s/0/370ef804-d62c-11e1-b547-00144feabdc0.html#axzz2e1G9htZN. Accessed 5 September 2013. For other analysts, however, RMB still remains substantially undervalued in the range of 15–30 %.

[27] *Ibid.*

countries (Asian and African).[28] While China imports raw materials from the latter countries, the surpluses generated by trade with Western countries are used in the acquisition of foreign exchange reserves, most of which are then invested in low-yield US Securities. For the World Bank, this reserve accumulation reflects "not only the country's growing role as the centre of a rapidly expanding and deepening East Asian production network but also a policy objective to strengthen the country's foreign exchange buffer against external shocks".[29]

To sum up China's growth model, it is an extensive growth model, reliant on high savings and high investment, which generates overproduction largely absorbed by exports.

5.1.2 Catching-Up or Paper Tiger?

China's miraculous economic growth, reminiscent of Japan's "catching-up" period during the 1960s and 1970s, has provoked many forecasts by the main international economic institutions.

Early projections were overly pessimistic: in 2003, for instance, it was forecast that the Chinese economy would have overtaken Japan's, ranking in second position in the world, by 2015, and outstrip the US economy by 2035. In fact, China overtook Japan already in 2010, and, considering the broad margin between China's outperformance and the continued slow growth of the advanced economies,[30] it goes without saying that catching-up with the United States will also be much earlier.

According to the IMF, China has already overtaken the United States in terms of purchasing power parity GDP in 2014[31] and by the next decade in terms of nominal GDP. Similar projections have been made by other institutions, such as the World Bank and OECD. The World Bank estimates that, with an average growth of 6.6 %, China will overtake the United States early in the next decade, while for the Paris-based OECD, China is bound to catch up with the United States sometime before 2030.

These forecasts are based on the assumption that Chinese growth will continue through the next decade by a similar rate as registered in recent years. Although it is true that more than 30 years of growth does not mean unending growth, even if China's growth does slow down, the final result might only be postponed. According to the World Bank: "even if China grows a third as slowly in the future

[28] Mazzei, Franco, Vittorio Volpi (2010). *La rivincita della Mano Visibile. Il Modello economico asiatico e l'Occidente*. Egea, Milano: p. 155.

[29] The World Bank and the Development Research Center of the State Council, P. R. China (2013). *China 2030: Building a Modern, Harmonious, and Creative Society*. The World Bank, Washington DC.

[30] The creation of this wide margin was accelerated by the global financial crisis started in 2008.

[31] International Monetary Fund (2014). World Economic Outlook Database. IMF Web. http://www.imf.org/external/pubs/ft/weo/2014/02/weodata/index.aspx.

compared with its past (6.6 percent a year on average compared with 9.9 percent over the past 30 years), it will become a high-income country sometime before 2030 and outstrip the United States in economic size".[32] This will be the first time in more than a century that the United States will not be the world's largest economy.

A large part of the literature agrees that China will surpass the United States not only in terms of economic size: many authors believe that by 2030 China's influence in the global economy could approach that of the United Kingdom in 1870 or the United States in 1945,[33] or even that of China before the "great divergence" (see Sect. 2.4.2). According to the Indian economist Subramanian, within the next 20 years, China is likely to be economically dominant, not only because "its market-based GDP is projected to equal that of the US and its PPP-based GDP by then to be twice that of the US", but mainly because "China's trade in goods will be nearly two times that of the US and Europe, and the renminbi stands a good chance of nipping at the heels of the dollar, if not eclipsing it, as the main reserve currency".[34] In addition, notwithstanding shrinkage in its trade surplus, China will be the world's largest creditor and will remain the largest manufacturer and exporter. In this impressive scenario, the shift of the geo-economic centre of gravity to the Asia-Pacific region, with China as its future hub in terms of trade and financial flows, will inevitably be accompanied by an equally important geopolitical and geostrategic shift.

Becoming the world's largest economy, however, does not automatically imply that China will be able to create an economically stable system. The extraordinary pace of growth of the past 35 years has created such huge imbalances and vulnerabilities that these may instead transform China into a "Paper Tiger". After all, as the Chinese proverb goes: *Wuji bifan*: as soon as a thing reaches its extremity, it reverses its course.

As indicated by the same institutions that forecast China's catching-up, the country's pattern of growth is no longer sustainable, and the many socio-economic distortions it has raised could soon serve, if uncorrected, as serious constraints on growth. The main contradictions and imbalances are presented below:

- First, economic activity in China has become too reliant on investment and credit,[35] while internal consumption has not been encouraged and remains too low.[36] Although the high level of investment has been one of the key factors in

---

[32] The World Bank and the Development Research Center of the State Council, P. R. China (2013). *China 2030: Building a Modern, Harmonious, and Creative Society*. The World Bank, Washington DC: p. 3.

[33] *Ibid.* p. 9.

[34] *Ibid.* p.10.

[35] International Monetary Fund (2013). "China: New Round of Reforms Needed for Continued Success". IMF Survey Magazine: Countries & Regions. Web. http://www.imf.org/external/pubs/ft/survey/so/2013/car071713a.htm. Accessed 18 September 2013.

[36] Berkofsky, Axel (2012). "L'economia cinese cresce velocemente. Ma sarà abbastanza?" Osservatorio di Politica Internazionale n. 62: p. 2.

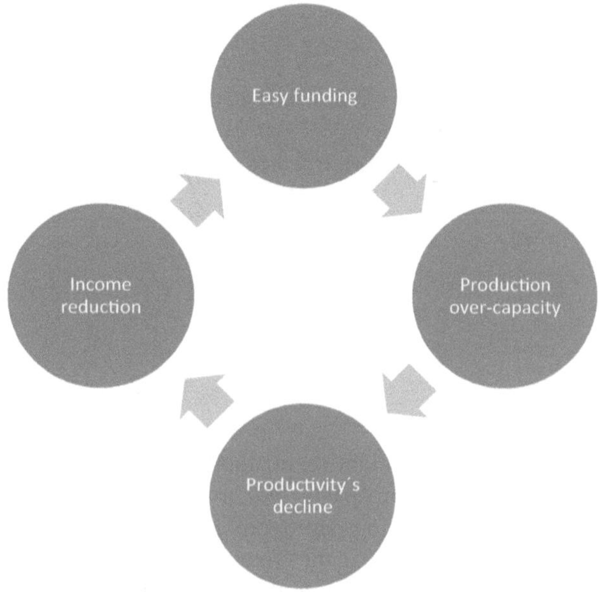

Fig. 5.4 The vicious circle of Chinese economy

the Chinese miracle, it has created a vicious circle, very well described by François Gipouloux.[37] The French sinologist emphasises that the lack of rigour in the credit system creates an irrational increase in investments and thus in production overcapacity; the overproduction in turn leads to a progressive decline in productivity. Ultimately, the decline of productivity, together with the continued accumulation of capital, is responsible for declining profitability for SOEs. This vicious circle is represented in Fig. 5.4.

As the IMF has also noted,[38] although investments have risen in recent years, growth has already started to moderate, with the return on investment declining to around 16 %, down from 25 % in the early 1990s. For the *Forbes* journalist Gordon Chang, returns on investment are so low that to add one yuan to China's GDP, some seven yuan of investment are required.[39]

In addition, given that the extensive growth model China has adopted has been too reliant on the input of a new workforce, once the growth rate of the labour force shrinks as a consequence of looming demographic changes, growth could suddenly decelerate, with dramatic repercussions both domestically and internationally.

[37] Gipouloux, François (2005). *La Chine du 21 siècle. Une nouvelle superpuissance.* Armand Collin, Paris.

[38] International Monetary Fund (2013). "People Republic of China Report". IMF Country Report No. 13/211: p. 21. Web. http://www.imf.org/external/pubs/ft/scr/2013/cr13211.pdf. Accessed 10 October 2013.

[39] Chang, Gordon G. (2012). "In morte del Miracolo Cinese". In: *USA contro Cina.* Limes, Rivista Italiana di Geopolitica No. 06/2012. December 2012.

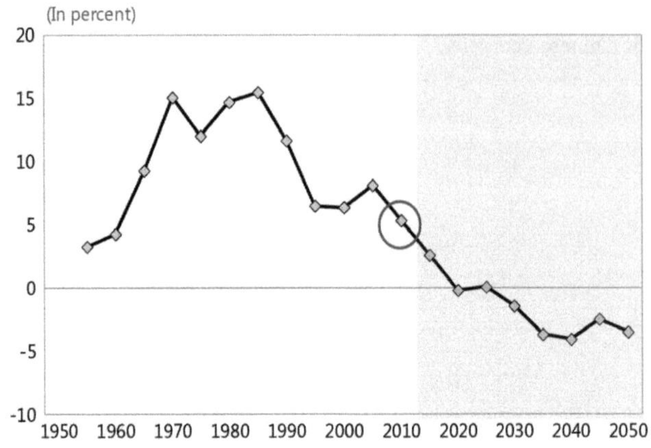

Fig. 5.5 Growth of working-age populations (IMF)

- Incredible as it might seem, another extremely critical issue for China is indeed its population prospects. For many international institutions, China is in the midst of wrenching demographic change.[40] The government's "one-child policy" and the subsequent "four-two-one problem" have made the ageing of the Chinese population and the related problem of workforce shrinkage a top priority issue, which could seriously threaten growth in the next few decades (Fig. 5.5).

  The IMF estimates that the working-age (15–64) population will start to fall in less than a decade, while the core industrial worker cohort (25–39 years old) will start shrinking as soon as 2015. "The dependency ratio (population younger than 15 and older than 64 as a share of the working-age population), which declined for decades, is projected to increase from 13.5 % in 2010 to around 30 % by 2030".[41]

  According to the Washington-based institution, "China will meet the Lewis Turning Point—where the supply of low-cost labour is exhausted—towards the end of the decade. And, as the surplus labour dwindles, labour costs will rise, which could affect prices, incomes and corporate profits in China, and would have implications for trade, employment and price developments in key trading partners"[42] For the IMF, this transformation makes it even more important to switch from an extensive to an intensive growth model. However, time is running out and, if not reformed, the current growth model could have dramatic

[40] International Monetary Fund (2013). "People Republic of China". IMF Country Report No. 13/211: p. 21. Web. http://www.imf.org/external/pubs/ft/scr/2013/cr13211.pdf. Accessed 10 October 2013.

[41] *Ibid.* p. 20.

[42] *Ibid.* p. 20.

consequences (e.g. bankruptcies and financial losses that could in turn ultimately make employment and growth stall).[43]

- External imbalances generated by China's growth model have to be considered as well. China has been too reliant on exports, and although globalisation and free trade will presumably remain irresistible trends in the future, its bilateral trade surpluses with key partners could give rise to protectionist measures. Alternatively, in the case of a decline in the advanced economies, China's exports (and thus its growth) could be compromised.

  The most relevant external imbalance China has to face has been termed its *chain-gang relations* with Washington, so called because Chinese overproduction and American overconsumption are "like the proverbial prisoners who seek to break free from one another but can't because they're chained together".[44]According to the sociologist Walden Bello, this peculiar relationship:

  > is increasingly taking the form of a vicious cycle. On the one hand, China's breakneck growth has increasingly depended on the ability of American consumers to continue their consumption of much of the output of China's production brought about by excessive investment. On the other hand, America's high consumption rate depends on Beijing's lending the US private and public sectors a significant portion of the trillion-plus dollars it has accumulated over the last decade from its yawning trade surplus with Washington.[45]

  There is broad consensus that this relationship is no longer sustainable for the actors. In addition, China, which uses its currency reserves as a buffer against external shocks, could ironically also face the risk of significant capital losses if the dollar weakens. The World Bank has stated that "efforts to export capital in the form of outward FDI, especially to secure raw material supplies, has met with suspicion in some receiving countries, and unless appropriate steps are taken to address these problems, such risks and friction could grow".[46]

- Another risk stemming directly from the previously described vertical take-off of the Chinese economy is the sharp increase in socio-economic inequality.

---

[43] The IMF has drawn a scenario beyond 2018 illustrating the consequences of continuing the current model of growth. IMF explains it as follows: The scenario assumes a further build-up of excess capacity and misallocation of resources. With demographic trends implying a decline in the labour force after 2015 and exhaustion of surplus labour around 2020, the returns on investment would be progressively lower than envisaged, which would cause bankruptcies and financial losses. [. . .] The outcomes could be costly not just in terms of direct fiscal cost of clean-up, but also because the financial losses and deleveraging would in turn generate an adverse feedback loop that hampers employment and growth. The convergence process would stall, with the economy slowing to around 4 %, and GDP per capita would remain about a quarter of that of the United States through 2030. *Ibid.* p. 20.

[44] Bello, Walden. "Chain-gang economics". Foreign-policy in Focus. 30 October 2006. Web. http://fpif.org/chain-gang_economics/. Accessed 7 October 2013.

[45] *Ibid.*

[46] The World Bank and the Development Research Center of the State Council, P. R. China (2013). *China 2030: Building a Modern, Harmonious, and Creative Society*. The World Bank, Washington DC.

Although the accelerated economic ascendancy has translated into a significant reduction in poverty, one of its main consequences has been extremely unequal development on a regional basis, which has fractured Chinese territory socially into "three Chinas": a densely populated "coastal China", rich and well integrated in the global economy; an "internal China" which acts as a supply of plentiful low-paid labour to coastal China but remains quite underdeveloped; and a "peripheral China", whose regions (Xinjiang, Tibet, and Inner Mongolia), populated by ethnic minorities, play the role of *buffer states*. Despite being rich in raw materials, these regions remain among the least developed in the country.[47]

The 11 coastal provinces alone contribute around 60 % of national GDP and account for over 90 % of import and export flows, while their per capita income is on average 2.9 times higher than in the rest of the country. This income ratio reaches the extreme peak of 13:1 when comparing the richest province of Shanghai with the poorest of Guizhou.[48] It is estimated that in internal China and peripheral China, more than 130 million people—about a tenth of China's population—still live below the $1.25 per-day international poverty line, while more than one third live on less than $2.[49] Large disparities exist not only between the coastal and inner regions but also between urban and rural areas. The urban–rural income ratio stands at an average of 3.2:1,[50] although many unofficial estimates show an even greater gap. Within the urban areas of rich coastal China as well, there is an increasing gap between *xingui* (the new rich) and *xiagang* (a term usually referring to redundant employees of the SOEs, the unemployed, or migrants), the latter also excluded from social protection.[51] In fact, besides per capita income, inequality has also climbed in terms of access to social welfare and quality public services.

During the pre-reform period, workers were guaranteed lifetime employment and a mechanism of social protection (*the iron/clay rice bowl*, which was linked to the working unit), assuring a package of benefits that included provisions for housing, medical care, childcare, schooling, and a pension for retired workers. With the liberalisation process and the rapid growth of the private sector, the

---

[47] Mazzei, Franco, Vittorio Volpi (2010). *La rivincita della Mano Visibile. Il Modello economico asiatico e l'Occidente*. Egea, Milano: pp. 135–136.

[48] Disegni, Simone (2012). "China's rise and the global economic downturn: threat to steady growth or opportunity to rebalance?". ISPI Analysis No. 107. Istituto per gli Studi di Politica Internazionale, Milano.

[49] *Ibid.*

[50] The World Bank and the Development Research Center of the State Council, P. R. China (2013). *China 2030: Building a Modern, Harmonious, and Creative Society*. The World Bank, Washington DC.

[51] The *hokou* system (the household registration) has prevented rural migrants to have access to welfare provisions and transformed them into second-class citizens in the cities. See Knight, John (2013). "Inequality in China: An Overview". The World Bank Policy Research Working Paper No. 6482. Washington DC: p. 8.

state had to drastically reform these mechanisms in order to maintain the competitiveness of the SOEs, ultimately causing the collapse of the welfare system.[52]

All these fractures and imbalances have already led to social tensions and public protests. These have hitherto been partially mitigated by the prospect of long-term collective wealth, which is implicitly promised to millions of rural and urban households by the country's continuous and burgeoning growth. However, should there be a sudden slowdown, the situation could easily get out of control. The British analyst George Friedman has shrewdly explained that, in the likely case of an economic downturn, where "the money stops rolling in, not only will the bank system spasm, but the entire fabric of Chinese society will shudder. In a country where poverty is endemic and unemployment widespread, the added pressure of an economic downturn will result in political instability".[53] Ultimately, this instability may even lead to political fragmentation.

To conclude on this point, it should be remembered that the numerous causes of social tension could—if unresolved—pose a real threat to growth and stability in the coming decades.

- Finally, there are many other non-negligible problems, such as a widespread sanitary degradation (SARS, avian influenza, etc.) and uncontrolled environmental deterioration, which could also become an impediment to the long-term development of the nation. China is the world's largest polluter: it is home to 16 of the world's most polluted cities, and, since 2005, it has been the largest emitter of greenhouse gases and sulphur dioxide, the latter exceeding that of the United States and EU combined.[54] Concerning water quality and supply, "one-third of major river systems, 85 % of lakes, and 57 % of underground water in monitoring sites are polluted. Nearly 300 million rural residents lack access to clean drinking water and some major rivers have become too polluted to supply drinking water".[55] Desertification, soil erosion, and grassland degradation are relevant problems as well and are causing the loss of a significant portion of grasslands every year. The environmental damage in 2010 cost the equivalent of 3.5 % of China's GDP[56] but, if uncorrected, will raise even more serious problems for growth in the future.

[52] Kochhar, Geeta (2008). "China's Urban Poor: An expanding Social Stratum". University of Nottingham—China Policy Institute, Discussion Paper No. 37.

[53] Friedman, George (2010). *The Next 100 Years. A Forecast for the 21st Century.* Anchor Books, New York: p. 96.

[54] See Thompson, Thomas N. "Choking China. The Superpower that Is Poisoning the World". Foreign Affairs Snapshots. 8 April 2013. Web. http://www.foreignaffairs.com/articles/139141/thomas-n-thompson/choking-on-china. Accessed 10 October 2013.

[55] International Monetary Fund (2013). "People Republic of China". IMF Country Report No. 13/211: p. 21. Web. http://www.imf.org/external/pubs/ft/scr/2013/cr13211.pdf. Accessed 10 October 2013.

[56] *Ibid.* p. 19.

Similarly, China's growth pattern has involved extremely intensive natural resources and energy use. As the World Bank puts it: "the rapid growth, growing urbanization, and structural change within manufacturing have combined to make China the world's largest energy user, outstripping the US in 2010". China is in addition the largest importer of concrete, coke, steel, aluminium, and nickel and the second largest importer of oil, after the United States. The country's energy dependency is growing rapidly, and, according to forecasts, by 2030 80 % of Chinese energy demand will have to be imported (thus probably transforming energy into China's *Achilles heel*). The energy issue will also contribute to making the external imbalances an even more critical factor for the future of the country.

5.1.2.1 Towards Reforms

All the above-described vulnerabilities and distortions pose a tangible threat to the country: especially if combined, they could act as a serious and non-manageable brake on growth, sapping its strength. A solution will largely depend on whether the Chinese leadership is able to introduce and implement the required structural reforms and move China from an investment-led, high-carbon growth model to a consumption-led, green growth model, less reliant on low-cost manufacturing and factor accumulation and more on technology, innovation, and productivity gains.[57]

Both the 11th and 12th Five-Year Plan outlined the necessity of reform in these areas, but timely and focused implementation will be crucial for their success. It is also anticipated that the adoption of structural reforms and transition policies will inevitably imply, as a side effect, a slowdown in China's economic performance. It is evident that the transition from an extensive to an intensive and more sustainable growth model and the attempt to move from middle-income to high-income status will be accompanied by slower growth.

5.1.3 Hard or Soft Landing?

There is a wide consensus among scholars and economists that China's growth will not continue at the same impressive rate in the coming years. However, there is still no telling what kind of slowdown—or landing—China will face: soft or hard?

While the former implies a moderate slowdown (hypothetically an annual GDP growth of 7 %), a hard landing would occur if the country experiences a serious

[57] The World Bank (2013). "People's Republic of China. Developing Skills for Economic Transformation and Social Harmony. Yunnan Province". The World Bank Repot No. ACS3321.

breakdown in its growth.[58] For most economists, this level is represented by an average annual growth of 6 % or less. Admittedly, neither scenario contemplates an abrupt stop to China's GDP growth. But each could cause the emergence of critical consequences for the country, though with different implications.

In an article in the *Financial Times*, Martin Wolf illustrated this underlying difference with an interesting analogy: China's economy is like a 747 jet; after a vertical take-off, "in recent years, a couple of engines have not been working well, and the pilot is now loath to keep straining the remaining good engines. He is allowing the plane to slow down, but if it slows too much, it will fall below stall speed and drop out of the sky".[59]

In other terms, a moderate slowdown—caused, for instance, by a macroeconomic shock or by the effects of the implementation of planned reforms—could still be managed by the government, because China's fiscal and debt position gives it the space to respond with counter-cyclical measures. A more severe slowdown (e.g. growth of 5 %), on the contrary, could produce catastrophic effects, both domestically and internationally. For the World Bank, such an eventuality will for sure "unmask inefficiencies and contingent liabilities in banks, enterprises, and different levels of government—heretofore hidden under the veil of rapid growth—and will precipitate a fiscal and financial crisis".[60] Ultimately, it will also lead to slower growth and possible economic stagnation, with implications for social stability and the global economy that are hard to predict.

In 2012 the growth rate fell below 8 % for the first time in the twenty-first century and a similar trend is expected in 2014 and 2015. This seems, to some extent, to confirm the emergence of a gradual and manageable soft landing. However, given the increasing number of negative economic indicators, some economists are beginning to ask whether the economy might face a much more difficult situation. They are starting to believe that the current slowdown is more structural than cyclical and that, even if the government wanted to, it might not be able to induce a big rebound in growth.[61]

It is perhaps too early to make accurate projections. What is sure is that the government is already implementing several measures aimed at supporting China's transition to more balanced and sustainable growth. China's 12th Five-Year Plan

[58] According to the *Financial Times* definition, a hard landing "occurs when an economy that has recorded a period of very rapid growth experiences a severe slowdown, normally due to overheating and an excessive policy response such as substantial credit tightening, a revaluation of the currency, etc." See "Lexicon: Hard-landing". Financial Times. Web. http://lexicon.ft.com/Term?term=hard-landing. Accessed 12 September 2013.

[59] Wolf, Martin. "Risks of a hard landing for China". Financial Times. 2 July 2013.

[60] The World Bank and the Development Research Center of the State Council, P. R. China (2013). *China 2030: Building a Modern, Harmonious, and Creative Society*. The World Bank, Washington DC: p. 10.

[61] Anderlini, Jamil. "Economist weight Chinese hard landing". Financial Times. 19 August 2012. Web. http://www.ft.com/intl/cms/s/0/cc05e828-e860-11e1-8ffc-00144feab49a.html#axzz2bVSNjZFc. Accessed 15 September 2013.

(2011–2015), for instance, already envisions measures to address environmental and social imbalances, to reorient the economy from an extensive growth model to an intensive one, to reduce its external imbalances by boosting domestic demand and rebalancing investments and consumption, and to overcome the regional and rural–urban divides. Its annual growth target of 7 % "clearly shows the intention to focus on quality of life, rather than pace of growth".[62]

To conclude, Beijing has started to see the risk of a dramatic downturn as a double opportunity for radically reshaping its growth model and for correcting its numerous and unsustainable socio-economic distortions.

5.1.4 Implications for the Lunar Programme

The complexity of the socio-economic framework, in which China's space programme operates, makes it hard to identify clear and precise implications for the country's manned lunar exploration programme. At least three "independent variables" need to be considered in assessing the future of China's lunar ambitions: the hypothesis of continued rapid growth, that of a hard landing and that of a soft landing. Needless to say, the three variables are bound to produce quite different scenarios for the programme. In addition, the existence of numerous dependent variables might prevent the establishment of an unequivocal relationship between independent variable and expected outcome. It remains useful to sketch the main possible trends, however.

If China succeeds in going forward with its impressive economic performance throughout the next decade, no significant obstacles to any Moon plans can be envisaged from an economic point of view. The emergence of the powerful and economically dominant China forecast by many would on the contrary be perfectly in line with the country's growing space aspirations and in particular with the target of a manned lunar landing. As already noted, given that the country would outstrip the United States in economic size, sending taikonauts Moon-wards could also imply the "passage of the relay baton" to China in the technological and geopolitical world hierarchy.

The "sinification" of the global space community that some authors have already started to discuss would eventually become a consolidated reality, with effects on a wide range of issues.[63] China could, for instance, have strong bargaining power in the battle to implement its norms and standards at the international level (e.g. concerning its BeiDou system and its Earth Observing System), while its

[62] The World Bank and the Development Research Center of the State Council, P. R. China (2013). *China 2030: Building a Modern, Harmonious, and Creative Society*. The World Bank, Washington DC.

[63] Solomone, Stacey (2013). *China's Strategy in Space*. Springer, New York: pp. 37–41.

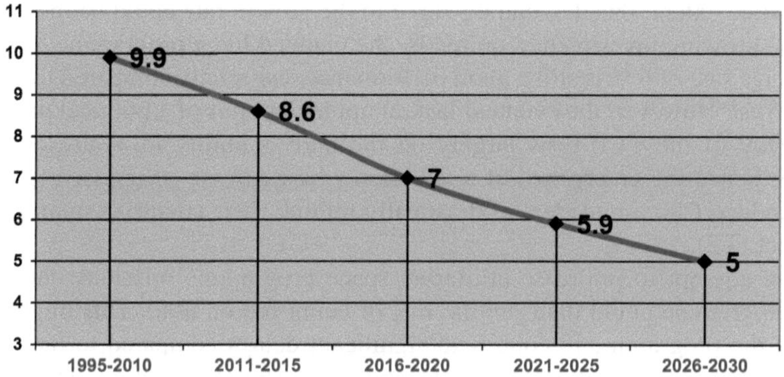

Fig. 5.6 China's growth rate 1995–2030 (*Source*: World Bank)

future space station could be used as "another bargaining chip in the international negotiations for manned access to space".[64]

However, the numerous vulnerabilities and challenges the country will face in the coming years may create serious obstacles not only to continued growth but also to the continuity of China's ambitions in space. If the country faces a hard landing, it is likely that any plan to get to the Moon will be seriously compromised. Hypothetically, under this scenario, not only would the manned lunar exploration programme be put on the backburner, but other significant segments of China's space programme would also have to be abandoned.

Fortunately, for the moment this dramatic possibility is still far from being imminent. Besides the fact that China still has several instruments with which to manage the forthcoming challenges, the government authorities have, as mentioned, already started to implement some of the structural reforms necessary to correct the numerous imbalances and vulnerabilities its growth has generated.[65]

In the less threatening case of a soft landing—which at the moment appears more likely, given China's ability to overcome difficulties and buffer shocks—growth will moderate, but the country will still succeed in becoming both a high-income economy and the world's largest. In this scenario, it can be expected that the continuity of the manned lunar programme will be at less risk since it could provide a success story deflecting attention away from the slowdown.

It must be emphasised, however, that even in this case such continuity cannot be taken for granted. In order to better assess this point, it is useful to look at the forecast descending curve of China's future GDP growth rate. As previously mentioned, by assuming a steady implementation of the reforms and no relevant international shocks, China's projected growth will inevitably still slow down (see Fig. 5.6).

[64] *Ibid.* p. 39.

[65] See, for instance, Feigenbaum, Evan A., and Damien Ma. "*The rise of China's reformers?*". Foreign Affairs Snapshots. 17 April 2013. Web. http://www.foreignaffairs.com/articles/139295/evan-a-feigenbaum-and-damien-ma/the-rise-of-chinas-reformers. Accessed 2 September 2013.

To some extent, this descending curve in the growth rate could be incompatible with the growing investment required by the manned lunar programme. Of course, an average rate of 6 % is still a good performance, especially compared to low rates in the West. However, the eventual lack of appreciation is of a political nature. The legitimacy of the CCP rests largely on the party's ability to assure continuous growth; hence the emergence of a situation where growth progressively declines could induce Chinese leaders to drastically rethink their country's main priorities (see Sect. 3.2).

In an attempt to prioritise utilitarian space programme missions, the manned lunar programme could then run the risk of being put on hold. This might in turn compel the programme to cope with significant delays compared to the roadmap presented by CAS, or alternatively, to undergo a substantive reconfiguration that would ultimately mean the need for international cooperative efforts. And yet even such a scenario would in the end rely on Chinese leaders' political will to continue pursuing a Moon programme.

5.2 Sociopolitical Conditions

Directly connected to the socio-economic variables, an equally relevant conditioning factor for China's manned lunar programme will be the attitude of the latest (2012) presidency, and in part that of Chinese civil society, towards this ambitious endeavour.

Over the years, manned spaceflight, responding more to political than scientific objectives, has witnessed a certain degree of political volatility. In China too, the history of the human spaceflight programme has clearly shown that, in a period of economic and sociopolitical turmoil, ambitious projects to send humans into space have had to be prioritised downwards.

Considering the current situation of pressing social, economic, and environmental problems and the fact that from a political-cycles point of view, the lunar programme is an inter-generational investment that will not primarily benefit the current policymakers (but only the CCP regime), it could be asked whether the new ruling class led by President Xi Jinping is willing to support such an ambitious programme wholeheartedly. The lack of a clear, strong commitment by the Chinese leadership suggests that at the moment a full consensus on the programme has not been reached.

The following sections will examine the possible "dilemmas" Chinese elites are facing in supporting the programme: the main schools of thought within the political debate surrounding the endeavour and the attitude of "civil society" towards its country's aspirations in space.

5.2.1 *Societal Needs Versus the Lunar Programme*

Despite its rapid economic ascendance and its impressive results in terms of poverty eradication and social development, China can still be considered a developing country. Deng Xiaoping's dream of a *xiaokang* (moderately well-off) society is still somewhat elusive, and this half success is now exacerbated by the sharp contrast between the excessive and sometimes ill-gotten wealth of the Chinese superclass and the poverty of the masses. With the second largest number of poor people in the world after India[66] and a gross national income per capita that in 2013 ranked only 107th in the world, the implementation of effective socially oriented reforms figures among the most relevant needs for China.[67]

Poverty eradication and enhancement of per capita income are just two of the societal challenges China will have to resolve. As already mentioned, the country is facing many other challenges that are dramatically affecting the living standards of its citizens. But above all, China's current leadership has to face the challenge of the so-called missing fifth modernisation.

The liberalisation process that started in 1978 with the launch of the *gaige kaifang* was not accompanied by any parallel political liberalisation. Deng Xiaoping made the point very clearly in 1987: "Democracy can develop only gradually, and we cannot copy Western systems. If we did, that would only make a mess of everything. Our socialist construction can only be carried out under leadership, in an orderly way and in an environment of stability and unity".[68]

This interpretation seems to a large extent to be in line with the democratic consolidation analyses of political scientists such as O'Donnell, Cappelli, and Carothers. These authors argue that "transition to democracy" risks being a universal catchword only: democracy cannot be seen as something that can "be crafted from scratch—at every latitude and in a short lapse of time—through a combination of constitutional design, shock-therapy market reforms, NGO promotion, etc."[69] In most such cases, the result has been a "destatification" that has plunged many

[66] According to the calculations of the WB and the IMF, there are more than 170 million people still living below the $1.25 international poverty line.

[67] "Gross national income per capita 2013, Atlas method and PPP". World Development Indicators database, World Bank (2014). Web. http://databank.worldbank.org/data/download/GNIPC.pdf. Accessed 20 December 2014.

[68] Deng Xiaoping's speech "We Must Carry out Socialist Construction in an Orderly Way under the Leadership of the Party". 8 March 1987. In: "The selected works of Deng Xiaoping. Modern Day Contributions to Marxism–Leninism". Web. http://dengxiaopingworks.wordpress.com/2013/03/18/we-must-carry-out-socialist-construction-in-an-orderly-way-under-the-leadership-of-the-party/. Accessed 10 October 2013.

[69] See in particular: O'Donnel, Guglielmo (1996). "The illusion of Consolidation". Journal of Democracy Vol. 7 (2): 34–51. Carothers, Thomas (2002). "The end of the Transition Paradigm" Journal of Democracy. Vol. 13 (1): 5–21. Cappelli, Ottorino (2010). "Premodern State Building in Post-Soviet Russia". In: Hill, Ronald J., and Ottorino Cappelli (eds). *Putin and Putinism.* Routledge, New York.

countries into political chaos, civil war, or even into the *reverse wave* of autocracy or sultanism.

However, it is also true that, 27 years later, the development of "democratic institutions" is still far from being initiated,[70] and the obsession with stability of the Deng era could also become self-defeating: it is indeed transforming itself into a source of new problems that may in the future pose real threats to the country's stability.[71] As a recent study by the World Bank and the P.R.C. Development Research Centre has shown, social tensions are not just expected to arise from the rampant inequality between the masses and the rich superclass or as a result of problems with ethnic minorities but also from other sources. "If the experience of other countries is of any guide—the *China 2030* research stresses—the rising ranks of the middle class and higher education levels will inevitably increase the demand for better social governance and greater opportunities for participation in public policy debate and implementation. Unmet, these demands could raise social tensions"[72] and result in public protests.

In stark contrast to the Chinese leaders' pledge for a "harmonious society", more than 100,000 "mass incidents" (riots, civil unrests, and protests involving more than 100 people in public space) are annually registered.[73] Moreover, the conditions for a new Tiananmen-style protest are for many scholars already present.[74] Since the Chinese leadership's biggest fear is that of social instability that could lead to the political fragmentation of the country, these events clearly do not bode well. As in imperial China, the spread of social instability is still considered a sign of the *geming*—the revocation of the celestial mandate—which calls for radical leadership changes.[75] Through the so-called Jasmine Revolutions,[76] Chinese leaders have

[70] The goal of democracy has been present only in the rhetoric of Chinese leaders, but it is clear that a first answer would lie in the replacement of this rhetoric with at least good political governance. See Van Wie Davis, Elisabeth (2009). "Governance in China in 2010". Asian Affairs. Vol. 35 (4): 195–211.

[71] As Sun Liping, of the Tsinghua University, writes: "The ultimate outcome of the rigid thinking of stability preservation and the massive stability preservation project is in fact the intensification of social tensions". Liping, Sun (2012). "The Wukan model and China's Democratic potentials". In: Leonard, Mark (ed). *China 3.0*. European Council on Foreign Relations, London: p.74.

[72] The World Bank and the Development Research Center of the State Council, P. R. China (2013). *China 2030: Building a Modern, Harmonious, and Creative Society*. The World Bank, Washington DC.

[73] For the majority of analysts, this number is conservative. Although dissimilar figures have been provided by foreign reports, there is no doubt that the annual number of so-called mass incidents has grown exponentially since the 1990s, passing from 8700 in 1993 to nearly 60,000 in 2003 to around 180,000 in 2008. See Tanner, Murray Scot (2014). "China Social Unrest Problem". Testimony before the US-China Economic and Security Review Commission. Web. http://www.uscc.gov/sites/default/files/Tanner_Written%20Testimony.pdf. Accessed 30 June 2014.

[74] Minxin, Pei. "Signs of a New Tiananmen". The Diplomat. 4 February 2012. Web. http://thediplomat.com/2012/04/04/signs-of-a-new-tiananmen-in-china/. Last accessed 27 September 2013.

[75] See Van Wie Davis, Elisabeth (2009). "Governance in China in 2010". Asian Affairs. Vol. 35 (4): 201.

[76] The term is used to indicate the succession of revolutionary waves that started in Tunisia in 2011 and then spreading to the area.

clearly been shown the need to implement social-oriented reforms, if they want to avoid the emergence of a similar situation in China.[77]

In a valuable contribution published by the European Council on Foreign Relations (ECFR), Mark Leonard has pointed out that "there is a widespread sense in Beijing that China has reached the end of an era. People are not just expecting new leaders but the end of a model of development that started with Deng Xiaoping's "opening and reforms" in 1979".[78] China 3.0 (the third era in its development since the 1949 takeover) will be defined by a quest for solutions to its affluence, stability, and power crises.[79]

In such a wrenching and rapidly changing situation, the political leadership installed in 2012 could face strong head winds in supporting China's most ambitious space endeavour. The dilemma can be summarised as follows: should China's space exploration ambitions find a leading place in the political agenda or would it be more useful to focus on the country's socio-economic development? Ultimately, why invest billions of *yuan* every year in the lunar programme, when China still needs to eradicate poverty, boost social development, implement much needed economic reforms, seize the opportunity of green development, and confront environmental pollution, water shortage, and sanitary degradation?

The question has partially found an answer in the analysis of China's rationales for the Moon, which appear to be beneficial for a wide range of domains, including economic development. In addition, it would be misleading to present the dilemma in too stark terms, because there is no direct antagonism between the lunar programme and the demand for social reforms. On the contrary, the former could also provide relevant benefits in the sociopolitical dimension, given the space programme's ability to boost national unity and social cohesiveness. It is also clear that the two objectives can be pursued simultaneously. Indeed, with an expanding economy, resources allocated to the space programme would not be taken away from other stakeholders with a sense of entitlement. Since the cake of the Chinese economy is getting bigger, there is very considerable unallocated revenue that could be allocated to the space programme without the fear that other slices (e.g. social reforms) will be directly affected.

Only in the case of a hard landing would the allocation of resources be more problematic and become an issue for political confrontation. Currently, China

---

[77] On this point, see Scisci, Francesco. "The Jasmine Lesson: Reform beats revolution". Asia Times. 31 July 2013. Web. http://www.atimes.com/atimes/China/CHIN-02-310713.html. Accessed 27 September 2013. See also Scisci, Francesco. "Lessons from Tahrir to Tiananmen". Asia Times. 17 February 2011. Web. http://www.atimes.com/atimes/China/MB17Ad01.html. Accessed 27 September 2013.

[78] Leonard, Mark (2012). "What does the new China think?". In: Leonard, Mark (ed). *China 3.0*. European Council on Foreign Relations, London: p.10.

[79] Mark Leonard and François Godement have explained that the Chinese like to think of history progressing in 30-year cycles. Chinese think of China 1.0 as the years of Mao Zedong (1949–1978), of China 2.0 as the years of the reforms and growth (1978–2008), and of China 3.0 as the forthcoming years of the recently established leadership. *Ibid.*

appears to be undergoing a gradual and manageable soft landing (an average annual growth under 8 %), which seems to be raising few problems in assuring an increasing (or at least a stable) budget for both the space programme and for social reforms.

However, since the country has reformed so little during a period of rapid expansion, many Chinese intellectuals are becoming cynical about reforms and fear that in a soft-landing scenario, these tasks will not be pursued. The Chinese scholar Deng Yuwen has even labelled the Hu Jintao–Wen Jiabao era a "lost decade", because the two leaders missed the opportunity offered by 10 % annual growth rate to implement needed socio-economic and democratic reforms.[80] With his kind of reasoning, one could expect that in the future the government might be in for a much rougher time in implementing reforms and thus also in maintaining its legitimacy.

Given its extremely high visibility, both domestically and internationally, a manned lunar exploration programme could risk being perceived as a mere "vanity project" at the expense of investment in public goods. Especially if China's leadership is unable to satisfy the demand for social reforms, the programme could become the target of numerous critics rather than a tool for supporting "national unity". In addition, two other relevant factors need to be considered.

First, as previously mentioned, from a political-cycles point of view, no direct benefit for the current ruling class is expected to be derived from the launch of a manned lunar exploration programme. Hypothetically, there will be at least two selection cycles—in 2018 and 2023—before Chinese taikonauts land on the Moon, and it can be envisaged that within the next two National Congresses of the CCP, the major portion of China's top leaders will retire.

Yet, a Moon endeavour will be highly beneficial for CCP legitimacy, and hence China has so far developed a three-step strategy for its long-term and visible space programmes (CLEP and the manned spaceflight programme) so as to make sure that every policy cycle can benefit from the accomplishments of long-standing space endeavours. The new leadership might, however, fear being judged more by the solutions it will be able give to the "three success traps" generated by the policies of the past 30 years, rather than by its achievements in space.

In addition, the current size of Chinese space programmes is also problematic: there are too many programmes in existence which, from a scientific, economic, and technological point of view, could be considered as domestically more important (e.g. BeiDou, communication satellites, earth observation satellites, new launcher fleet, etc.) and many existing goals still need to be accomplished. As Stacey Solomone has asked in this regard, "with mounting resource demands by the plethora of space programs, how will aerospace leadership be able to balance national security needs, develop new highly technical space programs, and open space to commercial use? It does not seem likely China will be able to accomplish all of these goals".[81] On the contrary, in a situation of gradual slowdown, it appears

[80] *Ibid.*

[81] *Cit.* Solomone, Stacey (2013). *China's Strategy in Space.* Springer, New York: p. 94.

more likely that China will face the need to prioritise its programme, something that will also cast doubts on the real need for a manned lunar landing.

In fact, Chinese politicians seem to have started believing that the scientific rationales and benefits of a manned programme are unsubstantiated and inadequate and that the strong political motivations for it are not sufficient to send taikonauts Moon-wards. This "role reversal" between the scientific and political community, although surprising, is not a new issue: also in the early 1990s, with the case of a first lunar probe, which was proposed to celebrate the return of Hong Kong to China, Chinese politicians cast serious doubts on the scientific nature of the project and eventually did not approve it.[82]

Although, as mentioned earlier, a full consensus on what to do with the programme has not yet been reached, it can be envisaged that all the likely trade-offs have already entered the political debate.

It must also be emphasised that the desired solutions to these trade-offs may vary significantly among the members of the ruling class, which is not embedded in a set of monolithic institutions, but on the contrary is a "melting pot" of often contrasting visions and thoughts.

The following section will therefore examine the main divides among Chinese policymakers and elites and assess how they could affect the manned lunar exploration programme.

5.2.2 *What Does the Chinese Leadership Think?*[83]

The existence in China of a one-party system and the lack of free elections and political transparency at international level have long obscured the fact that within the Chinese political and intellectual elites, many lively and articulate debates are taking place. Contrary to common belief, the Chinese regime is not a monolithic institution, but—at least in high-level circles—it allows dialectical thinking and open political confrontation, which ultimately, however, has to be re-channelled into a harmonious (*he*) and inclusive consensual policymaking process.

There are many schools of thought concerning political, economic, and foreign policy-related issues, and understanding these might prove useful exercise for sketching what the future of the manned lunar exploration programme might be.

As usefully documented by the European Council on Foreign Relations (ECFR), the current debate among Chinese officials and intellectual elites lies along some

[82] See Sect. 4.2.2.

[83] For the elaboration of this section, I have drawn heavily on the stimulating researches of Mark Leonard; see Leonard, Mark (2008). *What does China Think?* HarperCollins Publishers, London; see also Leonard, Mark (2012). "What does the new China think?". In: Leonard, Mark (ed). *China 3.0*. European Council on Foreign Relations, London: pp. 9–24.

Table 5.2 Chinese elite's schools of thought (*Source*: ECFR)

Economics	New Left	New Right
	Free-Market Egalitarians	
Politics	Neo-conservatives	Liberals
	Neo-Maoist	
Foreign policy	Globalists	Neo-Comms
	Defensive realists	

major "fault lines" in three main domains: economics, politics, and foreign policy. The debate can be graphically represented in Table 5.2:

(a) In the sphere of the economy, "the main divide is between a social Darwinist New Right that wants to unlock entrepreneurial energy by privatising all the state-owned companies and an egalitarian New Left that believes the next wave of growth will be stimulated by clever state planning".[84]

For the New Right representatives—who have Premier Li Keqiang as their champion—China should restart the interrupted privatisation of the SOEs, boost financial system liberalisation, and introduce a fundamental shift towards the *marketisation* of China's development model. Given these aims, it is clear that the target of their reforms is the supply side of the economy. On the other side, the New Left representatives believe that "bringing the state back in" to the economic processes is the only way to face China's looming economic difficulties. Admittedly, they do not reject the market but are strong supporters of the primacy of the "visible hand" (the state) over the "invisible hand" (the market) and therefore prefer "clever state planning" rather than wild and doubtful privatisation. Indeed, there is a widespread belief among the New Left representatives that the "current direction of economic liberalisation has led to a nexus between the CCP elites and business interests who have plundered the nation's assets under the cover of privatisation".[85]

Some New Left representatives, the so-called Free-Market Egalitarians, have a special focus on inequality: they emphasise that the government has been so obsessed by economic growth that it has completely forgotten everything else, with the result of creating a highly unequal society. For them, the problem can now only be solved by raising wages, reforming the *hukou* system, assuring access to social welfare (education, healthcare, pensions, etc.), seizing the opportunity of green development, and "socialising" capital.[86] If the New Right's focus is on the supply side, the reforms advocated by the New Left clearly focus on the demand side: an important goal for

[84] *Ibid.* p 11.

[85] Van Wie Davis, Elisabeth (2009). "Governance in China in 2010". Asian Affairs. Vol. 35 (4): 199.

[86] Leonard, Mark (2012). "What does the new China think?". In: Leonard, Mark (ed). *China 3.0*. European Council on Foreign Relations, London: p. 15.

stimulating the economy is to boost domestic consumption, rather than supporting exports though cheap labour.

(b) In the political domain, "the main divide is between political liberals who want to place limits on the power of the state, either through elections, the rule of law, or public participation, and neo-authoritarians who fear these measures will lead to a bureaucratised collective government that is unable to take tough decisions or challenge the vested interests of the corrupt, crony capitalist class".[87]

The core issue of this debate is the type of legitimacy the government should seek in a time of slower growth. While the neoconservatives would opt for the adoption of political solutions within the current institutional framework of authoritarianism, liberals would prefer to introduce institutional innovations. Although for liberals elections might eventually be essential in China, they are also aware of the risks involved (mainly their potential anti-systemic effects);[88] thus at the moment they prefer other formulas, such as strengthening the rule of law and integrating public consultation, referenda, and surveys as the central feature of decision-making.[89]

For Neo-conservatives and Neo-Maoists, on the other hand, these innovations would only lead to greater bureaucratisation of a state that has already been "captured" by corruption and powerful interest groups, thus making it incapable of making the radical choices that will be needed to legitimate it.[90] Ultimately, they are also convinced that only a charismatic leader like Mao or Deng—together with the party's power structures—can assure legitimacy and take the big, if sometimes painful, decisions for the good of the country. In short, in the political domain, Neo-Maoists want to see a much more political rather than democratic type of government, which includes among its priority tasks the development of more social programmes for the masses.

(c) In foreign policy, "the main divide is between defensive internationalists who want to play a role in the existing institutions of global governance or emphasise prudence and nationalists who want China to assert itself on the global stage".[91] This debate has largely been generated by the unbalanced, "low-profile" foreign policy model pursued since the Deng Xiaoping era. China's economic and political presence in every corner of the world, the

[87] *Ibid.* p. 11.

[88] See, for instance, Jun, Ma (2012). "Accountability without elections". In: Leonard, Mark (ed). *China 3.0*. European Council on Foreign Relations, London: pp. 80–87.

[89] It is worth underlining that features of this debate have become part of the political process: "intellectuals are for instance asked to brief the politburo in study sessions, prepare reports that feed into the party's five-year plans and advise on the government's white paper". Van Wie Davis, Elisabeth (2009). "Governance in China in 2010". Asian Affairs. Vol. 35 (4): 199.

[90] See, for instance, Wei, Pan and Shang Ying (2012). "A New Approach to Stability Preservation" In: Leonard, Mark (ed). *China 3.0*. European Council on Foreign Relations, London: pp. 88–94.

[91] Leonard, Mark (2012). "What does the new China think?". In: Leonard, Mark (ed). *China 3.0*. European Council on Foreign Relations, London: p 11.

size attained by its fast-rising economy, and its "fast-rising power" in the political and geostrategic domains have made this model difficult to sustain.

In particular, many nationalist officials and intellectuals (labelled Neo-Comms) have started to strongly criticise the too low-key position of the country concerning relevant issues on the international agenda and are now urging China to pursue a more assertive foreign policy and to translate its economic strength into political and military power. A continuation of the "prudent" approach in China's international relations is, on the contrary, supported by the so-called globalists and defensive realists. However, there are some differences among them: while the former would opt for an approach of "creative involvement", whose goal is to integrate China into the existing international order, the latter would prefer an even more cautious diplomacy. They strongly criticise China's excessive assertiveness in its territorial disputes, emphasising that this only creates the conditions for the revitalisation of American power in Asia. The recent adoption by the United States of the "pivot to Asia" strategy has only confirmed their fears. These fears are not, however, shared by Neo-Comms, who ultimately advocate a drastic "paradigm shift" in China's foreign policy: instead of looking for a multipolar world, the country should devote its energies to the forging of a new bipolar era.[92]

The various different groupings in these three main domains have through the years experienced (and will experience in the future) ups and downs in the battle to see their ideas approved and implemented as official policies. At regional government level, these debates have often translated into the adoption of different "experimental models", which correspond to the recipes proposed and supported by one or other schools of thought. The regions of Guangdong and Chongqing, for instance, are usually seen as the two best "competing archetypes" in the battle of ideas among Chinese elites.[93] At central government level, these debates have produced an iterative decision-making process, which is adjusted on a continuous basis, according to the relative power of the different groupings and to the evolution of the socio-economic environment. The fact that these debates are eventually synthesised in apparently stiff and coherent policymaking should not, however, overshadow their existence, the understanding of which is of extreme importance

[92] See, for instance, Xuetong, Yan (2012). "The weakening of the unipolar configuration". In: Leonard, Mark (ed). *China 3.0*. European Council on Foreign Relations, London: pp. 112–117.

[93] The difference between these two "archetypes" has been illustrated by M. Leonard as follows: Guangdong, a prosperous coastal region, has stood for a quest to move up the value chain economically while using a free media, civil society, and political openness to quell social tensions. Chongqing, by contrast, was about turning a backward inland province into a laboratory for egalitarian social policies and domestic consumption. *Cit*. Leonard, Mark (2012). "What does the new China think?". In: Leonard, Mark (ed). *China 3.0*. European Council on Foreign Relations, London: p 11.

Table 5.3 China's lunar exploration programme: the internal debate

Lunar programme	Pro	Con	Conditional support
Political reasons	Neo-Comms	Liberals	Globalists
	Nationalists		Neo-Maoists
Economic reasons	New Left	Left Egalitarians	New Right

for drawing conclusions on the country's space programme as well as for Western policymakers more generally.

5.2.2.1 So What About the Lunar Exploration Programme?

The existence of these debates strongly influences and permeates the environment in which the Chinese space programme operates and is bound to produce tangible implications for the manned lunar exploration programme. It could therefore be useful to try to extrapolate the main positions of the Chinese elites and project them on to the country's space programme, in order to identify the fault lines among decision-makers with regard to the lunar endeavour.

Using what has been discussed so far, it is possible to build up a table—similar to the one previously presented—which shows where the different schools of thought could be positioned in the debate surrounding the manned lunar exploration programme. Table 5.3 envisages the existence of two main parameters—economic and political—and contemplates three different positions relative to the quest for the Moon: favourable, against, and conditional support. Table 5.3 places the different schools of thought of Table 5.2 in this matrix.

Such a schematic is, of course, inherently inaccurate and simplistic, yet hopefully, still illustrative. With no trace of publicly available "position papers" by Chinese officials, the identified positions of the different schools of thought represent only Weberian-style "ideal types", which are considered for their salient characteristics. Ultimately, a wide range of considerations might be taken into consideration by the different schools of thought, thus making reality much more complex and fluid than captured by this table.

The aim of the schematic is not, however, to depict known facts but to give a sense of the possible debates taking place within Chinese elites with regard to the country's lunar endeavour. In doing this, it ultimately aims to emphasise the need for Western policymakers to change their mental maps and avoid a process of *reification* when dealing with China.

The Political Debate

From a political point of view, the main divide is assumed to be between Neo-Comms and nationalists on one side and the Liberals on the other.

(a) Nationalists and Neo-Comms can be regarded as the strongest supporters of the lunar programme. Nationalists are expected to emphasise the positive outcomes this endeavour could bring to the CCP and to Chinese citizens: legitimacy for the former, national pride and social cohesion for the latter. In nationalists' eyes, an eventual taikonaut landing on the Moon would not only strengthen the political and symbolic dimensions of Chinese identity[94] but would also be an essential and highly visible step towards the realisation of the *Zhongguo meng* (the "Chinese dream").

Neo-Comms are also expected to firmly support the programme, as they would consider this endeavour a very powerful tool for projecting the country's soft and hard power on the international arena. In particular, their support would be correlated with their efforts to win increased international prestige after the period of "low-profile presence" on the world stage during the Deng Xiaoping era. Beyond symbolising its regained superpower status, Neo-Comms would in addition consider this ambitious endeavour an indispensable component for enhancing China's comprehensive national power (CNP).

(b) Contrarily, it can be imagined that the Liberals, who are willing to put limits on the government's power, might argue that in a time of economic challenges, social reforms should have the priority in the regime agenda. They might argue in this regard that the lunar endeavour could be a self-defeating project for regime legitimacy: instead of boosting the sense of unity and fostering nationalism, it might exacerbate social tensions and thus act against the CCP's "*tianming*".

(c) Finally, in a position of conditioned support, one can expect to find Globalists and Neo-Maoists. For the former, China should embark on this complex endeavour, but only on the condition that it does so via international cooperation. The lunar programme could thus become the symbol of China's will to act as a responsible power in the global arena, as well as tangible proof of its intention to realise the recently adopted vision of a *Shijie meng* (the "world dream"). For their part, Neo-Maoists could argue that the lunar programme has the power to become the sign of a renewed Maoist-style spirit of collectivism: it would represent "symbolic collective esteem", the victory of state planning over counterproductive *marketisation* and the primacy of China's SOEs over private initiatives. The support these representatives would provide is, however, conditioned by their possible economic concerns.

---

[94] According to sinologists, Chinese identity comprises three dimensions. The first is "the political dimension of sovereignty, which means the full acceptance of the political and religious legitimacy of *Tianxia* [...]. The second is the symbolic dimension of civilisation, which means the sharing of common cultural practices through fully accepting the values of *Zhongguo*. The third is the [ethnic] genealogical dimension of family ramifications, which refers to the direct [...] descendants of Chinese ancestors". Quoted from: Mazzei, Franco (2003). "Intercultural Variables and Japanese Socio-Economic Performances". In: Lavagnino, Alessandra Cristina et al. (eds). *Reflections on Asia. Essays in honour of Enrica Collotti Pischel*. Franco Angeli, Milano.

Socio-economic Factors

From a socio-economic point of view, interestingly, the main divide could be among representatives of the Left itself: more precisely between the New Left and the Free-Market Egalitarians.

(a) New Left representatives can be expected to be in favour of the lunar programme. A successful such endeavour would in fact form part of the "smart state planning" measures they recommend to meet the looming economic difficulties: practically it could represent an effective way to spur technology innovation and boost the country's economic growth and ultimately improve the welfare level of the whole society. For the New Left, these targets can be reached only if the government is allowed to be more active in owning or directly controlling the country's major industries, in business regulation, and in directing scientific and technological advances through targeted initiatives such as the lunar exploration programme. More generally, the aerospace sector, with its hybrid connotations worldwide, represents itself a valid alternative to the free-market model of capitalism supported by the New Right.[95]

(b) On the other hand, Free-Market Egalitarians could be expected to strenuously oppose the implementation of the lunar exploration programme, simply because they think that China has focused too much on its ambitions of *grandeur* at the expense of other projects. They believe that, in a time of economic slowdown, the country should start to rethink its priorities and focus on the enhancement of quality welfare and the promotion of equal opportunity for everyone. In their eyes, if the CCP regime wants to avoid widespread and unmanageable social tensions, the pauperised masses should stop being "fed" with vanity projects like the lunar endeavour and instead be provided with tangible and more direct benefits (e.g. wage raises, quality state-backed healthcare, pensions, education, social protection for unregistered households, etc.).

(c) Finally, New Rightists could place themselves in a position of conditional support. In their opinion, China should perhaps embark on the lunar endeavour but on condition that the implementation of the programme does not become the pretext for avoiding or postponing the privatisation of the SOE sector, including CASC and its subsidiaries, and the commercialisation of space-related products and services. As with the dynamic of the telecommunication corporations, government involvement in the aerospace industry should ultimately be limited, and controlled market freedom for CASC provided in order to boost the spirit of Chinese space business and entrepreneurial culture.

[95] It should be noted that also in the United States and Europe, the aerospace sector is not a model of free-market capitalism and that industrial competitiveness is largely determined by nonmarket factors. For an assessment of the space sector in Europe, see Hayward, Keith (2011). "The Structure and Dynamics of the European Space Industry Base". ESPI Perspectives 55. European Space Policy Institute, Vienna: 3.

The above presented ideas and positions are ultimately just sketches of a possible debate, whose progression is bound to have different implications for China's lunar ambitions. When combining these debates with the economic variables presented in Sect. 3.1, it becomes even harder to predict what the most probable course of the Chinese lunar endeavour will be.

5.2.3 What Does Chinese "Civil Society" Want?

In most of the "Western" spacefaring nations, the successful realisation of a highly complex and socially visible endeavour such as the manned lunar exploration programme would not only rest on the establishment's political will but ultimately also on the support of what the democracies of the West label "civil society". In a country like the United States, one of the most paradigmatic cases, public support for the achievement of the nation's ambitions is a determining factor, and civil society plays a crucial role in deciding the direction of policymakers' choices.

The situation in this respect turns out to be problematic at best in the case of China. In a number of contributions, sinologists have claimed that China does not have a civil society in the normal sense. Besides the fact that the notion of *civis* is extraneous to China's sociopolitical tradition, the development of a bourgeois class (the linchpin of every civil society) has followed a quite different trajectory from its "Western counterpart".[96] It is true that the structure of society has significantly changed during the reform period and that new elites (social classes) have made their entry into the public sphere; however, as the French sinologist Marie-Claire Bèrgere has poignantly remarked, these new classes are at best expression of a *consenting bourgeoisie*, created by the regime through the transformation of former bureaucrats.[97] The Italian sinologist Franco Mazzei has noted that the recently settled economic elites have very little of the bourgeois: they are made up of managers and enterprise cadres—including aerospace professionals—legitimised by their economic and technical competence but sensitive to power inducements and devoid of own values. Significantly, almost 90 % of the Chinese new rich have family relationships with high-ranking officials and Party cadres. In addition, the dependence of professional business and middle class employees on the state for employment and resources makes them reluctant to oppose the state.[98]

This type of analysis is also now present in much of the current affairs journalism: after initially assuming that the emerging market economy would create a sizeable middle class that would then become the backbone of civil society and a driving force for Western-style democracy, journalists then retracted this position and even started to debate whether the concept of civil society was ultimately

[96] The bourgeoisie in China has not been the driving force for the emergence of capitalism, the development of which was instead managed by the state. See Sect. 5.1.1.

[97] Bergère, Marie-Claire (2000). *La Chine de 1949 à nos jours*. A. Colin, Paris.

[98] Mazzei, Franco, Vittorio Volpi (2006). *Asia al Centro*. Universitá Bocconi Editore, Milano.

relevant. For instance, they have argued that "Chinese citizens largely see themselves as submissive subjects looking to officials to protect and safeguard their interests"[99] rather than becoming promoters of their own interests, especially concerning sociopolitical rights.

This evidence does not explain, however, why the central government has put so much effort into feeding its citizens with their country's space achievements. Indeed, these efforts seem to demonstrate the government's need to legitimise its increasing expenditure on space and thus the relevance of the space programme for Chinese "civil society".

Ultimately, even if it is true that the issue of civil society in China is conceptually difficult, it should not be forgotten that even in China public support is a determining factor for success in terms of achieving national collective esteem. And it should not be forgotten that Deng Xiaoping's dream was to create a society in which "citizens would be comfortable enough to lift their eyes above the daily struggle of subsistence";[100] hence his success has effectively contributed to the creation of the conditions for the emergence of citizens who are more aware of the policies adopted by their government. In addition, the country's rapid growth, the improvements in education and living standards, and the accessibility of the Internet and social media (albeit with its problematic trimmings[101]) have combined to strengthen also the position of those social classes not directly linked to the political sphere. And, as previously described, this growing middle class will ultimately act as a "catalyst for improved governance, better delivery of public services, and the empowerment of civil society".[102]

Aware of the numerous ifs and buts of this issue, it is important to investigate what the civil society attitude towards the country's lunar ambitions might be, especially in the light of the previously described framework of social and economic challenges.

It can be argued that an enthusiastic space identity and culture have emerged within Chinese society, especially thanks to the successes of the manned spaceflight programme. Stacey Solomone has pointed out that Yang Liwei's flight was the main event that paved the way for the Chinese people to think of themselves once again as part of a true spacefaring nation since the historic 1970 launch of the

[99] Van Wie Davis, Elisabeth (2009). "Governance in China in 2010". Asian Affairs. Vol. 35 (4): 195–211.

[100] Deng Xiaoping statement on the realisation of a *Xiaokang* society (1979). Quoted from: Leonard, Mark. "China's Affluence Crisis". Reuters US Edition. 31 July 2012. Web. http://blogs.reuters.com/mark-leonard/2012/07/31/chinas-affluence-crisis/. Accessed 04 October 2013.

[101] If it is true that the Internet has been often used to promote social requests, in the Chinese case it is also true the other way round: the government often uses the Internet to shape the public opinion. In this regard, Michael Anti has underlined that the government's strategy of "blocking and cloning" social media sites could actually reinforce the one-party state rather than weaken it. See Anti, Michael (2012). "The Chinanet and Smart Censorship". In: Leonard, Mark (ed). *China 3.0*. European Council on Foreign Relations, London: pp. 100–105.

[102] The World Bank and the Development Research Center of the State Council, P. R. China (2013). *China 2030: Building a Modern, Harmonious, and Creative Society*. The World Bank, Washington DC: p. 9.

DFH-1 satellite. Each successful human spaceflight has strengthened popular enthusiasm and facilitated the diffusion of space-related themes into society. "The central government's efforts to publicize these events has allowed for all Chinese to share in the successes of the human space flights. Even the poor Chinese farmer can be proud of China's human space program despite not having much impact on his life in rural society".[103]

It is especially the emerging middle class, however, that has dipped into this "space-generated elation". The success of space-related products (from the food industry to the entertainment industry, passing through the creation of civic groups on space issues) can therefore be seen as a clear manifestation of the will of the middle class to support the country's space ambitions. For them, a future landing on the Moon would certainly represent an important achievement that celebrates their wealth and social status, as well as being the most effective symbol for magnifying their national pride. Indeed, for a country that understands its superpower status as having been granted by nature,[104] in the eyes of the people the realisation of the lunar endeavour would be perceived as the strongest benchmark for feeling proud of China's resurgence. As the Chinese scholar Lu Zhongwei argued in an article published by the *Beijing Review*, "the manned lunar landing is an event that will excite Chinese citizens nationwide. The manned moon landing would help the country to reclaim its glory and splendour; just as (or even more) the Olympic Games 2008 in Beijing, the 60th anniversary of China in 2009 and the World Expo in 2010 in Shanghai, it will be the evidence of China's rise to power".[105]

To conclude on this point, it can be argued that Chinese society would be likely to support their country's ambitions. The obvious precondition for this to happen, however, is that the pursuit of the country's space ambitions is at least accompanied by steady and continuous advances in the wealth levels of Chinese society.

The situation, however, is rapidly changing and so could the attitude of Chinese society towards the country's space programme. If the looming socio-economic difficulties eventually take hold, one could envisage the emergence of quite opposite feelings towards the programme. In a time of economic difficulties, the current enthusiasm for the space programme could dwindle or even vanish. If the government is not able to assure increasing wealth for its population, the benefits expected from a manned lunar exploration programme will inevitably be compromised. Put

---

[103] Solomone, Stacey (2013). *China's Strategy in Space*. Springer, New York: p. 69.

[104] As Yan Xuetong poignantly argued in a renowned essay: "the rise of China is granted by nature. The Chinese are very proud of their early achievements in the human history of civilization. In the last 2,000 years China has enjoyed superpower status several times, such as during the Han Dynasty, the Tang Dynasty, and the early Qing Dynasty. Even as recently as the 1820, just 20 years before the Opium War, China's GDP accounted for 30% of the world GDP. This history of superpower status makes the Chinese people very proud of their country on the one hand and very sad about China's current international status on the other hand. They believe China's decline is a historical mistake which they should correct". *Cit.* Xuetong, Yan (2001). "The Rise of China in Chinese Eyes". Journal of Contemporary China Vol. 10 (26): 33–39.

[105] Zhongwei, Lu. "China Shoots for the Moon". Beijing Review. 18 September 2003.

simply, the pursuit of a "*panem et circenses*" strategy for gaining public approval will not be effective if the government does not guarantee the former!

Under that scenario, not only could the lunar programme be perceived by Chinese society as a useless project that does not bring any concrete benefits but also as something that comes at the expense of the welfare of the people. Ironically, it could also act as a catalyst for fomenting popular discontent with government policies and the lack of social reforms, rather than as an instrument of greater legitimacy. The clear risk for the CCP is that the programme could be instrumentalised by the detractors of the regime. In that scenario, those elites against a lunar programme (e.g. Free-Market Egalitarians and Liberals) could find themselves in a position of relative power, making the realisation of this endeavour precarious.

5.3 Considerations on China's Technological Innovation

China's manned lunar exploration programme represents a very ambitious activity that not only requires the expenditure of many billions of RMB and strong political and public support; it also involves difficult technological feats, fraught with dangers of failure. As discussed in Sect. 4.3, China is investing heavily and steadily in the development of a whole set of technologies in order to pave the way for its lunar endeavour, but questions as to its technological capabilities, and in particular innovation capabilities, still abound.

In fact, although China has been extremely successful in integrating foreign technologies into its space systems, it is still far from becoming a source of independent innovation in products and processes, as well as in relevant space-related technological breakthroughs. Indeed, these types of innovation will ultimately prove to be key for the realisation of an "indigenous" manned lunar exploration programme. Only through a genuine process of innovation can the objective of a "red Moon" be readily reached.

If, on the one hand, innovation is a conditioning factor, it is, on the other hand, also the distillation of a multitude of other factors, which need to be better understood in order to determine why a former world leader in scientific and technological discovery is now struggling to become innovative. In the following sections, consideration of China's efforts to become a genuine source of technological innovation will be discussed. The points of strength on which the "space orgware" can rely and the weaknesses it will face will be investigated subsequently, and their impact on the lunar endeavour assessed.

5.3.1 Towards Genuine Independent Innovation?

For more than a decade now, in their official documents and public statements, China's policymakers have stated their intention to reshape the country's position as the world's manufacturer.[106] Despite the size of the national economy, they have gradually become aware that China is not an economic powerhouse, primarily because of its weak innovative capacity. To paraphrase the winner of the Nobel Prize in economics, Paul Krugman, the key to China's economic growth was its *perspiration* (manufacturing capacity) rather than its *inspiration* (technology innovation).[107] From an economic point of view and in terms of international prestige as well, this position can no longer be considered satisfying. As already noted in Chap. 3, policymakers' solution is to push for a "national innovation system" so as to break the country's dependence on foreign technology and move away from the "Made in China" paradigm towards a "Created in China" or "Innovated in China" paradigm.

Zizhu chuangxin, the Chinese term indicating indigenous innovation, was introduced in a 2006 state-issued report, entitled "Guidelines on National Medium- and Long-Term Program for Science and Technology Development".[108] As already mentioned, these guidelines aim to "lay... the foundation for China to become a science and technology power by the middle of the 21st century" and identify a number of ambitious targets to support the development of an indigenous innovation system. In particular, the document recommends that R&D expenditure reach 2.5 % of China' GDP by 2020, in order to reduce the country's reliance on foreign technology to 30 % or below.[109] In addition, it is expected that by the same year the number of patents granted to Chinese nationals will rank among the top five in the world, and China will have developed a number of breakthrough technologies, thanks to the implementation of 16 megaprojects in different sectors (one of which is specifically dedicated to manned spaceflight and lunar exploration).[110]

Besides these goals, a further objective is to "enhanc[e] original innovation through co-innovation and re-innovation based on the assimilation of imported

---

[106] See, for instance, the below reported Jiang Zemin speech in August 1999 and Hu Jintao speech of January 2006 on the role of technology innovation, in which he stressed: "in the face of international scientific development and increasing international competition, by seeing the development of science and technology as a central thread in the development strategy and actively committing to its progress, China can seize the opportunity for development".

[107] Krugman, Paul (1994). "The Myth of the Asian Miracle". Foreign Affairs. Vol. 73 (6).

[108] See The State Council of the People's Republic of China. The National Medium- and Long-Term Program for Science and Technology Development (2006–2020). An Outline. Beijing, China. 2006. Available at: http://sydney.edu.au/global-health/international-networks/National_Outline_for_Medium_and_Long_Term_ST_Development1.doc. See also Segal, Adam. "China's Innovation Wall. Beijing Push for Home-ground Technology". Foreign Affairs Snapshots. 28 September 2010. Web. http://www.foreignaffairs.com/articles/66753/adam-segal/chinas-innovation-wall. Accessed 19 October 2013.

[109] To compare the reliance on foreign technology, in 2006 it was estimated to be the 60 % and the 2006 gross expenditure on R&D was 1.3 % of China's GDP. See "China issues guidelines on sci-tech development program". Chinese Government's Official Web Portal. 9 September 2006. Web. http://www.gov.cn/english/2006-02/09/content_184426.htm. Accessed 23 October 2013.

[110] It has to be also underlined that while the Guidelines identified the goals and specific sector to focus, it was the 11th Five-Year Plan (2006–2010) for high-technology industries that formally detailed the 16 megaprojects. *Ibid.*

technologies".[111] Thus, the 16 megaprojects, as the "major carriers of uplifting indigenous innovation capacity", will be *assimilating and absorbing* advanced technologies imported from outside China so the country can "develop a range of major equipment and key products that possess proprietary intellectual property rights".[112]

This type of innovation, usually labelled "innovation with Chinese characteristics" is clearly worrisome. Indeed, it has provoked increasing criticism. European and US governments and business communities in particular inevitably label such a practice copying or plagiarism: they claim that Chinese officials have not only shown an intention to tolerate the practice of copying but have even encouraged it.[113] As a result of this industrial policy, an increasing number of disputes have been opened at the WTO, with unpleasant results for the academic and scientific community.[114]

It should, however, be pointed out that in China—and more general in the *Sinic world*—the practice of copying has deep cultural roots that will make it difficult to completely abandon this "exercise".[115] It is perhaps not a coincidence that other north-east Asian governments have raised the problem with much less intensity.[116] In addition, in China the practice of copying is ultimately considered an essential part of the creative process. Copying is not conceived of merely as a knockoff of certain technologies or products: it is the starting point of what can be defined as *recombinative innovation*, realised through importation, absorption, assimilation, and re-innovation. This *recombinative innovation* has ultimately proved highly

---

[111] Raustialia, Kal and Christopher Sprigman (2013). "Fake It Till You Make It. The Good News about China's Knockoff Economy". Foreign Affairs Vol. 92 (4): 25–30.

[112] *Cit.* McGregor, James (2010). "China's Drive for Indigenous Innovation. A Web of Industrial Policies". Global Regulatory Cooperation Project—US Chamber of Commerce. APCO worldwide. Web. http://www.uschamber.com/sites/default/files/reports/100728chinareport_0.pdf.

[113] Raustialia, Kal, and Christopher Sprigman (2013). "Fake It Till You Make It. The Good News about China's Knockoff Economy". Foreign Affairs Vol. 92 (4): 25–30.

[114] The US government has, for instance, banned Chinese scientists who were involved in research projects at NASA and advisory bodies in both countries from continuing their research in the United States and even from attending NASA conferences. Recently this ban has been partially reversed. See "Nasa reverses conference's ban on Chinese scientists". BBC News. 21 October 2013. Web. http://www.bbc.co.uk/news/world-asia-24618824. Accessed 21 October 2013.

[115] It is well known that *Sinic* countries (China, Japan, Taiwan, the two Koreas, and Singapore) may be considered "re-elaboration cultures", in the sense that the same "object or process" can be continuously copied or repeated, without losing its original value. In these countries, the concepts of original and copy carry quite different meanings than in "Western cultures". Especially in literary and artistic production, sinologists have explained that "original" means a sort of conformity to a primary source. A work can thus be regarded as original when it conforms to the primary source. For these cultures, this conformity is never perceived as either a mere slavish imitation nor as plagiarism but as something that possesses its own value, as it expresses deference or recognition for predecessors. In addition, what ultimately counts for these cultures is the final product, not its point of departure.

[116] See, for instance, the map of disputes between the WTO Members. "Map of Disputes between WTO Members". World Trade Organisation (2014). Web. http://www.wto.org/english/tratope/dispu_e/dispu_maps_e.htm?country_selected=CHN&sense=e. Accessed 5 January 2014.

successful for China's economic and technological catch-up as well as for advancing its global competitiveness.

In an interesting essay published in *Foreign Affairs*, Kal Raustiala and Christopher Sprigman have observed that the most successful innovative products copied by China, such as the iPhone (*Hiphone*), Twitter (*Weibo*), and YouTube (*Youku*) are not only "real fakes"—in the sense that these products do not try to hide their origins nor aim at being exchanged for the original—but add value to the product itself. *Youku*, for instance, which began as an undistinguished clone of YouTube, evolved into a platform also delivering original content and thus into a serious competitor to traditional broadcast TV in China, an important feat that the original service lacks.[117] In other words, China's *assimilating* and *absorbing* innovation could have *in se* the potential to become genuine innovation as well.

Similarly, in its space programme, China has become "quite adept at acquiring foreign aerospace technologies and creating new and improved models by recombining them in novel ways and *sinifying* them rather than developing new technologies from scratch".[118] Once acquired, these technologies have been successfully incorporated into the country's space systems. As Stacey Solomone has pointed out, on the occasion of the launch of the Chang'e-2 lunar orbiter, even President Hu Jintao recognised the successes of Chinese *recombinative innovation* capabilities.[119] The most evident example of Chinese ability in *recombinative innovation* in space is presented by the Shenzhou spacecraft, which, in spite of the numerous similarities with the Russian Soyuz, has undergone a process of significant transformation (see Sect. 4.2 for analysis of similarities and differences).

It can therefore be expected that China will to a large extent continue to make use of its *recombinative innovation* capabilities, which could ultimately serve as the basis for initiating a process of genuine innovation. However, what in the eyes of the Chinese leadership has emerged in parallel are the intrinsic limitations of this path to innovation and the need for China to transcend its longstanding game of technological catch-up. On the enterprise side, this need has been emphasised best by Li Guojie, the chairman of China's Dawning Corporation, who stated: "Our spirit of innovation is to avoid following the same route as the global industry leaders. We will never catch up with large multinationals if we follow their strategies. You can't leapfrog when you are following others in the same direction".[120] On the government side as well, Liu Yandong, the leading State Councillor responsible for technology policy, underlined this in 2007 by affirming: "the majority of the market is controlled by foreign companies, most core technology relies on imports, the situation is extremely grave as we are further pressured by

[117] Raustialia, Kal and Christopher Sprigman (2013). "Fake It Till You Make It. The Good News about China's Knockoff Economy". Foreign Affairs Vol. 92 (4): 25–30.

[118] Another interesting observation made is that China, "aware that with recombinative innovation comes both explicit and implicit knowledge, has been extremely successful in sinifying the explicit knowledge so that implicit influences are reduced". See Solomone, Stacey (2013). *China's Strategy in Space*. Springer, New York: p. 37.

[119] *Ibid.* p. 37.

[120] Quoted from: *Ibid.* p. 36.

developed countries who use blockades and technology controls—if we are not able to solve these problems we will forever be under the control of others".[121]

Especially for the successful realisation of a complex endeavour like the manned lunar exploration, the innovations introduced by an "assimilating and absorbing strategy" might not be sufficient, unless technology transfers from the United States take place, something that is highly unlikely.[122] Space programme policymakers have made it clear that "they cannot continuously leapfrog technologies because this means they will have to follow foreign strategies in the global space community. China does not want to continue to follow and play S&T catch -up, but is striving to learn to innovate and choose its own path to space".[123]

The need for a new direction in stimulating genuine innovation is already well shown by the particular mix of policies contained in the 2006 Guidelines. Beyond the "assimilating and absorbing" strategy, the document introduces new paths to foster technological innovation: it combines the traditional top-down, state-directed policies with a series of bottom-up measures, centred on university–industry collaboration, small start-ups, and venture capital, which seem to follow a Silicon Valley model.[124] The Guidelines particularly emphasise that the state's R&D results need to be shared with industry and that state-level engineering labs and various industrial engineering centres with joint R&D groups from companies, universities, and scientific institutes must be established.[125] In the long run, the objective is clearly to develop an ecosystem that stimulates broad-based creativity and innovation.

The cautious opening to these "Western models" clearly shows not only the pragmatism of the Chinese authorities in reaching their targets but also a certain historical mistrust in the unconditional acceptance of non-native ideas or models. It is in this regard noteworthy that during the period of institutional reforms launched by the Qing dynasty, an attempt to modernise the country's economic and industrial apparatus was pursued—at that time unsuccessfully—through the integration of Western-derived models with Chinese ones. *Zhong xue weiti xie xue weiyang* (Chinese knowledge as foundation, Western knowledge as an instrument) was a

---

[121] Quoted from: McGregor, James (2010). "China's Drive for Indigenous Innovation. A Web of Industrial Policies". Global Regulatory Cooperation Project—US Chamber of Commerce. APCO worldwide: p. 17. Web. http://www.uschamber.com/sites/default/files/reports/100728chinareport_0.pdf.

[122] Interestingly, even the United States have to a large extent lost their know-how and are thus now obliged to invest in *reverse engineering* and make up for lost ground.

[123] *Cit.* Solomone, Stacey (2013). *China's Strategy in Space*. Springer, New York.

[124] Segal, Adam. "China's Innovation Wall. Beijing Push for Home-ground Technology". Foreign Affairs. 28 September 2010. Web. http://www.foreignaffairs.com/articles/66753/adam-segal/chinas-innovation-wall. Accessed 19 October 2013.

[125] "China issues guidelines on sci-tech development program". Chinese Government's Official Web Portal. 9 September 2006. Web. http://www.gov.cn/english/2006-02/09/content_184426.htm. Accessed 23 October 2013.

popular slogan for this instrumental integration.[126] However, for many Chinese policymakers and intellectuals, such integration has ultimately been the symbol of the "destruction which the Chinese have inflicted upon themselves by importing—and misapplying—foreign ideas"[127]: a mistake that, for many, should be avoided in the future.

The attempt to diversify the sources and methodologies in innovation and to create more dynamism in the industrial–scientific environment was eventually emphasised in the 11th and 12th Five-Year Plans (2006–2010 and 2011–2015). The latest Plan identifies technology development as the sector—alongside health care and energy—that will receive the biggest boost over the duration of the Plan. In order to achieve the goal of enhancing indigenous innovation capabilities in the technology domain, the document identifies a number of tools that have been utilised consistently.

These tools include: (a) better utilisation of *R&D spending* (the document affirms that the government will invest heavily in science and technology R&D in order to bring about key breakthroughs in targeted technology sub-sectors, including life sciences, space, and nanotechnology); (b) *intellectual property* (China will expedite the implementation of a national strategy on IPR; it will create a legal environment in which IPR are respected and protected and will bring IPR management into the whole process of scientific and technological management); (c) *commercialisation* (the government will bring the research undertaken at government-sponsored universities and research institutions to the marketplace and will urge both large enterprises and SMEs to increase their R&D investments); (d) *administration* (China will strengthen fiscal and financial policies that support high-technology industry, including updating research funding management and venture capital investment systems)[128]; and (e) *education* (the 12th FYP emphasises that education reform in science and technology will be undertaken, and a human resources strategy for finding and nurturing talent will be developed. Initiatives will include improving the scientific achievement evaluations and rewards system, encouraging even more highly educated overseas Chinese to return to China to work and increasing investments in human capital).[129]

Although consistent, these policies might not be sufficient on their own to introduce the paradigmatic shift towards a genuine innovation system. Much will depend on how they are implemented and complemented and ultimately on how

---

[126] For an informed examination of the Qing dynasty's strategy for preserving the essence of Chinese culture while at the same time utilising elements of the West for practical purposes, see Wilkinson, Endymion (2013). Chinese History. A New Manual. Harvard University Press, Cambridge, MA: pp. 477–479.

[127] *Cit.* Leonard, Mark (2008). *What does China Think?* HarperCollins Publishers, London: pp.10–11.

[128] "China's 12th Five Year Plan. How it actually works and what's in store for the next five years". APCO Worldwide. 10 December 2010. Web. http://www.apcoworldwide.com/content/pdfs/chinas_12th_five-year_plan.pdf. Accessed 3 November 2013.

[129] *Ibid.*

China addresses its weaknesses and best benefit from its strengths. The following section will therefore elaborate further on these aspects and discuss whether China will be able to introduce the necessary independent technological innovation to support its lunar programme.

5.3.2 *Orgware Strengths and Weaknesses in Fostering Independent Innovation*

Technological innovation will significantly condition the development, timeframe, and direction of China's space programme. However, technological innovation is itself conditioned by numerous factors including, among others, government policies, the market environment, the education system, the composition of the business sector and its strategic orientation, the quantity and quality of R&D, national and international networking, cultural traditions, and the legal and political system.

In these domains, the institutions and authorities formed to plan and execute the space programme, or *orgware*, can rely on many points of strength; at the same time they need to address serious weaknesses and challenges in putting forward the development of an indigenous and innovative "hardware".[130]

- Starting with government policies, there are undoubtedly positive elements that might foster a genuine process of innovation, but various pitfalls exist. One problem is that, notwithstanding the introduction of the abovementioned bottom-up policies and incentives for enterprises, it is still the government which conducts the bulk of research and development (while effective innovation is usually driven by private enterprises and by their collaboration with universities and research institutes).[131] If it is true that the government has started to urge Chinese enterprises to set up R&D institutes and spend more on R&D and has encouraged them to share the state's R&D outputs,[132] in reality there are only weak incentives for Chinese firms to spend more on R&D. This is because competition is not stimulated and investment rarely yields commercially viable innovations. But without the pressures of free and fair competition—as the World Bank notes in its recent report—"the effects of all the other

[130] See Sect. 2.1.

[131] The United States example has, for instance, shown that innovation used to be an undertaking led by private companies rather than by the government. It then evolved in the 1980s into a collaborative exercise involving universities, research institutes, and government programmes. Block, Fred, and Matthew R. Keller (2008). "Where Do Innovations Come From? Transformations in the US National Innovation System, 1970–2006". The Information Technology and Innovation Foundation. Web. http://www.itif.org/files/Where_do_innovations_come_from.pdf. Accessed 5 November 2013.

[132] "China issues guidelines on sci-tech development program". Chinese Government's Official Web Portal. 9 September 2006. Web. http://www.gov.cn/english/2006-02/09/content_184426.htm. Accessed 23 October 2013.

policies aimed at encouraging innovation will unlikely have much effect".[133] Merely investing 2.5 % of China's GDP in R&D, although a positive step, might not be sufficient to foster innovation if complementary domestic reforms are not introduced and the quality of R&D is not significantly increased.[134]

In addition, the megaproject path, which represents the core of the government strategy for developing indigenous innovation, could have a negative impact on the innovation process. Even during the Guidelines drafting phase—as noted by a 2010 APCO report—many prominent scientists "opposed the idea focusing on mega-projects saying that innovation could only come from individuals or small teams working on projects that they were passionate about and had undergone rigorous examination. They argued that central planning and mega-projects would be wasteful exercises as thousands of people would have to come to consensus to move anything forward".[135] The obvious result of such policies could instead be to stifle competition among scientists and ultimately hamper the prospects for genuine innovation.

The community of Chinese scientists, both inside and outside China, ultimately criticised the plan for giving bureaucrats too much power and strongly recommended that the power of MOST over research directions and funding should be reduced, if not removed altogether.[136]

Nevertheless, the megaprojects path was eventually adopted, confirming once again the adage of "bureaucrats beating scientists". However, the relevance of this "conflict" between bureaucrats and scientists is mainly that it clearly shows China's greatest weakness in becoming innovative—*the excessive power exercised by the state bureaucracy*. This is a longstanding problem: one already documented by the prominent historian Joseph Needham in his monumental *Science and Civilisation in China*. China, once the undisputed leader of global innovation, completely lost its ability to innovate and invent between the fourteenth and fifteenth centuries because of the emergence of a strong "bureaucratic feudalism". As in the past, bureaucracy is again considered a cause of "China's research culture suffering problems of cronyism, mismanagement and ineffectiveness".[137] It has been argued that, with bureaucratic decision-making,

---

[133] The World Bank and the Development Research Center of the State Council, P. R. China (2013). *China 2030: Building a Modern, Harmonious, and Creative Society*. The World Bank, Washington DC: p. 21.

[134] *Ibid*. pp. 34–36.

[135] *Cit*. McGregor, James (2010). "China's Drive for Indigenous Innovation. A Web of Industrial Policies". Global Regulatory Cooperation Project—US Chamber of Commerce. APCO worldwide: pp. 11–12.

[136] *Ibid*. p. 12.

[137] See Shi, Yigong and Yi Rao. "China's Research Culture". Science. Vol. 329. 3 September 2010. Web. http://www.swissnexchina.org/resources/chinas-research-culture.pdf. Accessed 20 November 2013. See also Huang, Yangzhong. "The US is Quietly losing its Innovation Edge to China". The Diplomat. 27 October 2013. Web. http://thediplomat.com/china-power/the-us-is-quietly-losing-its-innovation-edge-to-china/. Accessed 20 November 2013.

"money would be allocated to mediocre projects based on personal connections instead of pursuing real science and technology development".[138] A popular ditty in Chinese scientific circles is "small grants, big review; medium grants, small review; big grants, no review"[139]; but to be effective, research should instead always be complemented by a stringent and disciplined process of refereeing and evaluating findings.[140]

Chinese policymakers and scientists are fully aware of this historical problem, but drastic changes are unlikely to be introduced.[141] It is then ironic—if not paradoxical—that the government's struggle to become innovative is ultimately hampered by its bureaucratic structures and by the government itself.

Where the space programme is concerned, one of the most evident weaknesses in creating an "ecosystem of innovation" is, ironically, the *space orgware* itself. Indeed, although it is the *orgware* that requires the speeding up of innovation in China's space programme, it also acts as a serious constraint on the development of the technology hardware. As noted earlier, the *orgware* of China's space programme is increasingly complex and often subject to reorganisation for political reasons rather than for achieving functional objectives.[142] Rigid bureaucratic policymaking is, at best, inappropriate in assisting the development of new technologies and for adapting to the changes that innovation often brings. In fact, a major problem is that the *orgware* has difficulty in keeping pace with and adapting to rapid technological change.[143] Quite to the contrary, it even tries to control the changes, but ultimately such attempts become detrimental to the possible introduction of disruptive innovation and to encourage much needed "creative destruction" in the technological domain.

---

[138] As underlined by Yigong Shi and Yi Rao, "a significant proportion of researchers in China spend too much time on building connections and not enough time attending seminars, discussing science, doing research, or training students (instead, using them as labourers in their laboratories). Most are too busy to be found in their own institutions. Some become part of the problem: They use connections to judge grant applicants and undervalue scientific merit". *Ibid.*

[139] McGregor, James (2010). "China's Drive for Indigenous Innovation. A Web of Industrial Policies". Global Regulatory Cooperation Project—US Chamber of Commerce. APCO worldwide: 12.

[140] The World Bank and the Development Research Center of the State Council, P. R. China (2013). *China 2030: Building a Modern, Harmonious, and Creative Society*. The World Bank, Washington DC.

[141] A noteworthy reason for this "inertia" can lie in the successful experience offered by the Japanese model. In the experience of China's neighbour, bureaucracy played a predominant role in guiding the innovation efforts, thanks to—among others—its high efficiency. This efficiency is however difficult to achieve in China at the moment. In addition the industrial and market environment was completely different.

[142] See Sect. 2.1. See also Solomone, Stacey (2013). *China's Strategy in Space*. Springer, New York: pp. 17–30.

[143] See Solomone, Stacey (2013). China's Strategy in Space. Springer, New York.

To conclude on these points, China's innovation policies risk failing to provide effective results, unless a general redefinition of the government's role in the national innovation system is undertaken. As the World Bank report has noted, the government should "shift away from targeted attempts at developing specific new technologies and move toward institutional development" (including a redefinition of the bureaucracy's role) and "an enabling environment that supports economy-wide innovation efforts within a competitive market system".[144]

- If it is true that China's bureaucracy and highly hierarchical structures hinder its innovation efforts, one should not forget that it is the *determination* and extreme *focus* of the same structures that offer China its strongest asset. Beijing's resolve to become the world's next technological innovation centre has in fact led the country's policymakers to build up physical, human, and financial capital that on its own is impressive and has few rivals.[145] Sooner or later these assets will produce surprising results. Just a few numbers can give a sense of the potential involved in this focus and determination.

  In the last few decades, research expenditure has been growing at a rate of more than 20 % a year, allowing the country to become the world's second largest R&D investor[146]; it will probably outstrip the United States as the global leader in R&D spending by 2023.[147] This expenditure has been flanked by the establishment of an increasing number of research institutes, clusters, and think tanks (and reinforcement of already existing bodies). CAS, for instance, the leading think tank for science and technology, now has 60,000 staff, 33 % of whom have PhD qualifications.[148] It gives lot of food for thought that today a single Chinese institution alone has more staff than the combined European national academies of sciences.[149] Every year Chinese universities provide

[144] The World Bank and the Development Research Center of the State Council, P. R. China (2013). *China 2030: Building a Modern, Harmonious, and Creative Society*. The World Bank, Washington DC: p. 35.

[145] See Zeng, Ming, and Peter Williamson (2007). *Dragons at your door: How Chinese cost innovation is disrupting global competition*. Harvard Business School Publishing, Boston: pp. 57–88.

[146] "China's R&D investment 2nd in the world: report". China Daily. 15 November 2011. Web. http://www.chinadaily.com.cn/china/2011-11/15/content_14093946.htm. Accessed 7 November 2013.

[147] Huang, Yangzhong. "The US is Quietly losing its Innovation Edge to China". The Diplomat. 27 October 2013. Web. http://thediplomat.com/china-power/the-us-is-quietly-losing-its-innovation-edge-to-china/. Accessed 20 November 2013.

[148] Chinese Academy of Sciences. "CAS statistical data". CAS fact-sheet. Web. http://english.cas.cn/ST/stsd/200909/P020120814348911674163.pdf. Accessed 20 November 2013.

[149] See Statistics of European Federation of Academies of Sciences and Humanities. European Federation of Academies of Sciences and Humanities. Web. http://www.allea.org/Pages/ALL/4/731.bGFuZz1FTkc.html. Accessed 20 November 2013.

"platoons of graduates", and by 2030, China is expected to have up to 200 million graduates: that will be more than the entire workforce of the United States![150]

Of course, these numbers do not automatically guarantee the success of Chinese innovation efforts, but it appears statistically unlikely that they will not introduce potentially surprising outcomes and the introduction—at least occasionally—of technological breakthroughs. There are, however, other relevant strengths and challenges that need to be considered.

- A fundamental element in creating an "ecosystem of innovation" deals with the efficient integration of *national research networks*. Indeed, as the World Bank notes, "the success of China's innovation policy will [also] depend on how effectively all branches of the research and innovation network (research institutes, universities, central and local governments, state and private enterprises) function together and how these efforts are leveraged internationally through global networks".[151]

  In this regard, China has undertaken the construction of eight regional innovation clusters. According to the CAS "Strategic Planning" document, these regional innovation clusters have "carried out actively various cooperative agreements, organized implementation of cooperative projects, built a number of research and development centres, and transfer and transformation centres to promote integration of industries, universities, and research institutes as well as transformation of technological achievements, which strengthened the CAS's relationship to regional innovation systems throughout the country".[152] The location of the eight innovation clusters is shown in Fig. 5.7.

  These efforts undoubtedly constitute positive steps towards the creation of an "ecosystem of innovation". However, it is clear that the national research networks are still far from being integrated countrywide or linked with the global R&D network. A key obstacle is that at national level there are no strong links and little mutually beneficial interplay among firms and between firms, the academic sector, and the government. Many firms located in inland cities are to a large extent isolated from those in the coastal areas and have not become integrated "research consortia" tasked with developing more complex technologies. Research has shown that the "complexity of new technologies [is] often

---

[150] The World Bank and the Development Research Center of the State Council, P.R.C. "China's Growth through Technological Convergence and Innovation" in *China 2030: Building a Modern, Harmonious, and Creative Society*. Washington, DC: World Bank. 2013. p. 156.

[151] The World Bank and the Development Research Center of the State Council, P.R.C. *China 2030: Building a Modern, Harmonious, and Creative Society*. Washington, DC. World Bank. 2013. p. 38.

[152] Chinese Academy of Sciences. "Strategic Planning". CAS Fact-sheet. Web. http://english.cas.cn/ST/spi2020/201106/P020110608576186252276.pdf. Accessed 20 November 2013.

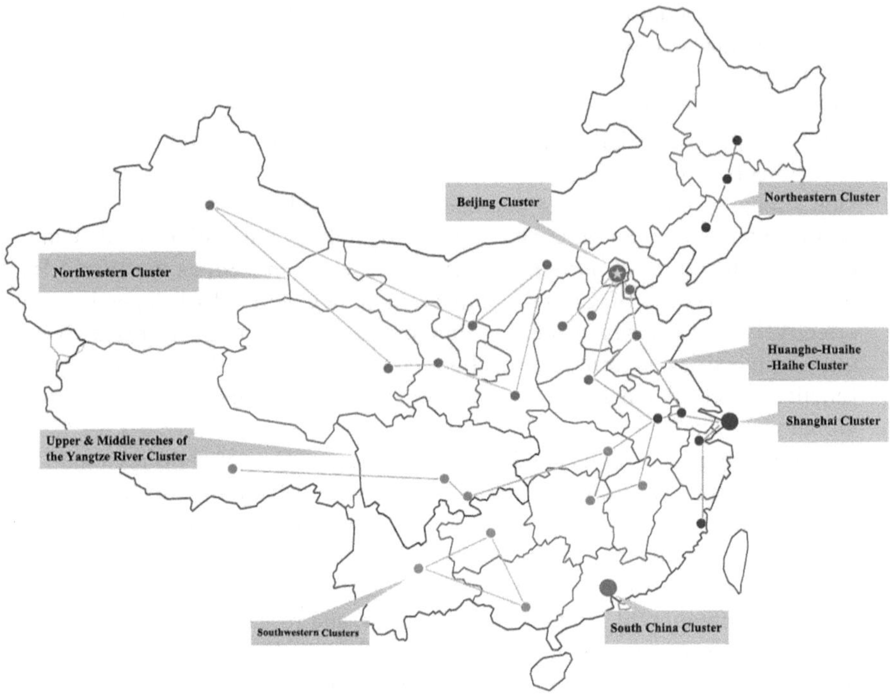

Fig. 5.7 China's research, innovation, and development clusters (*Source*: CAS)

beyond the internal capabilities of even very large companies".[153] The US example has shown that innovation has evolved from being a big company undertaking during the 1970s to a collaborative endeavour involving firms, universities, research institutes, start-ups, and government-funded programmes today.[154] Partnerships between firms or between firms and universities are therefore indispensable for advancing the necessary specialisation in frontier fields and for developing sophisticated new products or technologies.[155] Although the necessary expertise for introducing innovation may not be lacking, the linkage between laboratories/R&D centres, companies, and the market is

[153] Block, Fred, and Matthew R. Keller (2008). "Where Do Innovations Come From? Transformations in the US National Innovation System, 1970–2006". The Information Technology and Innovation Foundation. Web. http://www.itif.org/files/Where_do_innovations_come_from.pdf. Accessed 5 November 2013.

[154] McGregor, James (2010). "China's Drive for Indigenous Innovation. A Web of Industrial Policies". Global Regulatory Cooperation Project—US Chamber of Commerce. APCO worldwide: p. 36.

[155] The World Bank and the Development Research Center of the State Council, P. R. China (2013). *China 2030: Building a Modern, Harmonious, and Creative Society*. The World Bank, Washington DC: pp. 175–176.

fragmented, and China appears in addition to be ineffective at creating and managing spill-over effects from research findings and developed technologies.

The domestic research network also has difficulty being incorporated or efficiently linked to *global research networks*. It is hard for China to reach into the global network of science, collaboration, and research when it hunkers over the "techno-nationalism moat" at home.[156] Indeed, there is a widespread belief that China is pursuing a proactive—if not aggressive—policy of protection with regard to indigenous innovation.[157] Evident elements in this sense are the government's procurement policies, which often block products not designed and produced in China; its industrial and technology standards, used as market barriers to foreign technology; and the patent system, which often leads to the proliferation of junk patents used as weapons against foreign companies.[158]

For the World Bank, the "closed technology" strategy that China's policies have (un)intentionally induced may lead to "short-term gains but will be ultimately self-defeating", because Chinese companies and research institutes will often remain excluded from participation in research conducted in other parts of the world and from fruitful exchanges of know-how and new ideas, as well as from the important exchange of personnel and research staff and increased specialisation. An open innovation strategy, on the contrary, promises more sustained long-term rewards given that China could "benefit from participation in global R&D networks just as it has benefited from participation in global production networks".[159]

- The establishment of cooperative international research networks is also hampered by another fundamental weakness of China's innovation system: the *patenting system* and the protection of *intellectual property rights* (IPR). Official documents have emphasised that "China will further improve the national IPR system, create a legal environment in which intellectual property rights are respected and protected, increase people's awareness of IPR protection, and sternly crack down on IPR infringement". In practice, however, little has been done.

  The bottom line is not the introduction of new and stronger legal tools: China has already created the necessary laws to have respectable IPR protection. It has been part of the World Intellectual Property Organization (WIPO) since 1980

---

[156] McGregor, James (2010). "China's Drive for Indigenous Innovation. A Web of Industrial Policies". Global Regulatory Cooperation Project—US Chamber of Commerce. APCO worldwide: p. 6.

[157] It even seems that president Hu in 2009 modified Deng's famous slogan *Taoguang Yanghui, Yousuo Zuowei* (Keep a low profile and bide our time, while getting something accomplished) by adding the word *jiji* (actively), so to make the final phrase "while *actively* getting something accomplished. *Ibid.* p. 24.

[158] *Ibid.* p. 27.

[159] The World Bank and the Development Research Center of the State Council, P. R. China (2013). *China 2030: Building a Modern, Harmonious, and Creative Society.* The World Bank, Washington DC: p. 21.

and party to the main international agreements in the IPR domain[160]; it has adopted decent trademark laws, patent laws, and copyright laws (which have all been strengthened over the years); it has also established dedicated offices under the State Council for IPR administration (the State Intellectual Property Office and the State Administration for Industry and Commerce) as well as intellectual property courts in the main cities and provinces.[161] Thus, the problem is ultimately represented not by missing tools but rather by the lack of political will to pursue serious enforcement. There are both economic and political reasons for this inertia, but the point to underline is its negative impact on the country's innovation capabilities. Besides creating a climate of mistrust with possible international research partners, weak IPR protection is one of the main reasons for China's failure to bring home its top talent doing research abroad; it thus acts as another serious constraint on its innovation potential. To conclude on this point, only with the eventual enforcement of the IPR and patenting system will the growth of China's innovation capabilities be expedited.

- Education, the ultimate linchpin in the creation of any innovation system, is a source of both strength and weakness for China. The quality of university training has significantly improved, 11 Chinese universities are now in the top-ranked 200 universities of the world, and every year millions of graduates make their entry into the market (6.6 million in 2011 alone).[162] It goes without saying that this increasingly large supply of skilled graduates will help China's innovation efforts.

  In general terms, however, the productivity of university-level research and the quality of human resources are low: "the key role of universities so far centers not so much on cutting edge innovation but on adaptation and re-development of existing foreign technologies and products".[163] Furthermore, research management is not supportive of introducing the "Silicon Valley environment" envisaged in the 2006 Guidelines; and there are rigid boundaries that separate academic education from technical and vocational training, which ultimately hinders efficient university–industry linkages (UILs). For example, as Weiping Wu and You Zhou have explained, the number of firms that have

[160] The international IPR agreements signed by China include: the Paris Convention for the Protection of Industrial Property (1985); the Madrid Agreement for the International Registration of Trademarks (1989); the International Patent Cooperation Treaty (1994); and, with its WTO accession, the WTO Agreement on Trade-Related Aspects of Intellectual Property Rights (2001).

[161] McGregor, James (2010). "China's Drive for Indigenous Innovation. A Web of Industrial Policies". Global Regulatory Cooperation Project—US Chamber of Commerce. APCO worldwide: p. 25.

[162] The World Bank and the Development Research Center of the State Council, P. R. China (2013). *China 2030: Building a Modern, Harmonious, and Creative Society*. The World Bank, Washington DC: p. 21.

[163] *Cit.* Wu, Weiping, and Yu Zhou (2012). "The Third Mission Stalled? Universities in China's Technological Progress". Journal of Technology Transfer Vol. 37 (6).

Table 5.4 Share of manufacturing firms using universities as key R&D partners (in %)

	Product innovation (in %)	Process innovation (in %)
Size		
Large enterprises	3.9	5.3
Medium-sized enterprises	3.6	3.3
Small enterprises	3.7	2.4
Type		
State-owned enterprises	3.8	4.3
Foreign-invested enterprises	1.5	1.6
High-technology enterprises	4.7	4.1

actually sought out academia in pursuit of core technology is small: just 5 % (see Table 5.4).[164]

The World Bank has also stated that "universities need to develop innovative approaches to imparting knowledge and analytical skills and set up well-staffed specialised research institutes".[165] In this regard, China should encourage "leading foreign universities to set up campus in the country with domestic universities and impart modern governance standards, teaching methods and research management".[166] However, there are evident "political concerns" about adopting these policies that may limit effective implementation. And, even if such policies were to be adopted, it seems unlikely that Chinese people will start to think "American".

- Ultimately education is closely linked to another very influential factor of any process of innovation: culture. In line with the stimulating research conducted by the cultural psychologist Richard Nisbett,[167] one can argue that different cultures not only lead people to think different things but also via different mental processes; these processes are ultimately reflected in a number of different social and economic domains (including different innovation capabilities). In this light, the ultimate issue is to assess the influence of Chinese culture over China's innovation efforts.

  If one were to consider the United States as the ideal type of innovation culture, China appears to be its polar opposite. In fact, while individualism and personal assertiveness figure among the basic values of US culture, China (as well other *Sinic* cultures like Japan and Korea) has so far emphasised communitarianism and social conformism. Other relevant cultural differences are illustrated in Table 5.5.

---

[164] *Ibid.*

[165] The World Bank and the Development Research Center of the State Council, P. R. China (2013). *China 2030: Building a Modern, Harmonious, and Creative Society.* The World Bank, Washington DC: p. 23.

[166] *Ibid.* p. 24.

[167] Richard Nisbett, Richard (2003). *The Geography of Thought: How Asians and Westerners Think Differently... And Why.* Free Press, New York.

Table 5.5 US and Chinese culture in contrast

	United States	China
Basic values	Individual liberty	Group membership/consensus
	Independence	Group harmony
	Self-reliance	Communitarian sense
	Analytic	Holistic
	Legal	Trust
	Confrontation	Compromise
Organisation	Formal	Informal but hierarchical
	Competitive	Cooperative
	Fragmented	Generalist
Action	Short term	Long term
	Product focused	User focused
	Control	Human resources
Management style	Rationality	Relationships
	Structured	Flexible
	Directive	Adaptive
	Doing	Understanding

As interestingly pointed out by Franco Mazzei, while in the US performance can be considered individual, in China it is a collective endeavour: that is, it is an expression of the group as an operative entity. In short, in China the primary imperative of performance is to "support the group", and thus essential connotations are "conformism", "harmony", and "hierarchy". Conversely, in the United States, the primary objective is to "protect the individual", and essential characteristics are "action", "freedom", and "equality".[168]

Since innovation is often viewed as springing from individual talent and from the US cultural environment, it could therefore be expected that China's culture would act as an impediment that ultimately will not help the *orgware* spur either genuine innovation processes or, more importantly, the disruptive innovations that would be of great help for the space programme and the country.

However, analysing the incidence of culture over innovation processes through this limited prism may be too limiting, if not misleading. It is true that social conformism is perhaps not supportive of innovation—especially of disruptive innovation, which often arises in almost anarchic environments—yet this is not the end of the story.

Social conformism, for instance, should not be understood as automatically implying a mere amalgamation and dissolution of individual identity within the

[168] Mazzei, Franco (2003). "Intercultural Variables and Japanese Socio-Economic Performances". In: Lavagnino, Alessandra Cristina et al. (eds). *Reflections on Asia. Essays in honour of Enrica Collotti Pischel*. Franco Angeli, Milano.

group—on the contrary![169] Conformity to social norms requires the pursuit of harmony (*he*), which takes place within a person through a process of self-cultivation.[170] In turn, self-cultivation leads to an extreme level of focus, which itself is a potentially disrupting element. Taking Japanese experience as a guide, it was the extreme focus of Japanese society that ultimately spurred the introduction of a disruptive production system—or even a model of capitalism according to several economists: *Toyota-ism,* as opposed to *Fordism–Taylorism.* In short, Chinese "extreme focus" has per se the potential to spur, over time, the rise (or resurgence) of a genuine ecosystem of innovation.

Another cultural trait that might aid innovation can be found in the so-called non-linear logic that characterises the epistemological approach of *Sinic* cultures and has its roots in Taoist concepts of *yin* and *yang*. In China there is no dialectical (Hegelian) thinking based on the *logos* but a dialectic that can be labelled *relational* (or *yin-yang* type). As Edwin Reischauer wrote: "In the West the division was between good and evil, always in mortal combat with each other. In East Asia, the division of *ying* and *yang* was between night and day, male and female, light and darkness—that is between complementary forces that alternate with and balance each other. There is no strict good–bad dichotomy, but rather a sense of harmony and balance of forces".[171] In short, while in the West one lives under an "either...or..." (*tertium non datur*) type of logic, the *Sinic* mind is governed by a "both...and..." approach, where two seemingly opposite, or mutually exclusive, choices or qualities can coexist.[172]

This type of logic, combined with the ethical relativism of the *Sinic* value system,[173] may have a twofold positive reflection on Chinese innovation processes. First, it leads the Chinese to have a mind-set that is "more pragmatic than ideological, more relativist than absolutist, more intuitive than deductive, more contextual than structured";[174] second, it allows them to accept non-rational thinking in their mental processes.[175]

---

[169] As a famous Confucian analects (13) recites: "the wise man is looking for harmony without assimilation. The mean man is looking for assimilation without harmony". Zhou, Yuanxia (2008). "The Modern Significance of Confucianism". Asian Social Science Vol 4 (11): 12–16.

[170] In the Sinic world, the concept of harmony is not seen as something static, but as a process, a constant flux.

[171] *Cit*. Reischauer, Edwin, and M. B. Jansen (1995). *The Japanese Today—Change and Continuity*. Harvard University Press, Cambridge: p. 141.

[172] It could be, for instance, argued that it is ultimately this type of logic that allow China seeing itself as both a market and socialist economy without having a conflict between the two opposing states existing at the same time and place.

[173] See Mazzei, Franco, Vittorio Volpi (2006). *Asia al Centro*. Universitá Bocconi Editore, Milano.

[174] *Cit* Mazzei, Franco (2003). "Intercultural Variables and Japanese Socio-Economic Performances". In: Lavagnino, Alessandra Cristina et al. (eds). Reflections on Asia. Essays in honour of Enrica Collotti Pischel. Franco Angeli, Milano.

[175] See Richard Nisbett's book for more detailed analysis on China's rationality. These connotations leave much space for pragmatism that is reflected in a flexible and adaptive management

On the other hand, non-linear logic induces Chinese to adopt an iterative, holistic (as opposed to analytical) approach that examines all linkages between elements, incorporates the dynamics of change and contradictions, and embraces criticism and self-criticism. Although often leading to inaction and not "doing" (thus not being supportive of fostering rapid technological changes), it favours "understanding" and it is supportive of the emergence of processes of "creative destruction"—in economist Joseph Schumpeter's terms. Think Mao's continuing revolution... And from a past of permanent revolution to a future of permanent innovation, the step, after all, might not be too big!

It is beyond the remit of this book to provide a detailed and in-depth assessment of all the Chinese cultural traits influencing innovation processes, but it is clear that Chinese culture should not necessarily be seen as a hindrance towards the development of new ideas turning into innovative products or processes. Western countries could be surprised if China one day re-emerges as an important source of discovery and innovation on par with its Great Four Inventions.

5.3.3 Implications for the Moon Programme

In the light of the various strengths and weaknesses affecting Chinese innovation efforts, some considerations about their expected implications for China's manned lunar exploration programme are briefly provided below.

China has shown an uncanny ability to master all the competences necessary for independent human spaceflight and is now steadily investing in the development of the whole technological spectrum to build its own space station by the early 2020 and eventually implement a human lunar endeavour. The Chinese space community has also announced that some advanced, ground-breaking technologies (e.g. autonomous navigation and flight technology for spacecraft with high precision space positioning, as well as revolutionary nuclear-, solar sail-, and even antimatter-based propulsion systems[176]) are being actively explored in order to meet the requirements of its future deep space exploration programme.

Nevertheless, the development and mastery of all the necessary technologies and capabilities for embarking upon a manned lunar exploration will be both a lengthy and a challenging process, inherently fraught with difficulties. A bottom line

style. Pragmatism has eventually allowed accompanying the traditional consensual policymaking procedures with new, Western-derived managerial methods. In the telecommunications and aerospace industries, for instance, the new managers are pushing for innovation in managerial style and corporate culture to foment R&D breakthroughs while also keeping the CCP's role relevant in innovative managerial systems.

[176] Guo, Huadong, Ji Wu (eds) (2010). *Space Science and Technology in China: A Roadmap to 2050*. Chinese Academy of Sciences, Science Press (Springer), Beijing.

causing major difficulties probably lies in the fact that China's goals seem to be expanding faster than the capacity of the *orgware* to manage them.[177]

Indeed, the complex and highly bureaucratic institutional system that has been set up to unify, organise, and control the planning and development processes of China's space programme does not appear particularly suited to fostering and managing the technological innovation needed for a lunar endeavour. Too much power has been given to the state bureaucracy to direct and coordinate the development of technologies, causing China's research culture to suffer problems of the adequacy of professionalism and management.

The *orgware*'s difficulties in adopting functional change are manifest in the tendency to centralise the control processes, with consequent spiralling expenses, cost cuts, and compromised quality control. Cracks in quality control—which are also caused by the official policy of speeding up and intensifying concrete achievements—may in turn lead to malfunctions or even failures that will cause further delays. The mechanical malfunctions encountered on the recent Chang'e-3 mission, as well as the August 2011 launch failure and the operational problems in the Chinese satellites built for Venezuela and Nigeria, are for many analysts ultimately attributable to the inadequate institutional system set up to manage China's technological development. It can be expected that this type of problem will have a negative effect on the development of the technologies required for successful implementation of a manned lunar landing.

Limiting observations to rocket technology—which appears to be the most relevant technological issue—it can be noted that China has encountered considerable difficulties in the development of its new launcher fleet and has been forced to adjust the proposed schedule more than once. In particular, the decade of delays accumulated for the development of the LM-5 engines—which are still not operational—is a clear indicator of all the aforementioned difficulties and casts serious doubts on China's ability to complete the construction of the super-heavy-lift launcher LM-9 required for a manned lunar landing within the next decade. At the moment, "getting 3000 tons of thrust under a single rocket presents a formidable [if not insurmountable] technological challenge for China, whose largest engine so far, the YF-100, generates a thrust of just 120 tons".[178] To many analysts, completing the development of its two engines by 2020 and the overall construction of the LM-9 by 2025 appears too optimistic.

Difficulties and slowdowns could be also encountered in the development of other technologies and systems. However, China will be able (and prompted) to steadily invest huge human and financial resources and its determination should eventually secure the ability to reach the Moon.

[177] See Solomone, Stacey (2013). *China's Strategy in Space*. Springer, New York.

[178] *Cit.* Perrett, Bradley (2010). "Longer Marches". Aviation Week & Space Technology Vol. 172 (11): 22–23.

5.4 Concluding Remarks

The numerous economic, political, social, and technological conditioning factors assessed in the previous sections help us better understand the cautious and pragmatic approach that China has adopted in formulating its long-term objectives for human spaceflight. Apart from the statement of interest contained in the 2011 White Paper, Chinese leaders have so far avoided pledging a clear, strong commitment towards the implementation of a manned lunar landing programme. More importantly, the conditioning factors also help explain why China is visibly not interested in stimulating the confrontational stances involved in a space race, but is open to international cooperative undertakings in the area of human spaceflight (and might not want to go the Moon alone). This is demonstrated by the reiteration of its intention to open the Chinese Space Station to the global community.

To be sure, China's interest in cooperation is—and will be—essentially politically driven. However, alongside the political motivations provided in Chap. 3, cooperation is also expected to accrue a number of valuable benefits that could help China overcome the most challenging issues within the three macro-endogenous factors assessed in this chapter.

(a) In the economic realm, cooperation will serve as a precious instrument to offset the economic costs associated with the implementation of a lunar endeavour. Although it is acknowledged that international cooperation more often increases the total costs, these are spread among the partners. By coordinating the development of respective tasks, cooperation offers the opportunity to rationalise and optimise available resources and increase the utility of the programme. The *soft-landing* scenario that is projected for future Chinese economic development urges Chinese policy makers to be more pragmatic in implementing highly ambitious programmes, and cooperation clearly moves a programme in this direction. In addition, the large expenditure that will be required to launch and maintain in orbit the forthcoming CSS, combined with China's ever-increasing involvement in the whole spectrum of space activities, may also undermine the possibility of devoting a large, parallel investment to a lunar endeavour, if cooperation is not pursued. After all, NASA's experience with President G. W. Bush's Vision for Space Exploration has to a large extent shown that the cost of a solo human spaceflight undertaking is prohibitive when the country has to maintain a highly expensive infrastructure such as the ISS.

(b) In the political realm, cooperating with other partners could contribute to avoiding political volatility and indeed increasing stability in the implementation of the programme. Although China's past and present experiences demonstrate that the county is an extremely determined actor and would certainly accomplish its goals once committed, international cooperation will secure a much firmer commitment to the implementation of the programme by the forces that up to now have only provided limited support, namely, the New Right and the Globalists. For the latter, in particular, international cooperation represents the *conditio sine qua non* for embarking upon a manned lunar exploration programme.

In addition, cooperating with international partners offers the chance of minimising the negative impacts that a potential mission failure would generate. As mentioned, Chinese political leaders have a low tolerance of failure: they are fully aware that if space successes are a formidable instrument for boosting their political legitimacy, so too are space failures disastrous for their standing.[179] In a strategy of calculated risk, sharing with external partners the dangers entailed in high-risk activity such as human spaceflight would reduce the overall responsibility of the Chinese leadership.

(c) Finally, from the perspective of technological development too, international cooperation would further contribute to enhancing the robustness and performance of the programme and reducing the risk of failure. More broadly, cooperation offers the possibility of hastening China's breakthrough in key technology problems of human spaceflight and related areas, improving its research system capability, as well as the level of professionalism and project management in the field, and would provide access either directly or indirectly to new technologies and enrich the pool of scientific and technical capabilities.

In the light of these considerations, the next chapter will analyse the evolving international environment in which China's space programme is operating and identify a number of scenario alternatives at various junctures where cooperative ways to the Moon may emerge.

[179] Johnson-Freese, Joan (2004). "Space Weiqi. The Launch of Shenzhou V". Naval College Review Vol. 57 (2): 121–145.

Chapter 6
China, the Moon, and the World

China is a sleeping giant. Let her sleep, for when she wakes she will move the world.

Napoleon

The giant has eventually woken up and is indeed profoundly reshaping the panorama of the twenty-first-century geopolitical and economic realities. The impact of its rise (or resurgence) as a great power on the world stage has also become remarkable in the space arena, where its ambitious space programme is dramatically changing the institutional landscape of global space activities. The leading space powers are fully aware that Beijing's ascendancy as a space power represents a significant and potentially disruptive occurrence that can no longer be ignored.

With a nod to the renowned essay by John Ikenberry "The rise of China and the future of the West",[1] it can be argued that China's lunar ambitions will become one of the dramas to be faced by the global community in the twenty-first century. But exactly how this drama will play out remains an open question.

The aim of this chapter is to assess how Chinese determination to go to the Moon could affect the rest of the institutional landscape in the period leading up to its arrival there. It will provide an account of the posture the leading space powers could adopt vis-à-vis China and will accompany the analysis with suggestions of a limited number of alternative scenarios at the various junctures where more cooperative pathways might eventually become possible. In the process, it will also show that fears of an intra-Asian or China–US space race are, respectively, either unfounded or avoidable.

6.1 China's Rise, Western Decline?

China's arrival on the international space scene has seised global headlines, generating as much apprehension as positive expectation. As described above, although the space programme was initiated as early as 1956, it was not until the early days of

[1] Ikenberry, John (2008). "The rise of China and the future of the West". Foreign Affairs Vol. 87 (1): 23–37.

M. Aliberti, *When China Goes to the Moon...*, Studies in Space Policy 11,
DOI 10.1007/978-3-319-19473-8_6

the twenty-first century that it made a real "Great Leap Forward" and became an integral—indeed a highly visible—part of China's resurgence (*fuxin*) narrative.

Specifically, the resurgence occurred because by the beginning of the new century, not only had China fully demonstrated indigenous capabilities in manufacturing, launching, and controlling satellites, spanning both civilian and military requirements, it had also started the implementation of highly ambitious endeavours in the field of space exploration.

The impressive full-spectrum achievements reached by this relatively new actor have led some scholars to envisage the emergence of a "new space age". According to James Clay Moltz, this second space age symbolically began in October 2003, when China became the world's third space power to possess an independent human space flight capability.[2]

Since then, Chinese space activities have continued to advance at an astonishing rate. In the span of a decade, China has achieved success after success (suffice it to think of the 2008 spacewalk, the launch of the Tiangong orbital module and the subsequent rendezvous and docking operations, the acclaimed activities of China's first space teacher in 2012, and the landing of its first rover on the Moon in December 2013), showing a level of determination and ambition currently unmatched in other countries.[3] Indeed, what is perhaps more significant than these achievements is that in the meantime China seems to have succeeded in taking full advantage of the lull in the space activities of the major spacefaring nations.

The US space programme has in recent years undergone a "crisis of identity" with regard to its strategic direction and is now perceived to be a drift or, at best, in transition to an alternative paradigm for space activities, including private-orientated undertakings. As nicely captured by the National Research Council, which in 2012 was requested by a congressional directive to conduct a comprehensive independent assessment of NASA's strategic direction and agency management: "NASA is at a transitional point in its history and is facing a set of circumstances that it has not faced in combination before. The agency's budget, although level-funded in constant-year dollars, is under considerable stress, servicing increasingly expensive missions and a large, aging infrastructure established at the height of the Apollo program. Other than the long-range goal of sending humans to Mars, there is no strong, compelling national vision for the human spaceflight program, which is arguably the centerpiece of NASA's spectrum of mission areas. The lack of national consensus on NASA's most publicly visible mission, along with out-year budget uncertainty, has resulted in the lack of strategic focus

[2] Moltz, James Clay (2011). *Asia's Space Race: National Motivations, Regional Rivalries, and International Risks*. Columbia University Press, New York: p. 55.

[3] As the US National Academy of Sciences notes, China's achievements in space exploration are the latest in a program that "marches steadily and strategically toward what might eventually become a lead role among the nations in spaceflight". National Academy of Sciences (2014). *Pathways to Exploration: Rationales and Approaches for a US Program of Human Space Exploration*. The National Academy Press. Washington DC.

necessary for national agencies operating in today's budgetary reality. An effective agency response is vital, because at a time when the strategic importance of space is rising and the capabilities of other nations are increasing, U.S leadership is faltering".[4]

On its side Russia, formerly a rival to the United States, has since the early 1990s dramatically reduced its ambitions and liquidated a large part of its assets, including the trove of technological experience accumulated by generations of engineers.[5] In order to keep its industry alive while waiting for better times, it has redefined its role to become primarily a contractor to others (including India and China), while exploiting its multifaceted expertise for commercial and political purposes. Consequently, although still a technologically formidable space power, the country now appears to find significant difficulties in renewing its space industry and revitalising its space programme. Notwithstanding the ever stronger endorsement by President Putin, the programme still appears in a *stall*, as highlighted by the growing number of launch failures and accidents it has encountered in recent years.[6] The recent "Basic Russian Federation National Space Policy until 2030 and beyond", adopted in April 2012, subtly recognises these difficulties and calls for structural reform of the Russian space industry to make it more innovative, competitive, and commercially self-sustainable.[7]

As for Europe, it has undoubtedly reached a prominent position within the space hierarchy, now ranking first in commercial satellite production and launches and boasting a solid space science programme with significant achievements. However, given the fragmentation of its governance and the distinct strategies proposed by ESA, the EU, and the national agencies, it is struggling to establish a coherent long-term shared visions and strategy. As Peter Hulsroj poignantly remarks in the ESPI Yearbook on Space Policy, "the unresolved issues on authority stemming from the Lisbon Treaty, and the difficulties on agreeing on launcher strategies, ISS extension and GMES operational funding, [ultimately] play into the hands of those seeing emasculation rather than strengthening".[8]

[4] *Cit.* National Research Council (2012). *NASA Strategic Direction and the Need for a National Consensus*. The National Academy Press. Washington DC: p. 1.

[5] Blamont, Jacques (2012). "US Space Exploration Strategy: Is there a better way?" Space Policy No. 28 (4): 212–217.

[6] The last has occurred in May 2014. For a review of Russian recent launch failures, see "Russian rocket falls back to Earth with Super Satellite". Space Mart. 16 May 2014. Web. http://www.spacemart.com/reports/Russian_rocket_falls_back_to_Earth_with_super_satellite_999.html. Accessed 18 May 2014.

[7] "Russian Space-Based Activities' Development Strategy until 2030 and Beyond" Aviation Explorer. 27 April 2012. (Russian language source). Web. http://www.aex.ru/docs/8/2012/4/27/1561/. Accessed 18 May 2014.

[8] Hulsroj, Peter (2014). "The Psychology and Reality of the Financial Crisis in Terms of Space Cooperation". In Al-Ekabi, Cenan, Blandina Banares, Peter Hulsroj, Arne Lahcen (eds). *ESPI Yearbook on Space Policy 2011/2012. Space in Times of Financial Crisis*. Springer, Vienna: pp. 159–168. See also Sect. 7.3.1 for a more detailed analysis.

Finally, Japan, which by any number of measures has so far been the most accomplished space power in Asia since the late 1990s, has been experiencing a prolonged "space crisis" with regard to its strategic direction, management, and level of involvement in space activities.[9] Affected by a decade of economic stagnation and political instability known as *ushinawareta junen* (the lost decade), the space programme lost confidence and motivation: the space budget hardly rose at all, passing from ¥240 billion in 1997 to ¥250 billion 10 years later in 2007,[10] ambitious programmes were drastically downsized, and the space industry entered a period of profound confusion.[11] At the same time, the evolution of the post-Cold War security context required a difficult broadening of the activities pursued thus far, which ultimately put the country's space agenda at a crossroads. Notwithstanding that the 2008 reorganisation of its space mandate and management structure, as well as the recent increase in funding, is evidence of Japan's strong commitment to space activities, the country still appears challenged in maintaining its competitive edge and hesitant to implement ambitious programmes and discretely recognises that it will not be able to do it alone.

All in all, if compared to the determined Chinese juggernaut, the traditional space powers seem to have lost much of the wind in their sails. The recent financial crisis has intensified this perception. In fact, the broad margin between China's economic outperformance and the continued slow growth of the advanced economies is creating a degree of asymmetry not only in the ability and political will to fund space activities but also in the overall level of ambitions[12]: while Chinese ambitions are on the rise, those of the West appear to be on the slide, an image that for many commentators was well captured at the 2013 IAC, where the widening gulf between China's space missions and future plans, and the state of space missions in the United States, was described as dramatic. [13]

---

[9] For an analysis of the Japanese "space crisis", see among the others Moltz, James Clay (2011). *Asia's Space Race: National Motivations, Regional Rivalries, and International Risks*. Columbia University Press, New York: pp.55–59.

[10] Suzuki, Kazuto (2008). "Basic law for space activities: A new space policy for Japan for the 21st century". In Schrogl, Kai-Uwe, Charlotte Mathieu, Nicolas Peter (eds). *ESPI Yearbook 2006/ 2007: A new Impetus for Europe*. Springer, Vienna: pp. 226.

[11] As noted by Kazuto Suzuki, due to the budget decrease, the traditional contract arrangements through JAXA—a rotating prime contractors system and equal distribution of subcontracts—has become no longer affordable or effective. Thus, many space companies in Japan have shrunk the size of operations and several of them have exited the market (e.g. Toshiba). *Ibid.* pp. 227–228.

[12] It must be stressed that while the NASA, ESA, and JAXA budgets did not decrease during the financial crisis, ambitious programmes were not approved and some were terminated (e.g. the Constellation programme, the Aurora programme, and the HOPE programme in Japan). For an analysis of the impact of the financial crisis on decision-making, see Tegnér, Per (2014). "The Effect of the Financial Crisis in terms of Political Decision-Making". In Al-Ekabi, Cenan, Blandina Banares, Peter Hulsroj, Arne Lahcen (eds). *ESPI Yearbook on Space Policy 2011/ 2012. Space in Times of Financial Crisis*. Springer, Vienna: pp. 149–158.

[13] Freeman, Marsha. "Is Asia Taking the Lead In Space Exploration". EIR Science. 25 October 2013. Web. http://www.larouchepub.com/eiw/public/2013/eirv40n42-20131025/32-40_4042.pdf. Accessed 18 April 2014.

To be sure, the perceived decline remains, under the present circumstances, much more psychological than real,[14] as China is still far from becoming the pre-eminent space power in the short to medium term. It is emerging—or has already emerged—as a formidable spacefaring nation and can be now regarded as a peer of Russia, Europe, and Japan, but it seems unlikely that it will easily surpass the United States to become the undisputed master and the measure of all things space. In terms of both financial (see budgetary prospects, Sect. 2.2) and technological capabilities, parity is still a long way off.

However, self- and external perceptions remain a powerful discriminator in the effective ability to project leadership and power, and China's upcoming spaceflight endeavours will inevitably provide additional, tangible elements affecting such perceptions. The completion of the Chinese space station, at a time when a possible decommissioning of the ISS might make China the only country with a presence in LEO and, more importantly, the prospective Moon landing of a taikonaut, if not addressed in a timely manner, will only reinforce the perceived decline of the West.

As already stated, the latter achievement, in particular, could act as a powerful turning point for the creation of a new space consensus and produce—in the words of former NASA Administrator Michael Griffin—"...[an] enormous, and not fully predictable, effect on global perception of US leadership in the world". Further reflecting on this impact, it could be argued that for *all* the existing powers, such a landmark accomplishment could more broadly cause a profound "ontological shock" that would disrupt their *Weltanschauung* and undermine the sense of entitlement the West has so far asserted in shaping the global order and providing "the Rest" with a point of reference to admire and emulate. In short, the now tired, old powers of the West could be doomed to inexorable decay.

Predicting the cultural and hence political demise of the West as a result of a Chinese manned lunar landing might seem exaggerated—or even fallacious. Given that the United States achieved the same goal 45 years ago, it could be argued that the West can afford to react with indifference! Although this might represent an appropriate response, it is clear that human spaceflight achievements remain important status markers strongly shaping public perceptions, and for the billions of people raised in the post-Apollo era, the memory of that triumphant legacy is a very distant one and so perhaps is slowly becoming part of the mythology of the American era.

Furthermore, it should not be forgotten that two additional elements might well accompany China's lunar endeavour and maximise its impact: the landing of a *real Chang'e* (i.e. a woman) and the possible participation of emerging space nations. While further relativising the sociocultural primacy of the "tired, old West", these occurrences could have wide-ranging geopolitical implications, providing China

---

[14] See Hulsroj, Peter (2014). "The Psychology and Reality of the Financial Crisis in Terms of Space Cooperation". In Al-Ekabi, Cenan, Blandina Banares, Peter Hulsroj, Arne Lahcen (eds). *ESPI Yearbook on Space Policy 2011/2012. Space in Times of Financial Crisis*. Springer, Vienna: pp. 159–168.

with key elements to claim moral and political leadership, while casting serious doubt on the economic and political models of the West.

The already flickering "structural power" (i.e. what Susan Strange has defined as the power to shape and determine the structures of the global political economy, to set the international regimes of rules and customs governing international economic relations and, in short, the power to decide how things will be done[15]) so far maintained by the West could then receive a decisive blow eventually being replaced by a *sinification* of the international political global order.

This possibility—while far from inevitable—clearly represents an ominous prospect for the existing powers and demands a response, a response that would have to be shaped by the level of their political ambitions and economic possibilities. Doubtless this incipient awareness will stimulate—indeed for many it already has—the emergence of competitive dynamics, particularly in the United States, and imply that by default the global community could slide into a competition scenario, a "new space race". At the same time, however, different and unexplored pathways for the pursuit of human space exploration in a more cooperative framework could open up.

6.2 A Space Race, Again?

6.2.1 The Incontestable Prelude: Space Ambitions and Geopolitical Dynamics

The idea of a "new space race" has gained particular currency in the past decade, prompted by the reorientations that China's space achievements have precipitated in the space policies of a number of space powers.

In January 2004, shortly after the landmark flight of Yang Liwei, President George W. Bush announced the Vision for Space Exploration (VSE),[16] perhaps the "boldest US space policy decision since the launch of Apollo programme" by President J. F. Kennedy in May 1961.[17] VSE entailed a long-term space exploration plan calling for redirecting NASA's human exploration from low Earth orbit to the "Moon, Mars, and beyond".[18] Although the motivations behind the launch of VSE were manifestly manifold,[19] the overall impression was that the United States was

[15] See Strange, Susan (1988). *States and Market*. Pinter. London.

[16] Remarks by the President on US Space Policy. NASA Headquarters, Washington DC. 14 January 2014. Web. http://history.nasa.gov/Bush%20SEP.htm. Accessed 5 May 2014.

[17] Logsdon, John M (2008). "Why space exploration should be a global project". Space Policy 24 (1): pp. 3.

[18] Most of the funding for VSE was in fact to be redirected from other NASA activities, including terminating the space shuttle programme in 2010 and ending US participation in the ISS by 2016.

[19] For many commentators, it was the Space Shuttle Columbia accident that shook the nation and triggered an in-depth review of the purpose and goals of the human spaceflight programme.

directly reaffirming the will to maintain a flagship role on the world stage and defend its undisputed leadership in space, a leadership that Chinese incipient ascendancy was endangering. While President Bush did invite other countries to join, making clear that the VSE would have to be a "journey and not a race",[20] the parallel repeated refusal to allow China to join the ISS made it clear that the implementation of this new ambitious international endeavour would not be open to Chinese participation.

The United States was not the only nation for whom China's rise in space posed a direct challenge and raised major concerns. Beijing's neighbours also seemed affected and eager to react promptly. Just 1 week after the Shenzhou-5 mission, on 22 October 2003, Japan's Ministry of Education, Science and Technology (MEXT) decided to set up a commission to review Japanese space aims for the following 20 years, including long-term participation in the ISS and manned spaceflight.[21] The conclusions of the commission's review were issued in a policy proposal, released by Japan's Aerospace Exploration Agency (JAXA) in April 2005. Entitled *JAXA Vision 2025*, the document set forth ambitious plans for an autonomous manned spaceflight programme. Notably, it also included a long-term plan for a human landing on the Moon by 2025, a task to be achieved in close collaboration with the VSE.[22]

The following year (in November 2006), the Indian Space Research Organisation (ISRO), which in 2003 had already set up several study groups on space exploration, also produced a formal proposal for the development of a human spaceflight programme. The proposal was presented to Indian Prime Minister Mamohan Singh on 17 October 2006, and on the latter's advice, it was submitted by ISRO Chairman Gopalan Madhavan Nair to a cross section of the scientific community that met in a brainstorming session in Bangalore on 7 November 2006.[23] The conference agreed to immediately initiate a human spaceflight programme and to autonomously launch its first manned flight by 2014 and land an Indian astronaut on the Moon by 2020.[24] The decision represented a major

[20] As the President would articulate in his remarks: "We will invite other nations to share the challenges and opportunities of this new era of discovery. The vision I outline today is a journey, not a race, and I call on other nations to join us on this journey, in a spirit of cooperation and friendship". Remarks by the President on US Space Policy. NASA Headquarters, Washington DC. 14 January 2014. Web. http://history.nasa.gov/Bush%20SEP.htm. Accessed 5 May 2014.

[21] Sheehan, Michael (2007). *The International Politics of Space*. Routledge, New York: pp. 181–182.

[22] Japan Aerospace Exploration Agency. JAXA Vision—JAXA 2025. Tokyo, Japan. 31 March 2005. Available from: http://www.docstoc.com/docs/87745747/JAXA-Vision.

[23] Peter, Nicolas (2008). "Developments in space policies programmes and technologies throughout the world and Europe". In Schrogl, Kai-Uwe, Charlotte Mathieu, Nicolas Peter (eds). *ESPI Yearbook 2006/2007: A new Impetus for Europe*. Springer, Vienna: p.96.

[24] Harvey, Brian, Henk H.F. Smid, and Theo Pirard (2010). *Emerging Space Powers. The New Space Programs of Asia, the Middle East and South America*. Springer– Praxis Publishing, Chichester, UK: p. 238. See also Jayaraman, K.S. "ISRO Seeks Government Approval For Manned Spaceflight Program" Space News. 13 November 2006. Web. http://www.spacenews.com/article/isro-seeks-government-approval-manned-spaceflight-program. Accessed 16 May 2014.

change in Indian space policy which, since its inception (in 1972), had been entirely focused on developing space infrastructure for socio-economic purposes and had declined any potential involvement in human spaceflight. As many analysts have argued, although India would insist that its "manned programme was a logical next step for the Indian space programme, and not a reaction to the emergence of the Chinese manned programme, the shift in policy was so dramatic in comparison to the almost ideological opposition to manned flights, that the timing seems hardly coincidental".[25]

Understandably, all these developments have had a great impact on the global space community, leading a number of scholars and space policy analysts to envisage the emergence of a new race to the Moon. Two scenarios would be explored in this regard, that of an intra-Asian space race and that of Sino-American competition, both seen as logical fallouts of China's rise in space.[26]

Between 2007 and 2008, with the almost simultaneous launch of the Kaguya-1, Chang'e-1, and Chandrayaan-1 lunar orbiters by Japan, China, and India, it appeared that the Asian space race had already started. The proximity of the launching dates and the nature of the missions fed the impression that the three countries were competing against each other, a competition that would later extend to manned space capabilities and, quite likely, end up in a space arms race.

Besides these signals and the reorientation in the policies of a number of spacefaring nations following Chinese space achievements, the widespread belief in the emergence of a new space race rested on broader political considerations. After all, space activities, and human spaceflight in particular, have always been inherently linked to geopolitical dynamics, and when looking at China's relations with its neighbouring countries and with the United States, these are undoubtedly problematic at best.

In conventional wisdom, China is viewed as a revisionist power which, having nurtured deep grievances against the established global order, will inevitably come to blows with the regional status quo and the prevailing American hegemony in the international system. Of course, Chinese leaders have missed no occasion to insist that China's ascendancy will be peaceful and poses no threat to its neighbours or to the existing international political and economic setting. But mere reassurance—Henry Kissinger has aptly remarked—cannot "arrest the underlying dynamism. For were any nation determined to achieve dominance, would it not be offering

[25] Manned spaceflight and planetary exploration were exactly the goals that the space programme's founder, Vikram Sarabhai, had rejected in favour of a *developmental rationale*. As he insisted, the application of space technology to addressing development goals "is not to be confused with embarking on grandiose schemes" and specifically that "we do not have the fantasy of competing with the economically advanced nations in the exploration of the Moon or the planets, or manned spaceflight". Quoted from Sheehan, Michael (2007). *The International Politics of Space*. Routledge, New York: p. 156.

[26] Given the stable and cooperative relations, both in space and on Earth, between Japan, India, and the United States, the idea of a space race between the United States and Japan or India has quite obviously not been explored in the literature.

assurances of peaceful intent?"[27] And as noted by James Steinberg and Michael O' Hanlon, it should come as no surprise that "many members of the world community remain concerned and even sceptical, noting that history and international relations theory are replete with examples of conflict arising from clashes between a dominant and a rising power".[28]

In recent years, many have revisited the example of the twentieth-century Anglo-German rivalry as a prediction of what may ultimately await the United States and China in the twenty-first century.[29] As Henry Kissinger acknowledges while questioning whether history is bound to repeat itself, there are certainly a number of comparisons. Just like the Germany of Wilhelm II, China is a fast-rising power, seeking to increase its influence and its military might, while the United States, like Britain, is a declining hegemon trying to maintain the established balance of power. As historical analyses of Anglo-German rivalry have argued, regardless of German intentions, the rise of Germany "was incompatible with the existence of the British Empire. Formal assurances were meaningless and even if moderate German statesmen were to demonstrate their *bona fides*, moderate German foreign policy could at any stage merge into a conscious scheme for hegemony", thus ultimately precluding cooperation or even trust.[30] Similarly, no matter how much China commits itself to peaceful rise, confrontational stances are inherent in its rise, and any form of cooperation will simply give China scope to build its capacities for an eventual crisis. While Kissinger ultimately criticises the inescapability of such a logic, he also recognises that it provides the subtext of much current thought.

Interestingly, the utilisation of this "inevitable historical analogy" (as the US national security strategist Edward Luttwak recently defined it[31]) has not been limited to describing the potential evolution of Sino-American relations but has also been applied to the geopolitical context of East Asia and in particular to current Sino-Japanese relations (in which the possibility of a tragic collision is rapidly emerging). Here too many commentators have found deep historical parallels. Gideon Rachman has, for instance, argued: "as in the years before 1914—when a rising Germany confronted its neighbours, so now a rising China is in dispute with several neighbouring countries, above all Japan", providing all the ingredients to make the prospect of a "new Sarajevo" a plausible prospect.[32]

---

[27] Kissinger, Henry (2011). *On China*. Penguin Books, New York: p. 523.

[28] Steinberg, James, and Michael O' Hanlon (2014). "Keep Hope Alive. How to Prevent US-Chinese Relations From Blowing Up". Foreign Affairs Vol 93 (4): 107–117.

[29] Kissinger, Henry (2011). *On China*. Penguin Books, New York: p. 519.

[30] *Ibid*. p. 520.

[31] Luttwak, Edward (2012). *The rise of China vs. the logic of Strategy*, Harvard University Press, Cambridge, MA.

[32] Rachman, Gideon. "Times to think more about Sarajevo, less about Munich". Financial Times. 6 January 2014. See also Blumenthal, Daniel, Mike Green. "Japan and China: not yet 1914 but time to pay attention". Foreign Policy. 29 January. 2014. Web. http://shadow.foreignpolicy.com/posts/2014/01/28/japan_and_china_not_yet_1914_but_time_to_pay_attention. Accessed 15 March 2014.

Even Japanese Prime Minister Abe Shinzo drew the comparison, in a speech at the World Economic Forum in Davos in January 2014, ominously recognising that the sound and ever-increasing interdependence between China and Japan is not, on its own, sufficient to avoid the fate of Germany and Britain. After all, he explicitly remarked, "Britain and Germany were highly interdependent economically. They were the largest trade partners, but the war did break out". [33] The situation today is incontestably similar: East Asia is the economic engine that drives the world, and while the mutual economic ties of Tokyo and Beijing are deepening, political relations have recently worsened, exacerbated by the territorial row over the Senkaku/Diaoyu islands, Tokyo's mistrust of Beijing's military build-up, and the nationalistic grandstanding of Abe, whose visits to the Yasukuni shrine reopened the historical wounds from Japan's militaristic past.[34] Moreover, the lack of official channels of communication between the two governments and the unedifying behaviour endorsed by their respective diplomatic services further complicate the current state of affairs,[35] increasing the chance of miscalculation in an area that US Navy Admiral Samuel J. Locklear has recently labelled "the most militarized region of the world".[36]

Notwithstanding that the inexorable escalation and subsequent descent towards a "new Sarajevo" remains debateable for many,[37] this line of argument well illustrates the climate of strategic distrust that has surrounded the re-emergence of China as a great power and provides the political substratum on which the "space race inevitability" theory is inscribed. By considering space as a zero-sum game, China's ascendancy as a space power must represent a destabilising factor—if not a threat—to the current status quo, and confrontational stances will thus be inherent in the development of its space programme. For both Asian countries and the

[33] "Abe compares Japan-China tension to Britain, Germany before World War I". The Asahi Shinbun. 24 January 2014. Web. http://ajw.asahi.com/article/behind_news/politics/AJ201401240076/. Accessed 15 March 2014.

[34] The visits to the *Yasukuni shrine* by Japanese government's officials are part of a broader issue known as *rekishi mondai* (the problem of history) dealing with Japan's purported reluctance to atone for its militarism and occupation of East and South East Asia during the 1930s. Part of this problem is also the rejectionism contained in the textbooks adopted by the MEXT (the so-called kyokasho mondai) and by politicians of the Liberal Democratic Party with regard to issues like the Nanjing massacre or invasion of China.

[35] The so-called fumie gaiko (expiate diplomacy) pursued by the Chinese government to keep Japanese diplomatic outreach down is, for instance, not much appreciated by Tokyo, given that Japan is the only country in the world to have denied its own *ius ad bellum* and that China also continues to falsify "historical facts" in its textbooks.

[36] Roulo, Claudette. "Pacom Area of Responsibility Defined by Superlatives". American Forces Press Service. 16 January 2014. Web. http://www.defense.gov/news/newsarticle.aspx?id=121499. 16 March 2014.

[37] For many analysts, there are striking differences in the current geopolitical context that could make the analogy erroneous. For an analysis of this issue, see Gadi, Franz-Stefan. "Let's Drop the Anglo-German Historical Analogy Once and For All When Discussing China". China–US Focus. 7 February 2014. Web. http://www.chinausfocus.com/foreign-policy/lets-drop-the-anglo-german-historical-analogy-once-and-for-all-when-discussing-china/. Accessed 16 March 2014.

United States—which must exclude cooperation in order to avoid an even more rapid catch up by China—the "path of competition" is but the perceived policy reality they are facing, whether they like it or not.

Although the idea of a "new space race" has come to dominate the debate on the international politics of space,[38] there are scholars and analysts who strongly criticise these types of approaches, which in their view have dangerous policy implications.[39] Indeed, while too reductive a formula to describe the current and much more complex international dynamics in space, the US–Soviet space race analogy might eventually become a tragic self-fulfilling prophecy if its logic is endorsed. The existence of this "academic disquiet" shows how the theme remains open to divergent and even contending readings of the current international dynamics in space.

6.2.2 *Assessing the Prospect of an Intra-Asian Space Race*

The idea of a space race easily catches the imagination of the general public and the attention of policymakers for almost self-explanatory reasons. The triumphant legacy of Apollo has profoundly shaped—and some argue even distorted—the way the world looks at human spaceflight, including the role of cooperation and competition.[40] However, different interpretations can be offered.

6.2.2.1 How to Assess a Space Race: Parameters

With acknowledgement of the analysis offered in a contribution by Kazuto Suzuki,[41] a first step, when drawing a comparison between current intra-Asian or global space dynamics and the US–Soviet space race, is to undertake a more precise investigation of the parameters and constituencies that defined that first space race. In very general terms, the US–Soviet space race made can be seen as a proxy for geopolitical rivalries, aiming to demonstrate respective superiority in space capabilities, particularly manned spaceflight capabilities. The technological superiority required for such supremacy was seen by both actors as necessary for national

[38] Besides the growing number of articles in the media, the idea of an intra-Asian space race (Japan, China, and India) has been brilliantly described in the literature, by a number of eminent scholars, such as Ajey Lele or James Clay Moltz. Others, like Erich Seedhouse, have on the contrary argued that a new space race is bound to take place but that this will be between China and the United States.

[39] See, for instance, Zhang, Yongjin (2013). "The eagle eyes the dragon in space—A critique". Space Policy 29 (2):113–120.

[40] National Research Council (2012). *NASA Strategic Direction and the Need for a National Consensus*. The National Academy Press. Washington DC.

[41] Suzuki, Kazuto (2013). "The contest for leadership in East Asia: Japanese and Chinese approaches to outer space". Space Policy 29 (2): 99–106.

security and economic affluence and symbolic of ideological superiority. Three interconnected yet specific arenas of competition were thus to hand in the space race of the original antagonists.

First, there was a competition for international prestige or soft power, which substantiated itself in the pursuit of space firsts in the civilian space programmes. The launch of satellites, planetary probes, and astronauts, the race to the Moon, and the construction of space stations are all examples of this one-upmanship contest, which served not only to prove superpower status but also superiority over the other and therefore the superiority of the respective ideology.

The United States and the USSR also competed in the development of their respective hard power. Although neither country's manned space programme had any real military value (they were rather a surrogate for military capabilities), other space assets soon became precious instruments for advancing the military might of the two superpowers. Telecommunications, Earth observation, meteorology, and navigation were all considered necessary capabilities to gain strategic advantage over the competing bloc.

A third arena of competition was the provision of space-related services or public goods. Besides assisting other countries in the development of space capabilities (respectively, Western European countries and Japan and Eastern European countries and Central Asia), the United States and the USSR created, among other things, the Intelsat and Intersputnik programmes for satellite telecommunications and the Intercosmos and Freedom station programmes to offer flight opportunities to their allies.

If we now look at these three arenas of competition today, the hypothesis of an intra-Asian space race loses much of its cogency, instead revealing the existence of different playing fields and approaches to space activities. As the following sections will explain, the evolution of the policy postures of Japan and India demonstrate that current—and conceivably future—dynamics cannot truly be understood as heralding a space race.

6.2.2.2 Space Exploration and Manned Spaceflight

It should first be acknowledged that the dynamics of the past 5 years have drastically downsized the prospect of a one-upmanship contest in space exploration. It is true that Japan, China, and India each launched almost concomitant lunar probes between 2007 and 2008 and that both JAXA and ISRO announced ambitious plans for robotic and human space exploration, more or less openly, intended to counterbalance China's rising star. However, very little has followed in practical terms.

- In Japan, the lunar exploration programme has encountered political setbacks that have compromised its steady implementation. In spite of the resounding success of the Kaguya-1 mission, policymakers have in fact provided little political backing and financial commitment to the programme. Because of the climate of severe budgetary pressures affecting Japan's economy since 2007 on the one side and the termination of US plans on the other, the follow-on Kaguya-2 moon lander scheduled for launch in 2010 and tactically intended to anticipate the 2013 Chinese Chang'e-3 mission was suspended and has not yet received the final "green light" from the government.[42] The mission is now scheduled for 2018, i.e. possibly even after China's Chang'e-5 sample return mission.

Even more ill-fated was the proposal for the implementation of an autonomous human spaceflight programme contained in JAXA's *Vision 2025*. In spite of the mounting excitement among the Japanese public generated by the release of the document, as early as September 2008 (i.e. a few days after China's first space-walk and thus in very stark contrast to it), Kawamura Takeo, then Chief Secretary of the Japanese Cabinet and a leading figure within Japanese space policy planning, declared that no Japanese autonomous manned space programme was in sight, beyond the long-term involvement in the ISS programme.[43] Not only that, he also said that Japan was "perhaps reaching a moment in time where we may consider cooperating with China over space issues".[44]

The decision to refrain from manned spaceflight was clearly carried out in the space budget allocations of the following years. Since 2008, the Space Policy Commission has given the lowest priority to manned spaceflight in the budget plan, limiting funds to the completion flight of the 2009-introduced H-II Transfer Vehicle (HTV) and to the full utilisation of the Japanese experimental module KIBO on board the ISS (see Annex D for an overview of Japan's exploration programme).

More broadly, the country has continued to focus on "less noble but practical space activities"[45] that are in line with its traditional science and technology-oriented approach, thus apparently standing aside and allowing China to pass. Of course, Japan remains concerned about Chinese development and future plans. The real concerns however are not Chinese space capabilities per se but—as argued in the next section—utilisation of these achievements to pursue diplomatic objectives in the Asian region. The concern is the pursuit of soft power within the region.

- In India the manned spaceflight programme initially received stronger political backing than in Japan, as A. P. Abdul Kalam, then the country's president, had for 20 years served as a director of ISRO and was a strong supporter of the human spaceflight programme. Initial funding began in April 2007, with the

[42] "Moon Lander Selene 2". Japan Aerospace Exploration Agency. Web. http://www.jspec.jaxa.jp/e/activity/selene2.html. Accessed 5 May 2014. See also Tanaka, S. et al. (2013). "Present status of the Selene-2". 44th Lunar and Planetary Science Conference. Web. http://www.lpi.usra.edu/meetings/lpsc2013/pdf/1838.pdf. Accessed 5 May 2014.

[43] "Japanese Official Calls for Space Cooperation With China". Kyodo News Service. 25 September 2008. See Saiget, Robert J. "Shenzhou 7 Astronauts Brace For Space Walk". Space Daily. 26 September 2008. Web. http://www.spacedaily.com/reports/Shenzhou_7_Astronauts_Brace_For_Space_Walk_999.html. Accessed 6 May 2014.

[44] Quoted from *Ibid.*

[45] Suzuki, Kazuto (2013). "The contest for leadership in East Asia: Japanese and Chinese approaches to outer space". Space Policy 29 (2): 99–106.

objective of sending the first Indian astronaut to orbit by 2014.[46] To support this ambitious exploration programme, ISRO took the decision to human-rate its GeoSynchronous Launch Vehicle (GSLV III) and develop a two-person manned capsule.[47] Actions were initiated accordingly. By the middle of 2007, ISRO had already validated its re-entry technology with the successful recovery of a space capsule and had started to work on pre-projects, including long-lead items for human missions such as spacesuits and simulation facilities.[48]

Following a working group on space cooperation with Russia, an agreement was reached between the two countries in December 2008 for the Indian manned spacecraft to be built following the trusted Russian Soyuz design, thus echoing the Chinese path.[49] In that context India also considered sending one of its citizens into space on board a Russian spacecraft, to acquire the skills necessary for future manned space missions, but that plan never materialised.

But also for India early expectations had to be reconciled with reality. In spite of a sharp overall increase in the space budget, the financial commitment to human spaceflight has remained rather limited. ISRO had estimated that the project leading to a first manned flight would cost from US$2.5 to $3 billion a year (more than three times the agency's current annual budget!).[50] However, the funds so far allocated to the programme have not exceeded a few hundred million dollars.[51] In addition, cooperation with Russia in the development of a Soyuz-based manned capsule has not gone as planned, while the development of the enhanced GSLV Mark III launch vehicle, whose first flight was scheduled for 2013, has encountered considerable delays and is still under development.[52]

[46] Peter, Nicolas (2008). "Developments in space policies programmes and technologies throughout the world and Europe". In Schrogl, Kai-Uwe, Charlotte Mathieu, Nicolas Peter (eds). *ESPI Yearbook 2006/2007: A new Impetus for Europe*. Springer, Vienna: p. 96.

[47] "GSLV Mark III". Indian Space Research Organisation. Web. http://www.isro.org/Launchvehicles/GSLVMARKIII/mark3.aspx. Accessed 10 May 2014.

[48] Peter, Nicolas (2009). "Developments in space policies programmes and technologies throughout the world and Europe". In Schrogl, Kai-Uwe, Charlotte Mathieu, Nicolas Peter (eds). *ESPI Yearbook on Space Policy 2007/2008: From Policies to Programmes*. Springer, Vienna 2009: p. 81.

[49] Also echoing the Chinese approach, it was decided that the first mission would be for a day, while the second for a week. Harvey, Brian, Henk H.F. Smid, and Theo Pirard (2010). *Emerging Space Powers. The New Space Programs of Asia, the Middle East and South America*. Springer–Praxis Publishing, Chichester, UK: p. 238.

[50] Jayaraman, K.S. "ISRO Seeks Government Approval For Manned Spaceflight Program" Space News. 13 November 2006. Web. http://www.spacenews.com/article/isro-seeks-government-approval-manned-spaceflight-program/. Accessed 18 May 2014. En passant, it is noteworthy that apparently in order to motivate such large expenditure, the scientific community started to "convince" politicians about the long-term prospect of getting helium-3 from the lunar surface.

[51] See Euroconsult (2014). *Profiles of Government Space Programs, Analysis of Over 80 Countries & Agencies*. Euroconsult Profiles Series, Paris.

[52] The GSLV experienced problems in April 2010 when the main cryogenic engine on India's domestically produced third stage failed to ignite. See Moltz, James Clay (2011). *Asia's Space Race: National Motivations, Regional Rivalries, and International Risks*. Columbia University Press, New York: p. 125.

Probably as a result of these adverse developments, in August 2013 ISRO Chairman Koppillil Radhakrishnan announced that a human spaceflight mission was not an ISRO priority,[53] a position that had already been clear since the release of the 12th Five-Year Plan (2012–2017), in which human spaceflight did not figure among the list of projects to be implemented over the period.[54]

Although in his announcement Mr. Radhakrishnan was keen to emphasise that the country would not be starting from scratch should a decision for human spaceflight eventually be taken, he avoided stating a definite timeframe for the potential launch, limiting himself to specifying: "We are not going to see the human spaceflight as a programme in the 12th Five Year-Plan. *We will see maybe later*".[55] As a result, it is now very difficult to make the case for one-upmanship with China. Even in the event of an abrupt volte-face in the current Indian space agenda, it can be projected that the country could make its first steps only by the end of the current decade or early in the next. By that time China will already have started to assemble the core module of its space station in orbit and will likely have mastered the crucial technologies to reach the ultimate target of a moon landing.

Fully aware of its short-term inability to compete directly with China in this arena, India seems already to be looking for alternative paths to gain international prestige, including a much-heralded reorientation of its space programme towards Mars exploration, where it has more chances of achieving "space firsts", at least in comparison to China. The launch of a Mars orbiter mission in November 2013 could be interpreted in this light.[56]

---

[53] "Human space flight mission off ISRO priority list". NDTV. 16 August 2013. Web. http://www.ndtv.com/article/india/human-space-flight-mission-off-isro-priority-list-406551. Accessed 18 May 2014. The position was also clarified in a personal interview with Ajey Lele, where he stressed, "As of today, a human mission is not in our space agenda. We are in a very early phase of developing a few critical technologies required for realising a human mission". Lele, Ajey (2014). *Mission Mars. India's Quest for the Red Planet*. Springer, New Delhi: pp. 120–124.

[54] Planning Commission, Government of India. Twelfth five year plan (2012/2017). Faster, More Inclusive and Sustainable Growth. New Delhi, India. 2013: pp. 264–268.

[55] (Emphasis added). "Human space flight mission off ISRO priority list". NDTV. 16 August 2013. Web. http://www.ndtv.com/article/india/human-space-flight-mission-off-isro-priority-list-406551. Accessed 18 May 2014. In ISRO Space Vision 2025, the objective is limited to the development of a space vehicle capable of putting two humans into LEO and returning safe to Earth, something that China achieved as early as 2003.

[56] The mission has been a landmark achievement for India, demonstrating its technological prowess and providing the country with a significant source of national pride, as confirmed by the speech made by Prime Minister Modi in September 2014, immediately after the Mangalyaan spacecraft entered Mars orbit: "History has been created. We have dared to reach out into the unknown and have achieved the near impossible [...] the success of our space programme is a shining symbol of what we are capable of as a nation". Quoted from "India wins Asia's Mars race as spacecraft enters orbit". Mars Daily. 24 September 2014. Web. http://www.marsdaily.com/reports/India_wins_Asias_Mars_race_as_spacecraft_enters_orbit_999.html. Accessed 10 October 2014.

This strategy is also promoted by many prominent figures within the Indian space policy community. In an interview with Ajey Lele, Pallava Bagla, an authoritative commentator on the Indian space programme, advocated the benefits of such a strategy: "Now, there is an opportunity for India to become the first and the only Asian State (a few years ago Japan attempted a Mars mission but that also was a failure) to Mars, if its mission becomes successful. Also, this could provide an opportunity for India to demonstrate its capabilities well ahead of its regional rival China".[57]

All in all, although the inclusion of exploration in India's space portfolio suggests that prestige considerations have become an important factor,[58] it is unlikely that they will come to dominate the space programme at the expense of the development rationale. An examination of the priorities and goals listed in the 12th Five-Year Plan (2012–2017) shows a sharp contrast between the great attention devoted to the "societal use of space" (in particular tele-education and telemedicine projects and more broadly space applications through the use of the INSAT system) and the scant attention paid to *grandeur* projects. This indicates that the "needs-based approach" is still *the* overwhelming principle of Indian space policy.[59]

From this perspective, it appears evident that neither India nor Japan is competing directly with China in the pursuit of international prestige through human spaceflight achievements.

6.2.2.3 Military Space Competition

While Japan and India both appear extremely concerned about China's ever-growing military space capabilities, it is difficult—if not misleading—to make the case for a space arms race between the three Asian giants.

- With regard to Japan's posture, it should be that in July 2008, 1 year after the Chinese 2007 ASAT test, the country drastically revisited its "pacifist" space

[57] Quoted from Lele, Ajey (2014). *Mission Mars. India's Quest for the Red Planet.* Springer, New Delhi: p. 112.

[58] As also stressed by the physicist Barath Gopalalswamy, the space programme's broadening to space exploration indicates that "India no longer views space as only enhancing the living conditions of its citizens but also as a measure of global prestige". Quoted from Moltz, James Clay (2011). *Asia's Space Race: National Motivations, Regional Rivalries, and International Risks*. Columbia University Press, New York.

[59] As the document stresses, "significant developments have taken place in the area of societal applications of space technology. Some of the important ones are: (a) expansion of tele-education network to over 55.000 classrooms; (b) telemedicine facilities in over 382 hospitals". See Planning Commission, Government of India. Twelfth five year plan (2012/2017). Faster, More Inclusive and Sustainable Growth. New Delhi, India. 2013: pp. 264–268.

policy by approving the *Basic Law for Space Activities*,[60] which allowed a substantial reinterpretation of Japan's normative framework and, for the Self-Defence Force (SDF), opened up the possibility of developing, procuring, and operating space assets. At a superficial level, this policy revision can be read as a direct response to China's ASAT test and its increasing capabilities in military space assets. In fact, a closer look reveals that China was not the catalyst for this change. For one thing, the legislative process that saw the approval of this law was started in early 2005,[61] at a time when Sino-Japanese relations were showing signs of cautious improvement and 2 years before the ASAT test.[62] More importantly, the original decision—which has even longer roots—stemmed from the need to normalise Japan's cumbersome space policy, which, since the inception of the space programme, had unnaturally excluded the use of space technologies for security-related purposes.

The cumbersome situation was not only the result of the inflexible provisions of Article 9 of the Constitution, imposed after World War II,[63] but more specifically the result of a parliamentary resolution adopted by the Diet in 1969. The Resolution "Space Development for Exclusively Peaceful Purposes" banned any involvement in security-related space activities whatsoever and limited the development of space technologies to "exclusively peaceful purposes". As Suzuki explains, although the term "exclusively peaceful purposes" is not unique, as it appears in the Outer Space Treaty and the ESA Convention, the interpretation of this clause in Japan was very strict. The clause was

[60] The new "Basic Law for Space activities" introduced a series of major administrative and conceptual changes. In particular the switch of space planning from the Ministry of Education, Culture, Sports, Science and Technology (MEXT) to the Prime Minister's Cabinet underscores a shift in attitude concerning the strategic importance of space for national security and other areas. The new law also removes the ban on any defensive governmental uses of space. *Cit.* Peter, Nicolas (2009). "Developments in space policies programmes and technologies throughout the world and Europe". In Schrogl, Kai-Uwe, Charlotte Mathieu, Nicolas Peter (eds). *ESPI Yearbook on Space Policy 2007/2008: From Policies to Programmes*. Springer, Vienna 2009: p. 53.

[61] In that year the MEXT formed a study group called "Consultation Group for National Strategy for Space" that eventually proposed establishing a new law for space activities. Suzuki, Kazuto (2008). "Basic law for space activities: A new space policy for Japan for the 21st century". In Schrogl, Kai-Uwe, Charlotte Mathieu, Nicolas Peter (eds). *ESPI Yearbook 2006/2007: A new Impetus for Europe*. Springer, Vienna: pp. 225–238.

[62] It is, for instance, remarkable that Abe Shinzo, the future prime minister of Japan, published a programmatic book explaining the pro-cooperation posture Japanese government would adopt vis-à-vis China. Abe, Shinzo (2006). *Utsukushii Kuni he* (Towards a beautiful nation). Bungei Shunju, Tokyo: p. 152.

[63] The so-called no-war clause of Japan's constitution states: *Aspiring sincerely to an international peace based on justice and order, the Japanese people forever renounce war as a sovereign right of the nation and the threat or use of force as means of settling international disputes. In order to accomplish the aim of the preceding paragraph, land, sea, and air forces, as well as other war potential, will never be maintained. The right of belligerency of the state will not be recognized.* This restriction on the fundamental right of national self-defence is of course radical and can only be understood in the context of the post-World War II era during which the Japanese Constitution was written or imposed.

interpreted similarly to the peaceful use of atomic energy, with a consolidated belief that "space technology should not only be prohibited from being used for aggressive actions ('non-aggressive') but also that the military authorities should be prohibited from developing, owning, operating or using it ('non-military')".[64]

As such, the evolution reflected in the 2008 Basic Law was not intended as a step towards a "militarisation" of the space programme, but as part of the broader process of "normalisation" of Japan's international status. The political will to make of Japan a "normal country" (*futsu no kuni*)[65] had already become stronger in the aftermath of the Cold War, pushed by a deep-rooted desire to become a permanent member of the UN Security Council[66] and eventually accelerated by the need to face the brinkmanship strategy of North Korea, rather than by China's rise.[67]

Quite notably, the decision to initiate the Information Gathering Satellite (IGS) programme—Japan's first dual-use reconnaissance satellite—was taken immediately after the so-called Taepodong shock or North Korea's launch of a ballistic missile over Japanese territory in 1998. Implementation of the IGS programme would, however, prove to be particularly difficult. As a result of self-imposed legal constraints, the Japanese Defence Agency (now Ministry of Defence) was not allowed to participate in IGS development and operation. It was even difficult to mention the word "dual-use" because it implied the possibility of the participation of military authorities. Thus, although clearly the purpose of IGS was to monitor military activities, it had to be operated by a civilian authority and termed a "multipurpose satellite".[68]

It was the need to overcome these legal constraints and bring Japan's space activities in line with that of other countries, rather than the rise of China, that eventually resulted in the formation of a study group and the subsequent introduction of the 2008 Basic Space Law. The new law adopted the international standard interpretation of "peaceful uses of outer space" and thus recognises the possibility for the Ministry of Defence and the Self-Defence Forces to procure, own, and operate space systems. However, it specifies that any military use of space should be for defensive purposes in accordance with the 1967 Outer

[64] Suzuki, Kazuto (2013). "The contest for leadership in East Asia: Japanese and Chinese approaches to outer space". Space Policy 29 (2): 100.

[65] See Soeya, Yoshihide et al. (eds) (2011). *Japan as a 'Normal Country'? A Nation in Search of Its Place in the World.* University of Toronto Press, Toronto.

[66] Japan's anti-militaristic posture has gradually eroded since the end of the Cold War and in particular after the launch of the US Global War on Terrorism in 2001. Japan has become more involved in peacekeeping operations, and in 2007 Japan's Defence Agency was transformed into a fully fledged Ministry of Defence.

[67] See Pekkanen, Sadia M. and Paul Kallender-Umezu (2010). *In Defense of Japan. From the Market to the Military Space Policy.* Stanford University Press, Stanford, CA.

[68] Suzuki, Kazuto (2008). "Basic law for space activities: A new space policy for Japan for the 21st century". In Schrogl, Kai-Uwe, Charlotte Mathieu, Nicolas Peter (eds). *ESPI Yearbook 2006/2007: A new Impetus for Europe.* Springer, Vienna: pp. 225–238.

Space Treaty and with "the pacifist spirit of Japan's constitution".[69] It is thus implied that offensive space assets (e.g. ASAT capabilities or jamming technologies) cannot be developed. The objective is to tackle looming challenges, not to open up avenues for militarisation.

While this erosion of the "non-military" interpretation neither automatically inhibits the progressive erosion of the "non-aggressive" interpretation nor the parallel emergence of a space arms race with China, what counts is that Japan's Ministry of Defence (MoD) is not willing to invest in space capabilities—let alone compete with China in a space arms race. There are multiple reasons for this reluctance. First, antimilitarism remains a deep-rooted attitude within a number of Japan's parliamentary factions (including the Democratic Party) and within the nation as a whole.[70] Second, considering that the government does not want to invest more than 1 % of its GDP in defence and the current climate of budgetary constraints also affects the defence budget, "there is no luxury to increase spending for the unfamiliar domain of space".[71] Furthermore, the MoD has little experience and almost no staff or technical expertise in space technology. This makes it dependent on JAXA for expertise. As once again argued by Kazuto Suzuki, "Given the secretive nature of the MoD, it would not be acceptable to depend on a civilian agency to develop military sensitive technology. So instead of relying on JAXA the MoD has to date chosen not to invest so much in space".[72] It is, for instance, worth mentioning that satellite communications capabilities have continued to be procured from private companies (e.g. the Sky Perfect JSAT Corporation), rather than being developed through the autonomous efforts of the MoD.[73] Finally, many analysts have noted the importance of Japan's alliance with the United States and argued that these ties mean that Japan does not have to become a military space power (and compete in an arms race in space) in order to maintain its security.

That being said, one should not minimise the risks implied in Tokyo's strategic posture. As Henry Kissinger aptly points out in a description of the self-propelled international mechanisms that traditionally head towards collision, "when diplomacy no longer functions, relationships become increasingly concentrated on military strategy—first in the forms of arms races, then as a manoeuvring for strategic advantage even at risk of confrontation and, finally, in war itself".[74] In this light, it can be argued that, in its relationship with China,

---

[69] Article 2 of the Basic Space Law stipulates that the use of space systems for national and international security comply with both the framework of international agreements and a pacifist constitution. *Ibid.* p. 234.

[70] See Lanna, Noemi (2010). *Il Giappone e il Nuovo Ordine in Asia Orientale. L'altra faccia dell'ascesa della Cina.* Vita e Pensiero, Milano.

[71] Suzuki, Kazuto (2013). "The contest for leadership in East Asia: Japanese and Chinese approaches to outer space". Space Policy 29 (2): 101.

[72] *Ibid.* p. 101.

[73] *Ibid.* p. 103.

[74] Kissinger, Henry (2011). *On China.* Penguin Books, New York: p. 515.

Table 6.1 JACHINDIA: Asian powers in contrast (2013, US$ billion)

	Japan	China	India
GDP[a]	4901	9181	1870
Space budget[b]	3.69	3.43	1.25
Human spaceflight[b]	0.48	0.79	0.02
Military spending[c]	48.6	188	47.4

[a]IMF
[b]Euroconsult
[c]SIPRI

Japan has opted to skip the first step, moving steadily towards the succeeding one. Manifestly unwilling—and unable—to keep pace in an arms race with China, given the resource difficulty in competing with a giant (see Table 6.1) and the inherent antimilitarism of the Japanese population, the country is trying to gain strategic advantage by forging more or less formal mechanisms to contain the expansion of China's power. These manoeuvres have included an upgrading of the US–Japan security alliance (an upgrading that en passant impelled the country somewhat riskily to use the Senkaku/Diaoyu dispute as a test-bed for Washington's commitment)[75] and the proposal for the so-called Asia's Democratic Security Diamond, an informal alliance of Asia-Pacific democracies (Japan, India, Australia, and the United States) intended to hedge against China's power projection.[76]

Regardless of the result of these "earthly manoeuvres", it appears highly unlikely that space will become an arena of military competition between Japan and China. Admittedly China's steady increase in military spending and development of offensive space assets remain a major source of regional instability, but the Japanese government does not intend to acquire similar capabilities or to participate in a space arms race with China.

- As for India, over the last decade the country has dramatically increased its interest in the area of security-related space activities and has started to bring its

[75] For a detailed account on the evolution of the US–Japan Security Treaty, see Hook, Glenn D. et al. (2012). *Japan's International Relations*. Routledge, Oxon: pp. 126–152.

[76] This strategic imitative was already launched during Mr. Abe's first tenure as prime minister (September 2006–September 2007) and named the Quadrilateral Security Dialogue (Japan, the United States, India, and Australia) as a solution to the maritime disputes involving China. The new strategy is based on "three pillars: (1) reinvigorating the US-Japan alliance; (2) a reintroduction of the UK and France to Asia's international security realm; and (3) bolstering international cooperation between key democracies in the Indo-Pacific, such as India and Australia". Miller, J. Berkshire. "The Indian Piece of Abe's Security Diamond". The Diplomat. 29 May 2013. Web. http://thediplomat.com/2013/05/the-indian-piece-of-abes-security-diamond/. Accessed 10 May 2014. For Shinzo Abe's announcement on the Asia's Democratic Security Dialogue, see Abe, Shinzo. "Asia's Democratic Security Diamond". 27 December 2012. Web. http://www.project-syndicate.org/commentary/a-strategic-alliance-for-japan-and-india-by-shinzo-abe. Accessed 10 May 2014.

programme in line with the rationales typical of the other major space powers.[77] There are also indications that India might be making a transition from dual-use satellite technology to dedicated defence satellites.[78] In recent years, the Indian Air Force, the Indian Navy, and the Indian Army have been expressing their interest in acquiring dedicated satellites and advancing their expertise and capabilities in terms of space assets. However, the reality is that "today, the military continues to suffer from a lack of institutional support for space activity in comparison with ISRO and institutions under the civilian Department of Space" and is lamenting the risk of losing strategic ground vis-à-vis the development of other major powers, specifically China. [79]

Indian policymakers are, however, extremely wary of competing with China's military space rise in ways that they would ultimately find difficult to support. There are, first of all, structural impediments that work against the case for a military space race. For New Delhi's politicians, it is evident that over the next 10 years, only China (besides the United States) is likely to be able to spend more than $200 billion on defence.[80] More broadly, they are fully aware that China's "comprehensive national power" exceeds India's by a very large margin. With a GDP that is 1/5 that of China, as well as a space budget that is a small portion of the Chinese one (see Table 6.1), fomenting a space arms race is the least of the strategically plausible options for New Delhi.

Many believe that, even more than for China, time is for India an asset (perhaps the most precious one): it will allow the country to grow ever stronger and eventually catch up not only with the advanced economies of the West but also with the Central Kingdom. The more solid foundations of India's economy and its demographic prospects steadily march in New Delhi's favour.[81] According to the UN *World Population Prospect*, released in June 2013,

[77] For a detailed description of Indian involvement is security-related space activities, see Moltz, James Clay (2011). *Asia's Space Race: National Motivations, Regional Rivalries, and International Risks*. Columbia University Press, New York: pp. 127–131. See also Lele, Ajey (2013). *Asian Space Race: Rhetoric or Reality?* Springer, New Delhi: pp. 118–190.

[78] *Ibid*. pp. 188–189.

[79] *Cit*. Moltz, James Clay (2011). *Asia's Space Race: National Motivations, Regional Rivalries, and International Risks*, Columbia University Press, New York: p. 128.

[80] Xuetong, Yan (2012). "The weakening of the unipolar configuration". In Leonard, Mark (ed). *China 3.0*. European Council on Foreign Relations, London: p 11.pp. 118–123.

[81] The idea that India is bound to overtake China has been debated for more than a decade now. According to several scholars, India has adopted an Import-Substitution Industrialisation (ISI), not an Export-Oriented Industrialisation (EOI), strategy. Thus, unlike China, India is not dependent on export for its economic growth, having principally relied on internal consumption. According to the projections, by 2050 India's economy will rank second after that of China, and the demographic prospects, as stressed, will in the long-term play in New Delhi's favour. See Mazzei, Franco, Vittorio Volpi (2010). *La rivincita della Mano Visibile. Il Modello economico asiatico e l'Occidente*. Egea, Milano. See also Huang, Yasheng, and Tarun Khanna. "Can India Overtake China?". Foreign Policy. 1 July 2003. Web. http://foreignpolicy.com/2003/07/01/can-india-overtake-china/. Accessed 16 July 2014.

India's population will surpass China's by 2028.[82] Also noteworthy is that already in 2012 India's working age population grew by twelve million, while China's shrank by over three million.[83]

If India can ultimately leverage favourable demographics and the more solid foundations of its economy, it is believed that the country "will be in a position to close the strategic gap with China by mid-century".[84] However, for the immediate future China will remain a distant peer, and this gap compels India to keep a prudent approach vis-à-vis China, gradually and methodically building up its military capabilities, but in the meantime carefully postponing the strategic contest for another day.

In addition to these strategic considerations, it should be noted that the current Sino-Indian relations give rise for cautious optimism for the future. While their interplay remains enigmatic at best, both New Delhi and Beijing have in recent years adopted a more practical and positive stance towards each other. After all, as the leaders of the two countries have repeatedly affirmed, Asia is vast enough to allow the peaceful rise of both China and India, and it is in their interests to maintain a stable environment and a cooperative relationship. Today China and India are more "politically and economically engaged than at any time in recent history".[85]

Besides the ever-increasing economic exchanges and a maturing political dialogue, the Chinese and Indian armies have also started periodically to hold joint military exercises (they held their first joint exercises in 2007; these were followed by two more in 2008 and 2013), thus significantly expanding the level of cooperation.[86]

It is also significant that India is gently declining Tokyo's diplomatic offer of the formation of a Democratic Security Diamond.[87] Indeed, although willing to increase its cooperative relations with Tokyo as well as to gain strategic advantage vis-à-vis China, New Delhi is visibly reluctant to take part in initiatives that would limit the freedom of manoeuvre of its diplomacy and generate undesirable rifts in its maturing relationship with that country.

---

[82] United Nations' Department of Economic and Social Affairs. World Population Prospects. The 2012 Revision. United Nations. New York, 2013. Web. http://esa.un.org/wpp/documentation/pdf/WPP2012_%20KEY%20FINDINGS.pdf.

[83] Smith, Jeff M. "India and China: The End of Cold Peace?" The National Interest. 10 February 2014. Web. http://nationalinterest.org/commentary/india-china-the-end-cold-peace-9853. 28 May 2014.

[84] *Ibid.*

[85] *Ibid.*

[86] *Ibid.*

[87] See "India uncertain as Abe looks for anti-China Alliance". Global Times. 18 February 2014. Web. http://www.globaltimes.cn/content/843272.shtml. Accessed 28 May 2014. See also Miller, J. Berkshire. "The Indian Piece of Abe's Security Diamond". The Diplomat. 29 May 2013. Web. http://thediplomat.com/2013/05/the-indian-piece-of-abes-security-diamond/. Accessed 28 May 2014.

On the basis of these considerations, it becomes quite hard to make the case for a Sino-Indian space arms race.

6.2.2.4 Commercial/Diplomatic Space Competition

Unlike the first two arenas of competition (achievement of "space firsts" and competition over military capabilities), the third presents truly competitive dynamics, especially between Japan and China.

- The real Japanese concern is about the threat posed by China to its regional leadership in the provision of space-related services and more broadly to the efforts Japan has made in seeking to guide regional development and shaping its integration processes.[88] For many years Japan was the only country in the Asia-Pacific region able to play a leading role in the provision of space-related services and in the creation of cooperative undertakings. Its leadership in space was forged with the establishment of the Asia-Pacific Regional Space Agency Forum (APRSAF) in 1993, an informal mechanism of consultation intended to coordinate Asia-Pacific activities in space and enhance cooperation among the space agencies of various nations in the region.[89]

  In 2005, the transformation of the China-led AP-MSTA into APSCO (see Box 1, Chap. 1) eclipsed Japan's position, as the new organisation was a fully fledged international organisation providing very attractive programmes for emerging space countries in the Asia-Pacific. Many countries (including those targeted by Japan such as Bangladesh, Indonesia, Thailand, and Mongolia) greatly appreciated the initiative and became members of APSCO.

  In addition, the success of China's manned space programme had started to send a clear message that Chinese space technology was not only affordable but also proven and reliable. As Wolfgang Rathgeber noted in an early ESPI study, the message is strong and clear: "the barriers for developing countries to enter the space club are extremely high, but through cooperation with China it is possible to gain access to space technology and affordable launch service by the Long March".[90] In short, China's human spaceflight has in Japan's eyes gradually eroded its position as a technological leader in Asia and given relative advantage to the APSCO "diplomatic initiative" (note: the proposal to create APSCO was launched by China's Ministry of Foreign Affairs).

---

[88] Japan has been in the vanguard of the regional integration process in the past 20 years: suffice to think of its 1990 proposal for the creation of an East Asia Economic Caucus (EAEC) or the 1997 proposal about the establishment of an Asian Monetary Fund. While these initiatives have failed, thanks to the United States, others have been more successful, like the East Asia Community (EAC) meetings launched in 2008 and the leadership Japan exercises within the Asian Development Bank.

[89] More information on APRSAF available at the website: http://aprsaf.org.

[90] Rathgeber, Wolfgang (2007). "China Posture in Space. Implications for Europe". ESPI Report 3. European Space Policy Institute, Vienna: 54.

As a response to the competition introduced by China's emerging ambitions, JAXA and the Japanese Ministry of Education, Culture, Sports, Science and Technology (MEXT) initiated more ambitious programmes, gradually transforming APRSAF from a talking shop into a programme management structure. At the 15th APRSAF meeting, Sentinel Asia, a regional disaster-monitoring programme inspired by the European Copernicus programme, was launched. This was followed in 2008 by the Space Applications for Environment (SAFE) programme, which focuses on the development of space applications for analysing climate change. Mimicking the APSCO-led Small Multi-Mission Satellite (SMMS) programme, in 2009 Japan also launched the Satellite Technology for the Asia-Pacific Region (STAR), which focuses on small satellite development and technology transfer, and in 2011 the Asia Beneficial Collaboration through Kibo (KIBO-ABC) utilisation.[91] Japan's government calls Kibo the "gateway to ISS in Asia" and through this programme seeks to accommodate non-Japanese Asian participants in the full utilisation of this module.[92] A "Kibo Utilisation Office for Asia" was established in July 2010 and has already succeeded in gaining a dozen government-to-government contracts.[93]

As noted by Moltz, what is also notable about recent APRSAF meetings is "the gradual shift in venue away from Tokyo to other Asian localities where Japan can 'show the flag' and promote new forms of space cooperation".[94] In short, space is now used by Japan as a fully fledged diplomatic tool. The recently introduced practice of providing official development assistance (ODA) to developing countries in order to support their procurement of satellites or space services (notably, Japanese ones) confirms this trend.[95]

It is in this space arena—and only in this arena—that Japan and China are engaged in a strategic competition. This contest for leadership, however, should not be truly understood in the sense of a space race. In addition, it should not be forgotten that it has produced a number of positive outcomes in the broader sense.

Thus, many countries in the region have seen their policy options increase and have been increasingly provided with efficient and economic access to space applications, Earth observation data, and telecommunications capabilities. Sino-

[91] See Aliberti, Marco (2013). "Regionalisation of Space Activities in Asia?". ESPI Perspectives 66. European Space Policy Institute, Vienna.

[92] Robinson, Jana (2012). *Europe-Japan Strategic Partnership: the Space Dimension*. ESPI Report 40. European Space Policy Institute, Vienna: p. 36.

[93] For more information, see "Kibo Utilisation Office for Asia". Japan Aerospace Exploration Agency. Web. http://iss.jaxa.jp/en/kuoa/index.html. Accessed 3 June 2014.

[94] Moltz, James Clay (2011). *Asia's Space Race: National Motivations, Regional Rivalries, and International Risks*, Columbia University Press, New York: p. 68. Recent meetings have been held in Jakarta (Indonesia, 2006), Bangalore (India, 2007), Hanoi (Vietnam, 2008), Bangkok (Thailand, 2010), Melbourne (Australia, 2010) Singapore (2011), Kuala Lumpur (Malaysia, 2012), and Hanoi (Vietnam 2013).

[95] In 2011, for instance, the Japanese government provided about $1 billion of ODA to Vietnam, which was also intended to facilitate the purchasing of a Japanese Earth observation satellite (ASNARO). Suzuki, Kazuto (2013). "The contest for leadership in East Asia: Japanese and Chinese approaches to outer space". Space Policy 29 (2): 103.

Japanese competition has assisted them in the process of reaching the capability level of the international space community.[96]

- With respect to political influence over the region, the Indian position appears rather weak, as the country is not competing with Japan or China for leadership of regional cooperative undertakings: there is no Indian-led regional organisation that can be compared to the Japanese-led APRSAF or the Chinese-led APSCO. Admittedly, ISRO hosts the headquarters of the Centre for Space Science and Technology Education in Asia and the Pacific (CSSTEAP), which was established in 1995 in response to a resolution of the UN General Assembly recommending the creation of centres for space education in developing countries.[97] The CSSTEAP, which works under the auspices of the United Nations Economic and Social Commission for Asia and the Pacific (UNESCAP), has limited scope, however, and less ambitious targets than APSCO and APRSAF; it is unlikely to play a major role in giving political shape to cooperative undertakings in the region. CSSTEAP's core objectives are in fact limited to educational activities in space science and technology, a task that figures only among the marginal activities of APSCO and APRSAF.[98] Even by considering other initiatives, such as Modi's recent proposal to build a satellite within the framework of the South Asian Association for Regional Cooperation (SAARC), India's position remains frail in comparison to Japan and China.

  What is perhaps more interesting is that India is moving ahead in the commercial launch sector, an area where it has some competitive advantages vis-à-vis Japan, whose services are expensive,[99] and China, still strongly affected by the ITAR regime. Here too, however, Indian efforts are primarily oriented—like its broader political economy—towards internal consumption (achieving self-sufficiency in launching national satellites). Commercialisation efforts do not figure among the priorities of the Indian Space Transportation Systems during the 12th Five-Year Plan (2012–2017).[100]

---

[96] Aliberti, Marco (2013). "Regionalisation of Space Activities in Asia?". ESPI Perspectives 66. European Space Policy Institute, Vienna.

[97] UN General Assembly Resolution 45/72 of 11 December 1990. For more information on the process that led to the establishment of the centre, see the CSSTEAP website: http://www.cssteap.org/background. Accessed 6 June 2014.

[98] The main task of the centre is to develop the skills and knowledge of university educators, environmental research scientists, and project personnel in the design, development, and application of space science and technology for subsequent application in national and regional development and environment management. For more information, see "Objectives of the Centre". Centre for Space Science and Technology Education in Asia and the Pacific. Web http://www.cssteap.org/objectives-of-center. Accessed 6 June 2014.

[99] It should be noted that Japan has, in a sense, given up competing in the global market for commercial satellite launches, given the expanding, fierce competition of global providers and the high costs of its space services. Japan has indeed on some occasions used Indian launchers. See Moltz, James Clay (2011). *Asia's Space Race: National Motivations, Regional Rivalries, and International Risks*. Columbia University Press, New York: pp. 61–62.

[100] Planning Commission, Government of India. Twelfth five year plan (2012/2017). Faster, More Inclusive and Sustainable Growth. New Delhi, India. 2013: pp. 264–268.

Competition in this arena, however, is subject to global, rather than regional, dynamics and cannot therefore be labelled a space race. [101] If we think of it as a race, then, in the words of Ajey Lele, "it is a form of global space race, where every state is looking for a share".[102]

6.2.2.5 Conclusions

All in all, it appears that Asian space nations are pursuing different approaches to space activities and do not share the same "rules of engagement" as those that defined the first space race. In a nutshell, Japan does not compete with China and India in the first two arenas discussed; it only aims to maintain its regional leadership in the provision of space services, although, admittedly, also seeking soft power advantages from this. The Indian space programme, which remains mainly focused on "developing space infrastructure for domestic purposes, is less ambitious in respect to influencing the region"[103] and does not aim to challenge China with head-to-head endeavours (no manned spaceflight, for instance, but reorientation towards Mars exploration). As for China, it is clear that its efforts have higher targets than Japan and India.

There is, however, little doubt that Chinese lunar ambitions, once formalised, will greatly affect the posture of policymakers in both Tokyo and New Delhi. While arguably inducing a firmer commitment to autonomous human spaceflight capabilities (e.g. the Japanese decision to human-rate its HTV, which is expected in the coming years, and the Indian decision to complete development of a human-rated GSLV Mark III), it can be expected that JAXA and ISRO will seek to promote international cooperation and establish mutually complementary relationships with other partners.

In the case of India, the partner of choice is Russia[104] as well as the United States,[105] but cooperation with China (or even triangular cooperation

[101] The target countries of India's space launch services are not only Asian neighbours. Thanks to the reliability and relatively low cost of its services, India has launched satellites for Argentina, Belgium, Canada, Denmark, France, Germany, Indonesia, Israel, Japan, the Netherlands, Singapore, South Korea, Switzerland, and Turkey. See Moltz, James Clay (2011). *Asia's Space Race: National Motivations, Regional Rivalries, and International Risks*, Columbia University Press, New York: p. 126.

[102] Quoted from *ibid.* p. 126.

[103] Suzuki, Kazuto (2013). "The contest for leadership in East Asia: Japanese and Chinese approaches to outer space". Space Policy 29 (2): 99.

[104] For a detailed overview of India–Russia space cooperation, see Mathieu, Charlotte (2008). "Assessing Russia's space cooperation with China and India. Opportunities and Challenges for Europe". ESPI Report 12. European Space Policy Institute, Vienna.

[105] For more detailed information on the Indo-American space partnership, see Correll, Randal (2006). "US-India Space Partnership: the Jewel in the Crown". Astropolitics Vol. 4 (2): 159–177. For a more recent analysis, see Gopalalswamy, Barath (2011). "Indo-US Space Cooperation: Aiming Higher". CSIS Issue Perspectives Vol. 1 (2): 1–3.

Russia–China–India),[106] though unlikely, should not be excluded a priori. As for Japan, the most likely strategy is that the country will seek aggregation within the programmes of a third party, most notably the United States, and indeed use the prospect of a "Red Moon" as a bargaining chip to induce Americans to refocus their exploration programme from the asteroid-next scenario to a lunar landing plan, in which of course it will be eager to participate.[107]

6.2.3 *Alea Iacta Est? Introducing the Prospect of a Sino-American Space Race*

Whereas the application of the US–USSR space race analogy to describe current intra-Asian space dynamics appears too reductive and potentially misleading, the geopolitical positioning of the various Asian powers and the United States nevertheless superficially suggests that a space race similar to that between the United States and USSR could develop between China and the United States in the near future. In fact, all the ingredients seem already ingrained in current Sino-American interaction.

China, fulfilling its interpretation of its national destiny, will inexorably continue to pursue its interests on Earth as well in space,[108] marching steadily and strategically towards catching up with what the United States, the undisputed leader of space, has achieved. Despite China's declared interest in avoiding a space race or a strategic arms race,[109] formal assurances are not, on their own, sufficient to dispel threat perceptions. Quite to the contrary, the mere expansion of China's space ambitions could soon induce the United States to react, fuelling a self-propelling mechanism that will eventually lead to confrontation. It appears clear that, thanks to its triumphant history and current position in space, the United States is extremely

[106] According to Charlotte Mathieu, triangular cooperation India–China–Russia is unlikely, mainly because it might not be in Russia's interest to promote cooperation between those two countries and bring its two partners closer. In addition, at the moment India and China have neither particular needs nor real opportunities to work together on space projects. Mathieu, Charlotte (2008). "Assessing Russia's space cooperation with China and India. Opportunities and Challenges for Europe". ESPI Report 12. European Space Policy Institute, Vienna.

[107] The current US space exploration policy focuses on an Asteroid Retrieval Mission (ARM). See also Chap. 7 for further information.

[108] Following the line of argument expressed by Henry Kissinger's in his book on China. See Kissinger, Henry (2011). *On China*. Penguin Books, New York: p. 525.

[109] In the light of the considerations stated in Chap. 3 and Sect. 5.4, China's interest in dispelling the prospect of a space race appears consistent, both strategically and economically. See also Sect. 6.3.1 in this regard. In addition, as aptly noted by Joan Johnson-Freese, "China's slow, incremental program, juxtaposed with highly advanced, but largely politically stalled US efforts, creates a tortoise and hare scenario that China, as the underdog, significantly benefits from". *Cit.* Johnson-Freese, Joan (2006). "A New US-Sino Space relationship: Moving Toward Cooperation". Astropolitics Vol. 4 (2): 148.

sensitive to any potential challengers and alterations in the status quo. In a policy and defence culture that still tends towards Cold War models, the PRC simply looks to have replaced the USSR in American strategic calculations.[110] Therefore, any idea that the United States can afford to react with indifference to expanding Chinese ambitions is but wishful thinking. What is more plausible is that the United States' sense of its "manifest destiny" of greatness and China's belief in its celestial mandate to rule the *Tianxia* may sooner or later collide.

The geopolitical dynamics described in the previous sections seem to have already set the stage for a strategic confrontation between Beijing's rising power and Washington's declining hegemony and to have extended terrestrial rivalries into the space arena. Any possibility of cooperation is in fact currently frozen (see the next sections for a detailed account), and the separate space efforts of China and the United States are stimulating competition and increasing the likelihood of confrontation.

In the area of human spaceflight and exploration, things are clearly not boding well. The US political debate is populated by many advocates of competition, who "still point to Apollo as an example of what can be accomplished when external factors force the United States to strive harder".[111] Rising Chinese ambitions are often cited as a justification for increasing the NASA budget and thus initiating a new ambitious Apollo-type programme that would revitalise American space leadership. Similarly, in China, an increasing number of Neo-Comms scholars and officials are urging the government to pursue a more assertive foreign policy and to translate the techno-nationalist benefits stemming from the manned space programme into geopolitical influence.

If Chinese plans for a lunar endeavour eventually become a formal commitment, we can assume that the proposed ARM (see Appendix E) will be officially sanctioned and then used as a springboard to Mars exploration by the United States—presumably with partners like Japan and Europe—as an attempt to eclipse the Chinese lunar endeavour.

With regard to the acquisition of military space capabilities, competitive trends appear undeniable. The feeling in Washington—particularly within the Department of Defence (DoD)—is overwhelmingly anti-China. As pointed out by Yongjing Zhang of Bristol University, "the arrival of China as a great power has been under intense scrutiny by the US strategic gaze, especially since China's 2007 ASAT test. This strategic gaze extends US geopolitical envisioning of great power rivalries into space, the so-called fourth battlefield. This social inscription of space as a new battleground has produced an imaginary of China as threatening other, bent upon pursuing relentlessly a military space strategy aimed at a contest of supremacy in space with the USA".[112]

[110] See Hilborne, Mark (2013). "China's rise in space and US policy responses: a collision course?" Space Policy 29 (2): 121.

[111] *Cit.* National Research Council (2012). *NASA Strategic Direction and the Need for a National Consensus*. The National Academy Press. Washington DC: p.44.

[112] *Cit.* Zhang, Yongjin (2013). "The eagle eyes the dragon in space—A critique". Space Policy 29 (2): 120.

Thus, in response to the astonishing growth in Chinese military space capabilities—and as explicitly recommended by a number of reports and space policy analysts[113]—the United States has started to expend significant resources in military measures intended to defeat potential Chinese counter-space initiatives and reduce the risks of a "Pearl Harbor in space". For most commentators, "the potential for inadvertent escalation is real",[114] and sooner or later, the prisoner's dilemma in which the United States and China look trapped will inevitably trigger a space arms race.[115] It is just a matter of time.[116] The ultimate outcome—probably deleterious for both actors—remains uncertain.

Finally, with regard to the provision of international space services and public goods, new market realities have emerged, providing clear indications of increasing competition between the two juggernauts. China is gradually, yet continuously, gaining market segments at the expense of the traditional space powers, above all the United States, and its approach to cooperation is indirectly challenging the "rules of engagement" in the provision of commercial space services set by the West. This is visible not only in the Asian region, which could easily become the backyard of China's aerospace industries, but also in Africa and Latin America,[117] where Beijing has reached out to many international partners and customers.

The reasons for China's achievements in this domain are manifold, but two sets of circumstances seem particularly noteworthy. First, Chinese restrictions on and requirements for space technology transfers are low, while the space systems—

[113] Among others, see, for instance, Ashley Tellis' recommendations to the House Armed Services Subcommittees on Strategic Forces and Sea-power and Projection Forces of 28 January 2014. Tellis, Ashley (2014). "Does China Threaten the United States in Space?". Testimony to the House Armed Services Subcommittees on Strategic Forces and Sea-power and Projection Forces. Carnegie Endowment for International Peace. See also Pavelec, Sterling Michael (2012). "The Inevitability of the Weaponisation of Space: Technological Constructivism Versus Determinism". Astropolitics Vol. 10 (1): 39–48.

[114] *Cit.* Cheng, Dean (2011). "China's Space Program: A Growing Factor in US Security Planning". The Heritage Foundation, Backgrounder No. 2594.

[115] See, for instance, Lopez, Laura Delgado (2012). "Predicting an Arms Race in Space: Problematic Assumptions for Space Arms Control". Astropolitics Vol. 10 (1): 49–67. See also Krepon, Michael, and Julia Thompson (eds) (2013. *Anti-satellite Weapons, Deterrence and Sino-American Space Relations*. Stimson Center, Washington DC.

[116] Based on a new concept of technological development, S. M. Pavelec of the Air Command and Staff College has, for instance, strongly argued that "as technology advances, space weaponisation not only is likely, but indeed is inevitable in the near future". Interestingly, his core argument is that "the development of these weapons is inevitable and should therefore be accelerated in the United States, given the country's position as the lone superpower, to command and control the space commons. If the United States leads this drive for development, then in the end, as with thermonuclear weapons, space weapons will make the world more, not less, secure, and will contribute to the spread of democratic peace and globalized capitalism". *Cit.* Pavelec, Sterling Michael (2012). "The Inevitability of the Weaponisation of Space: Technological Constructivism Versus Determinism". Astropolitics Vol. 10 (1): 39–48.

[117] See Sect. 6.4 for a more detailed account of China's international outreach. For an interesting essay on Sino-Latin America cooperation, see Delgado-Lopez, Laura M (2012). "Sino-Latin American space cooperation: A smart move". Space Policy Vol. 28 (1): 7–14.

including the launch services—Beijing offers to its clients have been demonstrated to be not only affordable[118] but, thanks to the success of its manned space programme, also proven and reliable. Thus, for emerging space nations, thanks to China's assistance, the barriers associated with space activities are no longer insurmountable.[119]

Second, because of the severe governmental controls enforced by the US Congress since 1999, US commercial contracts with foreign entities have visibly diminished. This has created opportunities for cooperation that have been partly filled by China. Ironically, the ITAR regime set up by the United States to inhibit the development of dual-use technologies in non-allied countries has opened up new avenues and models of cooperation (e.g. the ITAR-free business model) that are beneficial for China and detrimental for US industries and governmental controls on international technology transfer. US frustration in this regard is palpable, and the reforms of the ITAR regime urged by the Obama Administration and by over 100 member companies of the Aerospace Industries Association clearly intend to remedy the loss of US competitive advantage.[120] Should the modifications proceed, further competition in this arena will be stimulated. Not only that, as emphasised by a former US official, possible US "over-reactions, which are read by Chinese public and its leaders as an expression of hostile intentions, [. . .] could turn China from an economic rival into an all-out enemy".[121]

In sum, in the three main arenas of competition that defined the first space race (competition for soft power, competition over the military capability of hard power, and competition about the provision of international services or public goods), Sino-American relations can easily be regarded as incipiently confrontational. They appear, in addition, to be already channelled along a trajectory similar to that taken by the two original space antagonists: the die, in short, seems cast.

Although it is perhaps true that—in the words of Mark Twain—*history does not repeat itself, but it does often rhyme*, the path of a Sino-American space race should be regarded as a predetermined dead end. Policy analysis, as Henry Kissinger urges, cannot confine itself to a mechanical application of historic analogies; it has an obligation to take account of the unprecedented elements of the contemporary scene.[122] The current state of play in US–China relations as well as the chessboard

[118] The total package China offers as global provider definitively undercuts both Western and Russian prices: 20 million € for the LM-2, 40 million € for the LM-3A, 50 million € for the LM-3C, 60 million € for the LM-3B, and 40 million € for the LM-4. See Harvey, Brian (2013). *China in Space. The Great Leap Forward.* Springer, New York.

[119] Rathgeber, Wolfgang (2007). "China Posture in Space. Implications for Europe". ESPI Report 3. European Space Policy Institute, Vienna. See also Suzuki, Kazuto (2013). "The contest for leadership in East Asia: Japanese and Chinese approaches to outer space". Space Policy 29 (2): 99–106.

[120] See Moltz, James Clay (2011). *Asia's Space Race: National Motivations, Regional Rivalries, and International Risks*, Columbia University Press, New York: p. 203.

[121] Shirk, Susan L. (2008). *China. Fragile Superpower.* Oxford University Press, New York: p. 268.

[122] See Kissinger, Henry (2011). *On China.* Penguin Books, New York: p. 534.

on which the two actors move is in fact far more multilayered and complex than those of the first space age, and if the competition path is one possible outcome, other trajectories might eventually be possible.

Before further elaborating on the posture the United States could adopt vis-à-vis Chinese lunar ambitions, it is, however, useful to reflect on the profound transformations in the geopolitical context of space activities. These transformations may not only make the space race analogy an obsolete paradigm but may induce the two actors to move completely beyond it.

6.3 Beyond the Space Race Paradigm...

The geopolitical context of international relations has changed dramatically since the Cold War and so have both the context in which space activities are embedded and the actors. The international system is no longer characterised by bipolarity, but multipolarity. As noted by many commentators, "this is a far cry from the us-versus-them world of the superpowers, whose leaders did not have to worry about the activities of third countries in space".[123]

Thus, in the current context there are several structural differences that offer reasonable prospects for avoiding a space race between the United States and China and indeed make a space race an anachronistic paradigm—a historical memory.

6.3.1 Chinese Cooperation Efforts and the Evolution of the Space Context

A first striking difference from the US–Soviet space race is the posture endorsed by Chinese policymakers and space officials. Contrary to conventional wisdom, it needs to be reiterated that China shows no interest in competition, but indeed appears open to cooperation in human spaceflight, especially with its alleged rival, the United States. This attitude is remarkably different from the posture adopted by the Soviet Union in the early 1960s, which steadily refused the overtures made by President Kennedy. As John M. Logsdon remarked: "racing Russia to the Moon was (in fact) John Kennedy's second priority. He would have preferred to see space as an area where international cooperation was the normal way to proceed".[124] This possibility was raised several times, including during Kennedy's face-to-face meeting with Mr. Khrushchev in Vienna on 3–4 June 1961, as well as

[123] *Cit.* Moltz, James Clay (2011). *Asia's Space Race: National Motivations, Regional Rivalries, and International Risks*. Columbia University Press, New York: p. 13.

[124] *Cit.* Logsdon, John M. (2008). "Why space exploration should be a global project". Space Policy 24 (1): 4.

in September 1963, when speaking before the General Assembly of the UN.[125] On both occasions, however, Khrushchev promptly discarded the option, and the death of Kennedy in November 1963 ended any possibility in this respect.

Unlike their Soviet counterparts of the 1960s, Chinese leaders appear fully aware that cooperation is preferable to competition. As argued in the previous chapters (see Sects. 3.2 and 5.4), space cooperation is not only a tactical manoeuvre intended to gain technological and operational know-how and experience. While enabling China to showcase its technological might and better tackle the costs related to a Moon landing programme, cooperation is more importantly a tool of China's foreign policy of *harmonious world*, which is based on the principles of no conflict, mutual respect, and win–win cooperation.

Space cooperation is but part of a broader strategy aimed at allaying the fears of the international community about the nature of China's rise and indeed at co-opting the West. It is also worth noting that Xi Jinping has spoken regularly about the need to realise a *shije meng* (*world dream*, to be built on the much discussed concept of *he*, harmony) and cooperation in an internationally visible space exploration programme would clearly support this perspective.

Moreover, cooperation would harvest valuable political paybacks. Cooperation with emerging space actors would help China claim the moral and political leadership of the developing world, and if cooperation with the major spacefaring nations (most prominently the United States) was also achieved, China's much-desired recognition as a great powerhouse (*yi xi zhi di*) would be sealed. In both cases cooperation would help China "achieve great power status within a system dominated by the United States, and increase its international influence without triggering a counterbalancing reaction".[126]

China's outreach in space has been spelled out on many occasions, perhaps most notably at the 64th IAC, where Wang Zhaoyao, director of the China Manned Space Agency, emphatically remarked:

> Sincere exchange and broad cooperation will contribute to development of the international human space. It has been China's consistent pursuit to carry out international cooperation and exchange on the principle of equality and mutual benefit, mutual respect and transparency. China has collaborated extensively with many other countries and regions in space technology, space medicine research, space science experiments and astronaut selection and training. In the construction and operating phase of our space station, we will seek for international cooperation in an even more open manner and willing to share space development accomplishments with other countries, especially developing countries.[127]

---

[125] During this historic speech the President said... "Why, therefore, should man's first flight to the moon be a matter of national competition? Surely we should explore whether the scientists and astronauts of our two countries—indeed of all the world—cannot work together in the conquest of space, sending someday to the Moon not the representatives of a single nation, but representatives of all our countries". *Ibid.* p. 4.

[126] Pollpeter, Kevin (2007). "Competing perceptions of the US and Chinese Space Programs". China Brief Vol. 7 (1): 4.

[127] Zhaoyao. Wang. "China Manned Space Program to Carry out International Cooperation in an Open Manner". China Manned Space Engineering Office. 25 September 2013. Web. http://en.cmse.gov.cn/show.php?contentid = 1350. Accessed 4 June 2014.

All these statements, which appear genuine, now place the ball in the court of spacefaring nations, particularly the United States. It will thus largely be Washington's political attitude that determines whether the Khrushchev–Kennedy positions have been reversed and whether or not a space race is eventually bound to play out.

It should be noted that, independent of the US posture, the dynamics of today's space environment allow space actors to move beyond either the paradigm of bilateral competition or cooperation of the first space race. This constitutes a second striking difference from the Cold War context.

The nature and scope of international relations in space have fundamentally changed in recent years, redefining the role of cooperation and competition and leading to new patterns in international space activities.[128] Three main dynamics define the current context in comparison to the first space race. These are the increasing institutionalisation, globalisation, and regionalisation of space activities.

(a) The first thing to note is that the institutional landscape of space activities, which was once dominated by two superpowers and a handful of second-tier space powers, has become less and less exclusive. The number of countries involved in space has increased tremendously: suffice it to mention that during the Cold War the number of space agencies did not exceed a dozen, while there are now more than 70 governmental entities having space agencies or the equivalent. The proliferation of new space agencies strengthened in the immediate aftermath of the Cold War,[129] accelerated in the late 1990s and still appears to be an ongoing process not limited to the advanced economies of the "North".[130]

    As a result of foreign direct investment and expansion of trade, combined with the broad diffusion of information and technology, the barriers to entry associated with space programmes have become lower. Consequently, countries previously unable to pursue space activities have seen greater opportunity to enter this arena. The phenomenon is not restricted to specific regions, but has been widespread: a number of Asian, Latin American, Eastern European, and African countries have all created their own space agencies (or alternatively small offices or inter-ministerial bodies) devoted to the management of space activities.[131]

---

[128] For a detailed analysis of the changing context of space activities, see two contributions of Nicolas Peter, respectively: Peter, Nicolas (2008). *Space Exploration 2025: Global Perspectives and Options for Europe*. ESPI Report 14. European Space Policy Institute, Vienna; and: Peter, Nicolas (2006). "The changing geopolitics of space activities" Space Policy 22 (2): 100–109.

[129] As noted by Nicolas Peter, the multiplication of space agencies, which is an ongoing process, can also be considered a direct result of the new geopolitical context that arose from the end of the Cold War and the subsequent break-up of the USSR, which led to the creation of national space agencies in Kazakhstan in 1991, in Ukraine in 1992, and Russia in 1992. This trend is still ongoing. *Cit*. Peter, Nicolas (2006). "The changing geopolitics of space activities" Space Policy 22 (2): 102.

[130] *Ibid*. p. 102.

[131] The most recent examples are the Belarus Space Agency (2010), Sri Lanka Space Agency (2010), Agencia Espacial Mexicana (2010), South African Space Agency (2011), and Turkmenistan National Space Agency (2011).

In short, roles and relationships in space are being dramatically redefined. Because of the multiplication of spacefaring countries and newcomers in space with increasing capabilities, the United States, Russia, and China are no longer the only players able to lead cooperative projects; indeed, there are an increasing number of possibilities for cooperation.[132] As recognised by the US National Research Council's (NRC) report on NASA's Strategic Direction, "countries that once depended on partnerships with the US to execute their space programmes, now have other choices, including going it alone".[133] In more detailed fashion, Nicolas Peter has noted:

> The number of potential partners with sophisticated government and industrial space technologies and capabilities has grown, and the majority of spacefaring countries have much more experience in working together in their space programmes. Thus, the nature and patterns of international relations in space are changing fundamentally and the multiplication of space countries with varying ranges of capabilities is leading to an increase in cooperative options that allow countries and space agencies to cooperate à-la-carte, rather than using the prior Cold-War era limited set of options. [134]

(b) An additional, closely related factor setting current space relations apart from the Cold War model is represented by globalisation tendencies. The impact of globalisation on the space environment has been significant and complex. Besides inducing profound transformations throughout the entire value chain of the space industry and subjecting the space sector to the economic laws that rule the global market, "the widening and speeding up of global interconnectedness"[135] has deepened the level of interdependence among space actors. In stark contrast to the economic and technological autonomy of the Cold War superpowers, countries and industries are now strongly dependent on each other. From the launch of commercial satellites to ambitious space exploration missions, the degree of autonomy countries have in the implementation of their space programmes has inexorably decreased.[136] Even the United States, which has so far been the most self-sufficient country in terms of space capabilities, today relies on Russian launchers to reach the ISS.

It has been observed that the interdependence introduced by globalisation raises the costs of any future space conflict. Hence, globalisation has the potential to transform competitive trends into cooperative ones. While the impact of globalisation may still be subject to different interpretations, it is

---

[132] Peter, Nicolas (2008). "Space Exploration 2025: Global Perspectives and Options for Europe". ESPI Report 14. European Space Policy Institute, Vienna: 31.

[133] *Cit.* National Research Council (2012). *NASA Strategic Direction and the Need for a National Consensus.* The National Academy Press. Washington DC: 2.

[134] *Cit. ibid.* p. 31.

[135] David Held's definition of Globalisation. See Held, David, Anthony McGrew, David Goldblatt, and Jonathan Perraton (eds) (1999). Global Transformations: Politics, Economics and Culture. Stanford University Press, Stanford: p. 2.

[136] For an extensive analysis on the accelerating processes of globalisation in the space sector, see Organisation for Economic Co-operation and Development (2014). The Space Economy at a Glance 2014. OECD Publishing, Paris.

true that states now rarely initiate and carry out a space programme without some element of foreign participation. Indeed, at commercial, industrial, and governmental levels, international cooperation has become the norm; both the United States and China are fully engaged in it! Furthermore, cooperation is no longer restricted to short-term programmes, but has evolved to encompass long-term agreements, as demonstrated by the increasing number of multiannual projects signed in recent years and, in a sense, also by the ISS.[137]

(c) Alongside globalisation tendencies, the post-bipolar world order has also witnessed a resurgence of regionalism in world politics.[138] The dramatic expansion of global trade and financial networks, though reinforcing transnationalism and a homogenisation of the international economy, has also given birth to regional economic blocs. The two phenomena have been blossoming almost symbiotically.

This regionalisation is clearly mirrored in the space arena, where the globalisation of space players has been accompanied by the strengthening and launching of new regional initiatives, which have consistently been gaining momentum since the mid-1990s. The surge in regional space initiatives has been experienced all around the globe, albeit with different levels of ambition and integration: suffice it to mention the APRSAF and APSCO frameworks in the Asia-Pacific, the ASEAN Sub-Committee on Space Technology and Applications (SCOSA) in South East Asia, the Space Conference of the Americas (CEA) in Central and Latin America,[139] and the Inter-Islamic Network for Space Technologies (ISNET) in the Middle East.[140] In Africa too, several countries are actively promoting the establishment of an African Space

---

[137] Peter, Nicolas (2006). "The changing geopolitics of space activities" Space Policy 22 (2): 108.

[138] "Resurgence" is the term used by Andrew Hurrell in a renowned essay to indicate a long-standing tendency initiated in the 1960s and then revived since the 1990s. As he interestingly observes: "the past decade has witnessed a resurgence of regionalism in world politics. Old regionalist organizations have been revived, new organizations formed, and regionalism and the call for strengthened regionalist arrangements have been central to many of the debates about the nature of the post-Cold War international order. The number, scope and diversity of regionalist schemes have grown significantly since the last major regionalist wave in the 1960s". *Cit.* Hurrell, Andrew (1995). "Explaining the resurgence of regions in world politics". Review of International Studies No. 21: 331–358.

[139] The *Confrencia Espacial de las Américas* (CEAs) was launched as an effort to facilitate cooperation among Latin American countries. Although these countries have not moved beyond the talking-shop level, there is increasing support for the creation of a pan-American cooperation structure, aimed at stimulating continental synergies in terms of space capabilities. See United Nations's website: http://www.un-spider.org/event-en/3747/2010-11-15/vi-space-conference-americas. More recently Brazil has taken the lead in such initiative. See also "Brazil Proposes Latin American Space Alliance". Parabolic Arc. 17 November 2013. Web. http://www.parabolicarc.com/2013/11/17/brazil-proposes-latin-american-space-alliance/. Accessed 15 June 2014.

[140] The ISNET is an interstate, non-political agency under the umbrella of the Organisation of the Islamic Conference (OIC). The purpose of ISNET is to promote space science, space technology, and their applications in OIC member countries (Bangladesh, Iraq, Indonesia, Morocco, Niger, Pakistan, Saudi Arabia, Tunisia and Turkey, Syria, Iran, Sudan, Azerbaijan, Senegal, Egypt, and Jordan). See ISNET website: http://www.isnet.org.pk/pages/membership.asp.

Agency (AfriSpace),[141] although it must be acknowledged that the political complexity and vastness of this region for a broad-based space agency presents a number of challenging issues that could undermine its realisation or even produce multiple solutions. This notwithstanding, the overall trend of regionalisation of space activities is not only an additional element demonstrating the importance attributed to cooperation in the post-Cold War era but also a factor bound to have a constructive impact on regional space governance and the sustainability of space activities.[142]

In sum, the growing institutionalisation, regionalisation, and globalisation of space activities, as major forces behind the transformation of the current institutional landscape, are leading to a new paradigm for international space activities in which cooperation plays the central role. Particularly in the field of space exploration and human spaceflight, all the above-mentioned transformations have been conducive to the gradual emergence of international mechanisms aimed at pooling resources and creating synergies in space exploration, while offsetting economic cost and avoiding expensive duplication of efforts. The result is to a large extent geopolitical non-confrontation. Thus, a stepping stone is provided for using space as a pathfinder for deeper and broader international cooperation.

The creation of these international mechanisms, initiated with the ISS programme, has in recent years acquired a more global nature. As early as 2006, 14 space agencies began a series of discussions on global interests in space exploration.[143] Together they took "the unprecedented step of elaborating a vision for peaceful robotic and human space exploration, focusing on destinations within the solar system where humans may one day live and work, and developed a common set of key space exploration themes".[144] This vision was articulated in the Global Exploration Strategy (GES): The Framework for

[141] There is no formal coordination yet of the space efforts of African countries, although many African countries are in favour of the constitution of an African Space Agency. A study is under way with the support of ITU, to evaluate current initiatives and study a possible joint African space policy. In the meantime a first pan-African telecommunications satellite (RascomStar) was contracted to Thales Alenia and launched by Ariane in 2010. In addition, in 2009, an agreement was signed in Algiers between Algeria, South Africa, Kenya, and Nigeria to create the African Resource Management Constellation, a constellation of satellites on the model of the SSLT's Disaster Monitoring Constellation, comprising four, potentially five, satellites. See Timberlake, Ian. "Africa eyes joint space agency". Phys.org. 4 September 2012. Web. http://phys.org/news/2012-09-ministers-african-space-agency.html. Accessed 25 July 2014.

[142] For an additional analysis in this regard with an eye to the Asian context, see Aliberti, Marco (2013). "Regionalisation of Space Activities in Asia?" ESPI Perspectives 66. European Space Policy Institute, Vienna.

[143] Members are ASI (Italy), CNES (France), CNSA (China), CSA (Canada), CSIRO (Australia), DLR (Germany), ESA (Europe), ISRO (India), JAXA (Japan), KARI (South Korea), NASA (USA), SSAU (Ukraine), Roscosmos (Russia), and UKSA (United Kingdom).

[144] *Cit.* "The Global Exploration Strategy. The Framework for Coordination". International Space Exploration Coordination Group. April 2007. Web. http://www.globalspaceexploration.org/c/document_library/get_file?uuid=119c14c4-6f68-49dd-94fa-af08ecb0c4f6&groupId=10812.

> Coordination, which was released in May 2007. The document outlining the GES Framework was a very important step towards the international recognition of human space exploration as a common goal of humankind. It argues that the future establishment of a formal—though non-binding and voluntary—coordination mechanism among interested space agencies could assist in the development and implementation of the GES.[145]

In response to this recommendation, in November 2007 the International Space Exploration Coordination Group (ISECG) was established. ISECG is an open, inclusive, and evolutionary platform, designed to coordinate a global space exploration programme by:

1. Providing a forum for participants to discuss their interests, objectives, and plans in space exploration
2. Promoting potential cooperation in space exploration activities for the purpose of (a) making use of all available resources, knowledge, and technological capabilities; (b) leveraging each agency's individual investment; (c) sharing lessons learned from national and international missions; (d) improving the safety of humans in space, for example, through the interoperability of life support systems; and (e) enhancing the overall robustness of global space exploration[146]

An additional pillar towards the development of cooperation in space exploration was added in 2011, when the International Space Exploration Forum (ISEF) was established, thanks to a diplomatic initiative led by the EU and ESA (see Sect. 7.3). The first session of this high-level meeting between the heads of world space agencies was held in Lucca in November 2011, while the United States hosted the second in Washington in January 2014.

GES, ISECG, and ISEF all make the compelling case that the next step in space exploration *must* be taken cooperatively, and hence, they all work towards achieving this goal. They remain at present, it is true, mechanisms of consultation and coordination, not cooperation. Thus, the implementation of a more committing international endeavour will be a lengthy and complex task, fraught with both political and operational issues. In this respect, it must be acknowledged that "collaboration on programmes with decadal timescales is intrinsically more problematic than on programs that can be accomplished within a few years, since longer timescales drastically increase the probability of diverging political realities".[147] Furthermore, there is no solid consensus over the next destination of space exploration.

---

[145] *Cit.* Dupas, Alain (2009). "International Cooperation in space exploration: Lessons from the past and perspective for the future". In Schrogl, Kai-Uwe, Charlotte Mathieu, Nicolas Peter (eds). *ESPI Yearbook on Space Policy 2007/2008: From Policies to Programmes*. Springer, Vienna 2009: p. 163.

[146] *Ibid.* p. 164.

[147] National Academy of Sciences (2014). *Pathways to Exploration: Rationales and Approaches for a US Program of Human Space Exploration*. The National Academy Press. Washington DC.

The importance of GES, ISECG, and ISEF lies in the fact that they create the foundations of a more solid international regime in space exploration that could transform potential trends of competition into cooperative ones. In accordance with international relations theory, regimes—defined by Stephen D. Krasner as a set of principles, norms, rules, and decision-making procedures around which actor's expectations converge on a given issue area[148]—positively affect the behaviour of actors after their formation. By promoting dialogue and mutual understanding, mechanisms of political transparency and confidence building are created and competitive impulses mitigated. Eventually, this generates a self-enforcing trend that can gradually channel international space relations towards new, more mature, and integrated forms of cooperation.[149]

What is also noteworthy about these mechanisms is that they are explicitly intended to be open and inclusive, making clear that "the flag-waving paradigm inherited from the Cold War and also applied to the ISS must be abandoned in favour of a resolute policy of *real* international cooperation involving *all* nations".[150] This is explicitly recognised by China's participation in the ISECG, alongside its alleged competitors, the United States, Japan, and India. In other words, spacefaring nations are not only all becoming aware that in a time of budgetary pressures, they must pool resources and create synergies but also that they must find ways to harvest the opportunities that China's space ascendancy is generating.

As Alanna Krolikowski argued in testimony to the US–China Economic and Security Review Commission, Beijing is an important potential partner for future large missions and, clearly, a costly one to exclude.[151] Leaving untapped the opportunities offered by its rise will inevitably prevent the programmes of the major spacefaring nations from being effective. Indeed, if the United States refuses to cooperate with China, it will just leave opportunities for collaboration open that will be filled by other spacefaring nations.

The attitude towards cooperation is more and more widespread and can be detected even within the US political debate. Indeed, the ultimate element that could set today's space context apart from the Cold War deals with the dilemmas faced by the current US administration.

---

[148] Krasner, Stephen D (1983). "Structural Causes and Regime Consequences: Regimes as Intervening Variables". In Krasner, Stephen D (eds). International Regimes. Cornell University Press, Ithaca, NY.

[149] The literature on iterated prisoner's dilemma shows that when the transparency level of the states' action increases and fears of defection are consequently reduced, incentives to cooperate inevitably increase. See Axelrod, Robert, and R.O. Keohane (1993). "Achieving Cooperation Under Anarchy: Strategies and Institutions". In Baldwin, D. A. (ed). *Neorealism and neoliberalism: the contemporary debate*. Columbia University Press, New York.

[150] Blamont, Jacques (2012). "US Space Exploration Strategy: Is there a better way?" Space Policy No. 28 (4): 214.

[151] *Cit.* Krowlikoski, Alanna (2011). "China's Civil and Commercial Space Activities and their Implications". In Hearing: The Implications of China's military and civil space programs. US—China Economic and Security Review Commission (2011). Web. http://www.uscc.gov/Hearings/hearing-implications-china's-military-and-civil-spaceprograms. Accessed 15 June 2014.

6.3.2 *The US Space Policy Dilemma: Engaging or Confronting China?*

6.3.2.1 The Political Background: Collision or G2?

As in the case of the US–Soviet relationship during the twentieth century, there is a widespread belief among analysts and scholars that the United States and China constitute the most important bilateral relationship of the twenty-first century. Yet, unlike the two superpowers contending for global leadership during the second half of the past century, there is still no saying how the precise nature of US–China relationship will develop. As pointed out by Minxin Pei in an article recently published in *Foreign Affairs*:

> The United States and China are clearly not allies. They share no overriding security interests or political values, and their conceptions of world order fundamentally clash. Whereas Beijing looks forward to a post-American, multipolar world, Washington is trying to preserve the liberal order it leads even as its relative power wanes. Meanwhile, numerous issues in East Asia, such as tensions over Taiwan and disputes between Beijing and Tokyo, are causing US and Chinese interests to collide more directly. Yet the two countries are not really adversaries, either. They do not see each other as implacable ideological or security threats. And the fact that their economies are so deeply intertwined makes both countries hell-bent on avoiding conflict.[152]

Given this premise, it comes as no surprise that views about the future evolution of the relationship and the grand strategy the United States should correspondingly adopt have often moved in diametrically opposed directions, generally depending on the theoretical lenses through which American scholars and policymakers have interpreted international relations and looked at the challenges (or opportunities) posed by China's rise. Broadly speaking, two main schools of thought can be identified.

Principally drawing on the ideas embedded within "hegemonic cycles theory", realist scholars are inclined to believe that China is inexorably emerging as a threat to the prevailing American hegemony and that confrontation—if not conflict—will be inherent in their future relationship, as exemplified by the aforementioned case of Anglo-German rivalry at the start of the twentieth century. Accordingly, a firm strategy of containment—the creation of an anti-China strategic bloc and the parallel preservation of nuclear and conventional military superiority—should be adopted by the United States in order to prevent Beijing acquiring enough power to challenge American primacy.

Conversely, liberals and exponents of social constructivism have generally taken a much more positive view about the future of US–China relations. They have often insisted that a country facing great domestic challenges like China (see Chap. 5) is not going to throw itself into strategic confrontation or a quest for world domination.[153] In addition, the proliferation of weapons of mass destruction and the huge

---

[152] *Cit.* Pei, Minxin (2014). "How China and America See Each Other". Foreign Affairs Vol. 93 (2): 143–147.

[153] Interestingly, this position is also endorsed by a realist thinker such as Henry Kissinger. See Kissinger, Henry (2011). *On China*. Penguin Books, New York: p. 512.

economic stakes between Washington and Beijing dictate cooperation. A clear and broad engagement policy towards China would greatly benefit the United States and give China a stake in the existing international order, thus increasing its global role as a status quo, rather than revisionist, power.[154] In the words of the former president of the World Bank, Robert Zoellick, engagement will help to eventually transform the country into a "responsible stakeholder".[155]

Relying on the sound and ever-growing economic interdependence of Washington and Beijing, some authors have even proposed the idea of a cooperative co-dominium over global economic governance and strategic issues. Usually synthesised in the formula of G2 or with the portmanteau word *Chimerica*,[156] the idea has gained momentum in the past decade, receiving support from very diverse observers. A number of prominent personalities within the US foreign policy debate have indeed advocated the crafting of such a symbiotic relationship.

As early as 2009, Zbigniew Brzezinski, former national security advisor, was saying: "we need an informal G-2... The relationship between the US and China has to be a comprehensive global partnership, paralleling our relations with Europe and Japan".[157] Likewise, Robert Zoellick has written that "without a strong G-2, the G-20 will disappoint".[158] Even a Machiavellian thinker like Henry Kissinger has often emphasised that "the Sino-American relationship needs to be taken to a new level"[159] and has advocated the creation of a "Pacific Community" between the United States and China.[160]

[154] Shambaugh, David (2002). "European and American Approaches to China: Different Bed, Same Dreams?" Asian Papers No 15. Sigur Center for Asian Studies: p.3.

[155] For an interesting analysis on the concept of "responsible power", see Shirk, Susan L. (2008). *China. Fragile Superpower*. Oxford University Press, New York: pp. 105–139.

[156] *Chimerica* was coined by the historian Niall Fergusson to describe the symbiotic nature of current and future relations between China and the United States. See Fergusson, Niall (2008). *The Ascent of Money: A Financial History of the World*. Penguin Press, New York. On the idea of a Sino-American G2, see Bergsten, Fred C (2008). "A partnership of equals". Foreign Affairs Vol. 87 (4): 57–69.

[157] Brzezinski, Zbigniew. "Remarks at a Seminar Commemorating the 30th Anniversary of the Establishment of Diplomatic Relations between China and the United States". National Committee on US–China Relations. 12 January 2009. Web. http://www.ncuscr.org/files/Zbigniew%20Brzezinski.pdf. Accessed 18 July 2014.

[158] Zoellick, Robert B., and Justin Yifu Lin. "Recovery: A Job for China and the US". The Washington Post. 6 March 2009. Web. http://www.washingtonpost.com/wp-dyn/content/article/2009/03/05/AR2009030502887.html. Accessed 18 July 2014.

[159] Kissinger, Henry. "The World must forge a new world order or retreat to chaos". The Independent. 20 January 2009. Web. http://www.independent.co.uk/voices/commentators/henry-kissinger-the-world-must-forge-a-new-order-or-retreat-to-chaos-1451416.html. Accessed 18 July 2014.

[160] The "Pacific Community" is defined by Kissinger as a region to which the United States, China, and other states all belong and in whose peaceful development they participate. Sino-American cooperation will enable other major countries to participate in the construction of a system perceived as joint rather than polarised between Chinese and American blocs. Kissinger, Henry (2011). *On China*. Penguin Books, New York: p. 523.

For containment proponents, however, "elevating the Sino-US bilateral relationship is not the solution. It will raise expectations of a level of partnership that cannot be met [because of mismatched interests, values, and capabilities] and exacerbate the very real differences that still exist between Washington and Beijing".[161] Only by constraining Chinese power from becoming more assertive and "structuring" (i.e. capable of shaping the structure of the international system) will the United States be able to preserve American primacy and thus a stable world order over the coming decades.[162]

What is perhaps more interesting than this academic debate over US foreign policy is that the two contending strategies have been simultaneously endorsed by the US administrations that have taken office since the late 1990s. Over the past 15 years, US–China policy has in fact been guided by a mix of containment and engagement (the so-called congagement[163]). More precisely—as former US National Security Advisor Edward Luttwak notes, the United States has in fact concurrently adopted three different China policies, two of which are moving in diametrically opposed directions.[164]

The first one is that of the US Treasury, which is vigorously promoting China's economic growth and technological advancement. This policy "focuses entirely on the benefits to US public finance of cheap capital from China's huge foreign currency reserve, and to US consumers and business alike of having unconstrained access to the cheapest possible imports of manufactured goods as well as raw materials, because they in effect increase the US standard of living without income increases, and reduce inputs costs to business".[165]

The second China policy is that promoted by the Department of State, which explicitly aims to confront China. Such a policy recognises the importance of engaging China in the economic spheres and certainly values cooperation with China whenever it is forthcoming, but at the same time remains carefully on guard against the rise of a strategic competitor. In short, unlike the US Treasury, the State Department does not see good reasons for pandering to China and its expanding ambitions. Accordingly, it has generally adopted different lines of conduct to reconcile the various US interests, which for many commentators has translated into a policy of "hedging". Typified by Obama's *Pivot to Asia* strategy, this policy could be defined as a softer, gentler form of containment. As applied to China today, it implies "building alliances with other Asia-Pacific countries, such as Australia, India, and Japan, which are not designed specifically to oppose China

[161] *Cit.* Economy, Elizabeth C., and Adam Segal (2009). "The G-2 Mirage". Foreign Affairs Vol. 88 (3): 14–23.

[162] See, for instance, Wesley, Clark. "Getting real about China" The New York Times. 10 October 2014.

[163] See one of the first articles on the Congagement strategy: Khalilzad, Zalmay (1999). "Congage China". RAND Issue Paper. Web. http://www.rand.org/pubs/issue_papers/IP187.html.

[164] Luttwak, Edward (2012). *The rise of China vs. the logic of strategy*. Harvard University Press, Cambridge, MA: pp. 213–247.

[165] *Cit. Ibid.* p. 214.

but could be useful in any future confrontation with it".[166] It also means restricting transfers of highly sensitive technologies and adopting "polite but insistent ideological warfare" against the authoritarian Chinese regime.

Finally, the third US–China policy is that endorsed by the Department of Defense (DoD), which clearly aims to contain China. As several security analysts have noted, the Pentagon's China posture is a multifaceted strategy that, inter alia, includes deploying American military forces all around China's periphery, maintaining a wide network of defence relationships with China's neighbours (the so-called strategic ring of encirclement), breaking China's "string of pearls" (also through political destabilisation in the Indochinese peninsula)[167] and even—at least according to Chinese complaints—undermining the regime and acquiring the means to cut if off from imports of oil and other raw materials.[168]

All in all, US policymakers seem to have so far avoided giving a precise shape to their China policy by adopting a hybrid strategy that draws on both realist and liberal ideas and combines elements of containment and engagement. To be sure, although conflicting, these strategies are not mutually exclusive, and there are many who continue to advocate a balanced combination of "stick and carrot". But as Chinese power continues to grow, maintaining such a balance will become harder than ever. Whether this implies moving towards a more resolute containment policy or elevating the relationship to a strategic global partnership forms the bottom line of US foreign policy dilemmas.

It is clear that the "unipolar moment"—as John Ikenberry puts it—will eventually pass. US dominance will eventually end. US grand strategy, accordingly, should be driven by one key question: what kind of international order would the United States like to see in place when it is less powerful?[169] If China and the United States manage to ease fears on both sides and build a cooperative strategic partnership, they can do a lot to create the "Pacific Community" mentioned by Kissinger.[170] But if their relations become focused on a zero-sum game logic, the

[166] *Cit.* Barysch, Katina, Charles Grant, and Mark Leonard (2005). "Embracing the Dragon: The EU's Partnership with China". Centre for European Reform, London.

[167] The expression "string of pearls" is generally used to indicate China's maritime strategy. It entails the establishment of a series of military and economic nodes around China. Simply put, every node is a "pearl", a sphere of influence that is secured and maintained through the use of economic, geopolitical, diplomatic, or military means. See "String of Pearls. Meaning. Policy. Implications". The Barrel. 18 December 2012. Web. http://thebarrel.in/string-of-pearls-meaning-policy-implications/#Several_things_included_in_Chinese_String_of_Pearls. Accessed 20 September 2014. For an analysis on China's strings of pearls, see also Kim, Shee Poon (2011). "An Anatomy of China's 'String of Pearls' Strategy". The Hikone Ronso No.387.

[168] Engdahl, William F. "Obama's Geopolitical China 'Pivot': The Pentagon Targets China". Global Research. 2 February 2013. Web. http://www.globalresearch.ca/obama-s-geopolitical-china-pivot-the-pentagon-targets-china/32474. Accessed 20 September 2014. See also Nathan, Andrew J. and Andrew Scobell (2012). "How China sees America". Foreign Affairs. Vol. 91 (5): 32–58.

[169] Ikenberry, John (2008). "The rise of China and the future of the West". Foreign Affairs Vol. 87 (1): 23–37.

[170] Kissinger, Henry (2011). *On China*. Penguin Books. New York, 2011: p. 523.

scenario of an international system polarised between China and America will become more probable, as well as the self-fulfilling prophecy of a new version of the Anglo-German rivalry that started WWI.

Interestingly, these trade-offs are fully and dramatically reproduced within the space arena, which might thus provide the first answer to solve this strategic equation. In order to shed some light on the complexities of Sino-American space relations and the dilemmas faced by the current US space policy vis-à-vis China, an overview of its evolution is now presented.

6.3.2.2 The Space Background: The Ups and Downs of US–China Space Relations

Strongly influenced by earthly dynamics, US–China space relations have gone through many ups and downs over the years.

After more than 20 years of almost no formal contact, in 1972 President Nixon made his historic overture to China, and scientific and technological cooperation soon became a tool of US foreign and security policy, used to create a front against the Soviet Union. Cooperation in space activities was symbolically launched in May 1978—when US National Security Advisor Zbigniew Brzezinski handed one gram of Moon rock to CCP Chairman Hua Guofeng as a gesture of goodwill. It became institutionalised on 31 January 1979 with the signature (by President Jimmy Carter and Deng Xiaoping) of the first formal cooperative agreement in science and technology.[171] Known as the "Umbrella Agreement", it incorporated a memorandum of understanding on "Cooperation in the Field of Space Technology", negotiated during the July 1978 visit of NASA Administrator Robert A. Frosch to China.

A Joint Commission on Scientific and Technological Cooperation and its related working groups was set up in 1980, with meetings held on a regular basis. In the same year, the administration decided to move China down the ladder of technology export control,[172]and President Reagan pledged additional liberalisation efforts in the US technology transfer policy in July 1983.[173] Relations in the space arena further improved after the 1986 *Challenger* disaster and temporary suspension of space shuttle launches, when the Reagan Administration agreed to let US satellite

[171] Note: it was just a few weeks after the "Joint Communique on the Establishment of Diplomatic Relations" of 1 January 1979.

[172] The decision was valuable as it opened up the possibility of selling the country high technology at a better rate than that sold to the USSR and to transfer selected "dual-use" technology.

[173] In his message to the Congress on 11 July 1983, Reagan stressed: "It is in our fundamental interest to advance our relations with China. Science and Technology are an essential part of that relationship and I have taken steps recently to ensure that China has improved access to US technology it needs for its economic modernisation goals". Quoted from Myrrhe, Jacqueline (2013). "Hop-on and hop-off, ...but where do you go ? US-China space cooperation -An attempt to achieve the impossible". In Lan, Chen, and Jacqueline Myrrhe (eds). *Go Taikonauts. All about China's space programme.* Issue 11. February 2014: 47–63.

manufacturers contract space launch services from China.[174] Thus, together with political considerations, economic interests became another important driver of US–China space relations. In January 1989 a memorandum of understanding was eventually signed, granting China the permission to launch nine US commercial satellites over the next 5 years.[175]

However, the Tiananmen crackdown of 4 June 1989 had a considerable impact on Sino-American relations, abruptly stifling political dialogue and all but severing the expanding ties in space. In the immediate aftermath of the notorious events, the Congress imposed sanctions prohibiting the agreed launches and exports of space and nuclear technology. High-level contacts between officials on both sides were also suspended, as well as efforts to liberalise controls on technology export. The restrictions, which are formally still in place, were, however, circumvented as early as April 1990, with the permission to launch Asiasat-1. Between 1990 and 1992, President George H. W. Bush granted a number of waivers of the sanctions, allowing US companies to use Chinese launches. In addition, two small Chinese chemical and materials experiments were allowed to fly on the space shuttle (mission STS-42) in January 1992.[176]

The practice of granting waivers continued during the Clinton presidency, which even simplified the waiver-approval process and elegantly used space technology transfer as an incentive for China to strengthen its export regime and honour its agreement not to sell missile technology to Iran and Syria. Overall, the years of the Clinton tenure were marked by increasing overtures towards China, as demonstrated by the considerable number of launches approved between 1993 and 1999,[177] and the decision to fly a Chinese alpha magnetic spectrometer on the space shuttle *Discovery* mission to the Mir space station in June 1998.[178] According to US–China space policy expert Gregory Kulacki, cooperation in commercial satellite launches played a role in China's decision not to build its own space shuttle.

The political climate rapidly changed after the Hughes/Loral launcher information scandal and the subsequent release of the Cox Commission's report in 1999. [179]

---

[174] Kulacki, Gregory (2011). "US and China need contact, not cold war". Nature. Vol. 474: 444.

[175] Harvey, Brian (2013). *China in Space. The Great Leap Forward.* Springer, New York: p.100.

[176] *Ibid.* p. 100.

[177] Overall 18 launches were contracted over this period, of which 15 were successful.

[178] Harvey, Brian (2013). *China in Space. The Great Leap Forward.* Springer, New York: p.15.

[179] A commission (the Cox Commission) was set up after the LM launch failure of 14 February 1996 and in particular following American allegations of Chinese technology espionage and the deliberate transfer of sensitive technology to China by US corporations such as Hughes and Lockheed Martin. Hughes and Loral workers were accused of having handed over to Chinese officials responsible for the accident investigation sensitive technical information with potential military uses. This in turn led to an additional investigation. According to Cox, the Chinese had "over decades and in a systematic way, [used] fair means and foul, neutral scientific conferences, licensing arrangements, dual use military-civilian technologies, and straightforward spying to ferret out information on nuclear technology, computers, rockets, submarines, and atomic bombs". *Ibid.* p.158.

The commission's investigation, which served as a proxy for a broader debate on the US posture vis-à-vis China, openly accused Beijing of a systematic and decade-old policy of espionage and theft[180] and recommended severing technical ties with China, particularly in the space domain. As a consequence, various administrative and legislative reforms were initiated, including a reclassification of satellite technology as munitions and the transfer of responsibility for their licencing from the Department of Commerce to the Department of State, so as to ensure that defence considerations were taken into proper account.[181]

In the years of severe satellite export restrictions that followed the release of the Cox report, not only did the United States adopt a strict policy of no cooperation in space activities that endures to the present, it also started to regard any Chinese effort in space with the utmost suspicion. In political parlance, China quickly became a rival—if not a threat to national security. The Pentagon's annual reports on the military power of the PRC, or the 2001 Rumsfeld Commission report warning of a "Pearl Harbor in space", give a clear picture of the predominant line of thinking during the George Bush Jr. Administration. The infamous 2007 ASAT test further increased tensions and stifled any effort to cooperate.

In addition, concerns about the challenge to US leadership posed by China's ambitions in space exploration fuelled the political debate, as clearly demonstrated during the March 2006 hearing of the House Appropriations Committee, when Congressman Tom DeLay concluded in alarmist fashion, "We have a space race [with China] going on right now and the American people are totally unaware of this".[182] The venomous statements contained in the 2006 National Space Policy and the repeated refusal to allow China to become a partner on the ISS only served to reinforce Chinese belief that the US goal was to contain China's rise as a space power and to constrain its development of space technology and even access to space. Whether or not this was the *real* intention of the United States can be debated, but—as Joan Johnson-Freese notes—"in one regard, the message of the United States to China has been crystal clear—the United States is not interested in cooperative space programs with China. Period".[183] Cooperation, if any, could at best occur in the form concisely described by former NASA Administrator Michael Griffin: "each nation will build its own highway to the moon and then they will cooperate when they get there".[184]

Against this hawkish posture held throughout the Bush tenure, the Obama Administration immediately appeared—at least in the rhetoric of official

---

[180] *Ibid.* p.156.

[181] See Rathgeber, Wolfgang (2007). "China's Posture in Space. Implications for Europe". ESPI Report 3. European Space Policy Institute, Vienna: p. 40.

[182] Wheeler, Larry. "US Losing Unofficial Space Race, Congressmen Say". Space.com. 31 March 2006. Web. http://www.space.com/1232-losing-unofficial-space-race-congressmen.html. Accessed 25 July 2014.

[183] Johnson-Freese, Joan (2006). "A New US-Sino Space relationship: Moving Toward Cooperation". Astropolitics Vol. 4 (2): 134.

[184] Quoted from Rathgeber, Wolfgang (2007). "China's Posture in Space. Implications for Europe". ESPI Report 3. European Space Policy Institute, Vienna: 40.

statements—keen on reshaping US–China policy and on reopening a window for dialogue and cooperation. As early as November 2009, Presidents Obama and Hu Jintao met in Beijing and called for the "initiation of a joint dialogue on (space science) human spaceflight and space exploration".[185] The director of the White House Office of Science and Technology Policy (OSTP) subsequently travelled to China three times to discuss potential areas for US–Chinese S&T cooperation, including space activities, and NASA Administrator Charles Bolden visited Chinese launch facilities in October 2010. Although "no concrete programmes arose from these travels, a second joint statement was signed in January 2011, committing the two countries to 'deepen dialogue and exchanges in the field of space' and to 'continue discussions on opportunities for practical future cooperation in the space arena, based on principles of transparency, reciprocity, and mutual benefit'".[186] In the same year, the Obama Administration sent proposals to the Congress for a unified licencing regime to operate through the Department of Commerce, so as to make it possible for commercial satellites to fly on the Chinese Long March rockets.[187]

In spite of this remarkable change in posture—which reveals an emerging awareness of the dangers involved in deliberately excluding China from American-led space activities—these efforts have ultimately not succeeded, thwarted by strong opposition within the US Congress. In particular, Republican Congressman Frank Wolf, chairman of the House Appropriations Subcommittee which funds NASA and the OSTP, led a successful campaign to prevent the Obama Administration cooperating with China. In April 2011 this resulted in an amendment inserted into the federal budget. Known as the "Wolf Amendment", it stipulates that no appropriated funds may be used by NASA or the OSTP "to develop, design, plan, promulgate, implement, or execute a bilateral policy, program, order, or contract of any kind to participate, collaborate, or coordinate bilaterally in any way with China".[188] Although this legal barrier does not prohibit Sino-US cooperation through multilateral mechanisms—as clarified by the eventual admission of Chinese scientists to an international conference on NASA's Kepler space telescope programme in October 2013[189]—it is clear that the

---

[185] See "China/US Should Cooperate In Space: Astronaut". Red Orbit. 30 April 2011. Web. http://www.redorbit.com/news/space/2038588/chinaus_should_cooperate_in_space_astronaut/. Accessed 25 July 2014.

[186] Quoted from Kulacki, Gregory (2011). "US and China need contact, not cold war". Nature. Vol. 474: 444.

[187] Harvey, Brian (2013). *China in Space. The Great Leap Forward.* Springer, New York.

[188] Quoted from Mervis, Jeffery "Spending Bill Prohibits US-China Collaborations". Science. 21 April 2011. Web. http://news.sciencemag.org/technology/2011/04/spending-bill-prohibits-US-china-collaborations. Accessed 26 July 2014.

[189] Chinese space officials were in fact initially prohibited from attending the conference, but Senator Frank Wolf then clarified that the law did not ban Chinese officials participating in multilateral fora or events. See "Wolf Letter To NASA's Bolden Correcting Record On Restrictions Involving Chinese Nationals". Press Release. 8 October 2013. Web. http://wolf.house.gov/press-releases/wolf-letter-to-nasas-bolden-correcting-record-on-restrictions-involving-chinese-nationals/. Accessed 21 July 2014.

Congress has made serious efforts to prevent the United States and China working together.

What is perhaps more surprising than firm opposition in the Congress to the prospect of US–China collaboration is the counterreaction the enactment of this bill generated. Both NASA and the OSTP immediately took a clear stance against the policy imposed by the Congress.

In particular, President Obama's science adviser, John Holdren, resolutely opposed the ban, emphasising that Obama's decision to promote S&T cooperation with China was part of the President's constitutional prerogatives and the best way to advance US interests. It was even reported that, when speaking before the House Commerce, Justice, Science Appropriations Subcommittee in May 2011, Holdren affirmed that the administration did not intend to comply with the statutory prohibition.[190] NASA Chief Charles Bolden pursued the same line in testimony to the Congress in November 2011, openly stating that "some level of engagement with China in space-related areas [is] consistent with the national interests of both our countries", thus subtly inferring that NASA would in any case look for ways to engage China, "when based on the principles of transparency, reciprocity and mutual benefit". [191] Indeed, Bolden met CAS President Bai Chunli and with CNSA Administrator Xu Dazhe on the occasion of the 64th IAC and the 2014 ISEF,[192] later stating that: "We are looking for ways in time to find different ways we can be a partner to them".[193] Furthermore, as recently as November 2014, Bolden visited the China Manned Space Agency (CMSA) office and had a bilateral meeting with Wang Zhaoyao, CMSA director, for sharing information on air traffic management and exchanging opinions on cooperation in human spaceflight.[194]

The existence of this difference of opinion between the Congress and the OSTP–NASA deserves particular attention—also by European policymakers—as it reveals the lack of a firm political consensus over the current China policy in space, which could potentially open up new avenues for future US–China space relations. A more detailed examination of the arguments of the two sides is provided below.

---

[190] US–China Economic and Security Review Commission. Hearing: The Implications of China's military and civil space programs. 2011. Web. http://www.uscc.gov/Hearings/hearing-implications-china's-military-and-civil-spaceprograms. Accessed 18 July 2014.

[191] Gallo, William. "Could US work with China on Space Issues?" Voice of America News. 11 June 2013. http://www.voanews.com/content/could-us-work-with-china-on-spaceissues/1679451.html?goback = %2Egde_139815_member_249152883. Accessed 21 July 2014.

[192] Johnson, Andrew (2014). "An Agreement to Disagree". In Lan, Chen, and Jacqueline Myrrhe (eds). *Go Taikonauts. All about China's space programme.* Issue 12. May 2014: pp. 21–26.

[193] Morring, Frank "NASA China meet on possible cooperation". Aviation Week & Space Technology. 16 January 2014. Web. http://aviationweek.com/space/nasa-china-meet-possible-cooperation. Accessed 21 July 2014.

[194] Morring, Frank. "Bolden Meets Human-Spaceflight Chief During China Visit". Aviation Week & Space Technology. 3 December 2014. Web. http://aviationweek.com/space/bolden-meets-human-spaceflight-chief-during-china-visit. Accessed 5 December 2014.

6.3.2.3 The Pro-China Versus the Anti-China Front

The first thing to note is that there is an ideological divide between the Congress and the military on the one side and parts of the administration, including NASA and the OSTP, on the other, as just demonstrated. While the latter (and more broadly all the advocates of cooperation) are accused of "naively helping China to win a new cold war",[195] the anti-China chorus is condemned for living in an obsolete Cold War mentality, whereas international space cooperation could provide enduring solutions to the emergence of adverse trends.[196]

Advocates of space cooperation criticise the non-cooperation policy so far adopted by the United States for being not only unnecessary but also ineffective. Space policy analysts such as Gregory Kulacki, manager of the China Project for the Union of Concerned Scientists, and James Clay Moltz have opined that the progress of Chinese space activity during the previous US administration suggests that restrictions have not achieved their aims and arguably have been counterproductive. For one thing sanctions have severed the links between the two countries and made a new generation of Chinese intellectuals resentful and suspicious of the United States.[197] In addition, China's resolve to develop its space programme has ultimately been strengthened.[198]

Furthermore, the freeze in US–China space cooperation is seriously hampering the effectiveness of the US space programme. While the current law is impeding the nation's ability to collaborate with China where it would provide benefit, traditional US international partners have not subjected themselves to such restrictions. Indeed, many observers have pointed out that the "containment policies" have failed to prevent—and in some instances have stimulated—Sino-European cooperation in space (see Chap. 7), "leading to the growth of an ITAR-free business model in both Europe and China, to the detriment of the US space industry" and of NASA international outreach.[199] In short, the current freeze in US–China space relations is creating opportunities for collaboration that are—and will be—filled by other spacefaring nations.

For advocates of cooperation, the Congress' concerns about the PLA's involvement in the Chinese space programme and thus the fears over national defence and

[195] Kulacki, Gregory (2011). "US and China need contact, not cold war". Nature. Vol. 474: 444.

[196] As recognised as early as 2004 by a Defence Science Board report, "the US government is reflexively inclined toward Cold-War-style responses to the new threat(s), without a thought or a care as to whether these were the best responses to a very different strategic situation". Quoted from Johnson-Freese, Joan (2006). "A New US-Sino Space relationship: Moving Toward Cooperation". Astropolitics Vol. 4 (2): 148.

[197] Kulacki, Gregory (2011). "US and China need contact, not cold war". Nature. Vol. 474: 444.

[198] Moltz, James Clay (2011). "China's Space Technology: International Dynamics and Implications for the United States". In Hearing: The Implications of China's military and civil space programs. US-China Economic and Security Review Commission (2011). Web. http://www.uscc.gov/Hearings/hearing-implications-china's-military-and-civil-spaceprograms. Accessed 15 June 2014.

[199] Hitchens, Teresa, and David Chen (2008). "Forging a Sino-US 'grand bargain' in space". Space Policy Vol 24 (3): 128–131.

inadvertent transfer of sensitive technology—though understandable—are overstated, distorted, and misperceived. While China's space programme is connected to the PLA, "these connections do not justify a refusal to cooperate together",[200] not least because cooperation would positively strengthen China's civilian space sector. As Alanna Krolikowski has pointed out in a testimony to a Congress subcommittee, "the discussion of US-China space cooperation should recognize that every form of international cooperation has domestic effects. Any form of cooperation or non-cooperation with the United States will empower some actors within the Chinese space establishment at the expense of others. Premised on the right conditions, international cooperation projects can make civilian actors more prominent and influential within the Chinese space sector". [201]

In addition, "if any interaction with the Chinese space programme is assumed to involve an associated interaction with the Chinese military,—Andrew Johnson has ironically remarked—then many of the United States' closest allies have already taken a step in that direction".[202] The US restrictions intended to limit (or even control) Chinese access to space technology have pushed China towards other global suppliers, which are increasingly available and disposed towards cooperation. Ironically, this has afforded less control over technology transfer than previously. As Joan Johnson-Freese puts it, "if the goal of the U.S is to assure that Chinese space development occurs in a peaceful, non-threatening manner, then the US must consider that it could actually be more effective in influencing Chinese programmatic direction and in determining what globally-available technology reaches China through engagement rather than detachment, and by stressing innovation and staying ahead, versus trying to constrain China".[203]

Thus, cooperation may introduce an important benefit by offsetting the need for China to develop unilaterally. Collaborating with China—instead of isolating it—could keep the country reliant on US technology rather than forcing it to develop technologies alone or purchase them on the global market. This could in turn give the United States leverage in other areas of the relationship and be more broadly conducive to an improvement in political relations.[204]

---

[200] Position expressed by former astronaut Leroy Chiao in a testimony for the US–China Economic and Security Review Commission in 2013. Quoted from Johnson, Andrew (2014). "An Agreement to Disagree". In Lan, Chen, and Jacqueline Myrrhe (eds). *Go Taikonauts. All about China's space programme*. Issue 12. May 2014: pp. 21–26.

[201] *Cit*. Krolikowski, Alanna (2011). "China's Civil and Commercial Space Activities and their Implications". In Hearing: The Implications of China's military and civil space programs. US-China Economic and Security Review Commission (2011). Web. http://www.uscc.gov/Hearings/hearing-implications-china's-military-and-civil-spaceprograms. Accessed 15 June 2014.

[202] *Cit*. Johnson, Andrew (2014). "An Agreement to Disagree". In Lan, Chen, and Jacqueline Myrrhe (eds). *Go Taikonauts. All about China's space programme*. Issue 12. May 2014: pp. 21–26.

[203] *Cit*. Johnson-Freese, Joan (2006). "A New US-Sino Space relationship: Moving Toward Cooperation". Astropolitics Vol. 4 (2): 134.

[204] *Cit*. Logan, Jeffrey (2007). "China's Space Program: Options for US-China Cooperation". Congressional Research Service Report for Congress, Washington DC.

Another potential positive payback stemming from cooperating with China would be greater US insight into China's space programme, technical capabilities, and intentions. While there is currently uncertainty and lack of transparency over China's space goals, resulting in the need for worst-case planning, regular dialogue and exchange of information could help the two nations understand each other's intentions more clearly, overcoming mutual mistrust and ambiguity. Over the long term, dialogue and cooperation could potentially give way to strengthen confidence and assurance of intentions and concerns about space and help address national security concerns while increasing transparency across the board.[205]

Equally importantly, cooperative undertakings could be an important way to maintain and in a sense renew US leadership. Many space policy experts believe that NASA is losing its appeal as trailblazer of the international space community's efforts. However, by leveraging the fact that the United States has already accomplished a manned lunar landing, embarking upon a cooperative programme with China (as well as other spacefaring nations) could generate the public perception of the United States assisting other nations to go beyond Earth, in a true spirit of leadership. As the National Research Council notes, the underlying issue is that the "U.S can advance its national goals in space by sharing the responsibility on a global scale—making the U.S a *real* leader among a host of nations contributing to space exploration and reaping the benefits", rather than excluding them. Such a posture would provide an important impetus to allaying the fears of the international community about the alleged US intention of pursuing space dominance.

Finally, advocates of cooperation highlight a fact that is too often overlooked: the alternative to cooperating with China could be a descent into an unpromising space race (also at the strategic threat level), bringing unaffordable financial and political burdens for the United States. The United States increased NASA's budget by 89 % in the months following Kennedy's 1961 Moon speech,[206] and NASA's expenditure peaked at 5.3 % of the federal budget in 1965.[207] This is unimaginable today, given the severe budget constraints faced by NASA and the fact that the United States is not a rapidly expanding but a plateauing economy. It thus behoves the United States to discard a space race scenario and consider opportunities for cooperation. Indeed, a gradual increase in cooperation with China would "make sense because it would reduce the cost of US space exploration, enabling both countries to continue gaining scientific knowledge" and also improving relations to a degree.

---

[205] Chambers, Rob (2009). "China space program: a new tool for PRC "soft power" in international relations?". Dissertation, Naval Postgraduate School, Monterey, CA.

[206] Logsdon, John M (2008). "Why space exploration should be a global project". Space Policy 24 (1): 3–5.

[207] According to official records (NASA website http://history.nasa.gov/Apollomon/Apollo.html), the NASA budget increased from $500 million in 1960 to a high point of $5.2 billion in 1965. In that year NASA funding level represented 5.3 % of the federal budget. In total, between 1959 and 1973 NASA spent $23.6 billion on human spaceflight, exclusive of infrastructure and support, of which nearly $20 billion was for Apollo. This expenditure would amount to $110 billion in 2010 terms!.

If the path of cooperation is pursued, it could not only avoid an ominous destabilisation of the US–China political relationship but, equally importantly, could prevent—or at least inhibit—the formation of adversarial blocs, including a strengthening of the Sino-Russian axis. While the latter relationship remains an "axis of convenience" and not a strategic alliance, the US's China exclusion policy might further cement their cooperation and eventually spur the emergence of two competing ideological blocs. [208] It is in Washington's interest not to make this happen!

On the other side of the spectrum, opponents of cooperation judge these rationales as inconsistent at best. They have argued that it is highly unlikely that cooperation in space will translate into friendly relations on Earth. Space cooperation will "never help overthrow old tensions and distrust between Washington and Beijing, no matter how many astronauts and *taikonauts* hug each other in LEO, because diplomatic progress always comes first".[209] The concrete possibility of diverging policy realities makes the prospect of multiyear cooperation projects intrinsically unfeasible, and thus, the United States should be extremely wary of any cooperation with China at least until the trajectory of Beijing's rise becomes clearer.

The likelihood of offsetting financial cost through cooperation is also considered over-optimistic or even the wrong way round. International missions—they claim—can cost as much or more than the equivalent mission would have cost if done by the United States alone.[210] Erich Seedhouse, for instance, has quipped, "Washington has learned from bitter experience that major international projects almost always end up costing more, taking longer, and delivering less than a national programme. From a financial perspective, the US-Russian cooperation experience (on the ISS) is one that the Americans will not want to repeat by collaborating with the Chinese".[211]

For detractors of cooperation, it is also clear that, in terms of relative gains, cooperation would benefit Chinese stakeholders more than vice versa. And a gain for the one side will spell a loss for the other in a zero-sum game. China does not possess any convincing capability that NASA lacks. Given its vast superiority over all the other space powers, there is little that the United States can harvest from cooperating with a second-tier space programme such as that of Beijing. As Congressman Frank Wolf bluntly puts it: "we don't want to give them the

---

[208] See Lo, Bobo (2008). *Axis of convenience: Moscow, Beijing and the new geopolitics.* Brookings Institution Press, Washington DC.

[209] *Cit.* Seedhouse, Erich (2010). *The New Space Race. China vs the United States*, Springer—Praxis Publishing, Chichester, UK: p. 212.

[210] In addition, US laws and regulations, such as the International Traffic in Arms Regulations, can greatly complicate international cooperation. For modest projects, the added complexity can easily make things more expensive and slower to complete. National Research Council (2012). *NASA Strategic Direction and the Need for a National Consensus.* The National Academy Press. Washington DC: p. 44.

[211] Seedhouse, Erich (2010). *The New Space Race. China vs the United States*, Springer – Praxis Publishing, Chichester, UK: p. 212.

opportunity to take advantage of our technology, and we have nothing to gain from dealing with them".[212] Using this logic, it becomes easy to understand the logic behind the pathway to space exploration envisioned by Michael Griffin when he spoke about the "two highways to the Moon".

In addition, regardless of US purposes, China's ambitious programme is in any case designed to compete with the United States in both civil and military arenas of space exploration and space utilisation, and thus, competition is a de facto policy reality, whether the United States likes it or not.[213]

All the opponents of cooperation agree that the real reason Sino-US cooperation is out of the question stems from the probability that China could use the partnership to gain direct or indirect access to sensitive technologies. After all, as the Cox Commission made clear, China has a long track record of technology acquisition by improper means that goes back to the days of Tsien Hsue Chien, the father of the Chinese space programme.[214] Although it is now recognised that the isolation policy has only slowed, but not completely stopped China's technological development, the risk of technology leaks or inadvertent assistance remains high and has potentially wide-ranging implications. Technology leaks and inadvertent assistance would both lead to China becoming a more formidable space power (reducing the gap with the United States) and ultimately jeopardise national security. Given the PLA's involvement in space activities and the current policy encouraging civil–military integration (CMI) of technologies endorsed by the MIIT as well as CASC and CASIC,[215] cooperation would certainly be used by the military to exploit dual-use applications. According to the 2013 Annual Report of the US–China Economic and Security Review Commission, "even ostensibly civilian projects, such as the Shenzhou missions and the Tiangong-series space labs, support the development of PLA space, counter space and conventional capabilities".[216]

---

[212] Wolf, Frank. "US Should Not Cooperate With People's Liberation Army to Help Develop China's Space Program". Press Release. 2 November 2011. Web. http://wolf.house.gov/media-center/press-releases/wolf-us-should-not-cooperate-with-peoples-liberation-army-to-help#.U6yH39zLTj0. Accessed 28 July 2014.

[213] See Seedhouse, Erich (2010). *The New Space Race. China vs the United States*, Springer—Praxis Publishing, Chichester, UK: p. 207.

[214] Tsien Hsue Chien was accused of spying and of communist sympathies. For a more detailed account, see Harvey, Brian (2013). *China in Space. The Great Leap Forward.* Springer, New York.

[215] Known in Chinese as *Yujun Yumin* (locating military potential in civilian capabilities), the CMI policy recommends efforts to leverage investments made in the civilian sector by finding an application with military potential.

[216] US–China Economic and Security Review Commission. 2013 Annual Report to Congress. US Government Printing Office, Washington DC, United States. November 2013: p. 230. This view has been well reflected in a letter written by Senator Frank???? Wolf to James Clapper in December 2013, which states: "Given that China does not separate civilian and military space programs, its purportedly civilian space accomplishments suggest concomitant military applications as well". Quoted from Johnson, Andrew (2014). "An Agreement to Disagree". In Lan, Chen, and Jacqueline Myrrhe (eds). *Go Taikonauts. All about China's space programme.* Issue 12. May 2014: pp. 21–26.

Many see in addition a moral argument behind the current freeze in US–China space relations. As once again argued by Congressman Wolf, "Frankly it boils down to a moral issue... would you have a bilateral programme with Stalin? There will come a day when the Chinese communist government will fall, repressive, totalitarian regimes always do. And when that day comes, books will be written about who helped sustain this government in their final days. Will US companies feature in that narrative? Will the US government?"

Such argumentation—which undoubtedly has traction with US politicians given the almost missionary zeal of the US sociopolitical tradition—has been used to silence cooperation proponents and relativise potential benefits. Put simply, any cooperation that would inevitably improve the moral standing of an authoritarian regime is simply unacceptable—whatever the benefits it might bring.

All in all, the arguments supported by the pro-camp and its critics are many and multifaceted. In order to shed some light on the debate, Table 6.2 summarises the main positions of the two camps.

6.3.2.4 Towards Reconciliation?

Up to now, the result of this political disagreement between engagement and containment proponents has translated into a *status quo policy* that has played into the hands of those opposing cooperation. There are, however, signs that things could change in favour of a more pragmatic policy.

The aforementioned pro-cooperative stances endorsed by NASA and the OSTP, and more broadly by the Obama Administration, appear to be accompanied by an increasing consensus within academia and the scientific community on the need to cooperate with China. Further, relevant stakeholders like the National Research Council and the National Academy of Sciences have started to take a clear posture in this regard, as recent developments demonstrate.

In response to a congressional directive, the National Academy of Sciences (NAS) was requested to conduct a comprehensive "review of the goals, core capabilities, and direction of human space flight". Its report, which was released in May 2014 and entitled, "Pathways to Exploration: Rationales and Approaches for a US Program of Human Space Exploration", is clear. It not only criticises the

Table 6.2 The US dilemmas/rationales

Anti-China camp	Pro-China camp
Jeopardises national security	Offsets economic costs
Inadvertent technology transfer	Avoids competitive trends
Reinforces China's relative position in the space hierarchy	Avoids the burden of a space race/space arms race
Cooperation more expensive than national programme	Improves relationships
	Improves transparency
Improves the moral standing of Chinese leaders	Avoids the formation of adversary blocs
Space diplomacy has no effect on Earth	Keeps leadership

posture adopted by the United States so far but also recommends a drastic volte-face in US–China policy.[217] As the document highlights, "the prohibition on NASA speaking to Chinese space authorities has left open opportunities for collaboration that are being filled by other spacefaring nations [. . ..] It is also evident that given the rapid development of China's capabilities in space, it is in the best interests of the United States to be open to its inclusion in future international partnerships. In particular, current federal law preventing NASA from participating in bilateral activities with the Chinese serves only to hinder US ability to bring China into its sphere of international partnerships and reduces substantially the potential international capability that might be pooled to reach Mars. Also, given the scale of the endeavour of a mission to Mars, contributions by international partners would have to be of unprecedented magnitude to defray a significant portion of the cost".[218]

What is also noteworthy about the NAS report is the comprehensive dismissal of NASA's approach to international cooperation and the recognition that the flag-waving model inherited from the Cold War era should be abandoned in favour of a resolute policy of real international cooperation "open to the inclusion of China and potentially other emerging space powers, as well as traditional international partners". The report emphasises that "future collaborations on major new endeavours should seek to incorporate: (a) A level of overall cost sharing appropriate to the true partnerships that will be necessary to pursue pathways beyond LEO. (b) Shared decision making with partners. This should include a detailed analysis, in concert with international partners, of the implications for human exploration of continuing the International Space Station beyond 2024".[219]

For NASA—which has generally envisioned "international cooperation as the acceptance by other partners of a programme conceived, planned, and directed by NASA"[220]—endorsing such recommendations would be a radical paradigm shift. To be sure, the recommendations are not intended to induce the United States to renounce its leadership—quite the contrary. As the aforementioned NRC report also notes, "the role of the United States remains to lead. But a new paradigm for leadership is required where partners are given a more equal voice and a more substantive role in key areas critical to mission success. To lead is not necessarily to command and it is possible to establish international partnerships where all the members take part in major decisions and their interests are clearly aired and considered".[221] The adoption of such model could help overcome the doldrums in which the US space programme has languished since the end of the Cold War.

[217] National Academy of Sciences (2014). *Pathways to Exploration: Rationales and Approaches for a US Program of Human Space Exploration.* The National Academy Press. Washington DC.

[218] *Ibid.*

[219] *Ibid.*

[220] *Cit.* Blamont, Jacques (2012). "US Space Exploration Strategy: Is there a better way?" Space Policy No. 28 (4): 213.

[221] *Cit.* National Research Council (2012). *NASA Strategic Direction and the Need for a National Consensus.* The National Academy Press. Washington DC: p. 44.

While these recommendations are undoubtedly bold in their conclusions and have been formally welcomed by NASA,[222] it remains to be seen whether US policymakers will be able and willing to adopt such a "double paradigm shift" (i.e. inclusion of China in the US international cooperative scheme and a revision of the US approach to cooperation). In the meantime, however, concrete steps towards reconciliation seem to have been taken. In January 2014, the NAS and the CAS announced the establishment of a CAS–NAS Forum for New Leaders in Space Science. According to the NAS Space Studies Board, the forum has a threefold aim:

- To identify and highlight research achievements of the best and brightest young scientists currently working at the frontiers of their scientific disciplines
- To build informal bridges between the space science communities in China and the United States
- To enhance the diffusion of insights gained from participation in the forum to the larger space science communities in China and the United States.[223]

  Although this initiative does not by any stretch of the imagination amount to explicit government-to-government cooperation—NAS being a non-governmental institution—it can nonetheless be regarded as a valuable "cooperation-enabling mechanism" that could establish informal channels of communication, increase confidence, and, in time, lead to the establishment of formal space cooperation. The recent meetings of Charles Bolden with CAS President Bai Chunli and CNSA Administrator Xu Dazhe also seem to have gone in that direction.

In addition to the increasing commitment to dialogue and cooperation, two other developments are worth highlighting. The first is the retirement of tireless China opponent Frank Wolf from the Congress.[224] The second is the announced intention of the DoD "to expand and deepen its engagement with the Chinese military in non-sensitive areas of mutual interest".[225] While the former could at least potentially lead to a congressional relaxation of posture,[226] the latter reveals a gradual change in approach by the DoD, which has so far been the strongest supporter of the

---

[222] After the release of report NASA issued a statement welcoming the report: "There is a consensus that our horizon goal should be a human mission to Mars", the US space agency said, adding "the stepping stone and pathways thrust of the NRC report complements NASA's on-going approach". See "US Needs to reexamine NASA's China exclusion policy": report. Xinhua News. 5 June 2014. Web. http://news.xinhuanet.com/english/sci/2014-06/05/c_133384132.htm. Accessed 15 July 2014.

[223] "Forum for New Leaders in Science". National Academy of Science—Space Studies Board (2014). Washington DC. Web. http://sites.nationalacademies.org/SSB/SSB_086017.htm/. Accessed 15 July 2014.

[224] Leone, Dan. "Frank Wolf, House's Top NASA Appropriator, Retiring in 2015". Space News. 10 April 2014. Web. http://www.spacenews.com/article/civil-space/38756frank-wolf-house's-top-nasa-appropriator-retiring-in-2015. Accessed 15 July 2014.

[225] US–China Economic and Security Review Commission. 2013 Annual Report to Congress. US Government Printing Office, Washington DC, United States. November 2013.

[226] In this regard the eventual congressional approval to share information on air traffic management with China that in November 2014 enabled Charles Bolden's meeting with Wang Zhaoyao, director of the CMSEO, is remarkable.

"containment policy". The US–China Economic Security Review Commission has reported that the DoD now praises the benefits of "military-to-military engagement, [since it] reduces the risks of conflict through accidents and miscalculations; builds lines of communications at strategic level, contributes to better overall bilateral relations, and creates opportunities to obtain greater contributions from China to international security".[227] China's participation in the "Cobra Gold 2014" military exercise can be regarded as an early manifestation of this gradual change.[228]

6.3.2.5 Some Conclusions

All in all, prospects for a gradual Sino-American rapprochement in space seem to have brightened in the past few years. Considering the unappetising alternative implied in not cooperating with China, NASA's statement of interest appears genuine and consistent, both strategically and economically. It is also clear, however, that President Obama's overtures currently remain more virtual than real and that promoters of cooperation are some way from getting the upper hand in the political debate. As Bolden recognised at the end of ISEF 2014, "Human spaceflight is not something that's going to happen with US [and] China in the foreseeable future, because we are forbidden from doing that by law, so let's just get that out there . . . That's not going to change; not today, anyway".[229]

However, given the declared interest in "finding alternative ways to be a partner with them" and also noting that current law prohibits only bilateral contacts,[230] the United States might eventually seek to engage in a multilateral cooperation scheme (including JAXA and ESA, if not *all* interested countries). Such multilateral cooperation could present a way to overcome opposition and legal restrictions and move the relationship in the direction that is increasingly advocated by the global space community.

Internal considerations of both the relative strength of the United States over the next 10 years and the likely evolution of Sino-US political relations will weigh on the feasibility of such prospects. And two important extraneous factors will influence the outcome: the eventual success of China's efforts to find alternative partners to the United States and the European posture.

---

[227] US–China Economic and Security Review Commission. 2013 Annual Report to Congress. US Government Printing Office, Washington DC, United States. November 2013: p. 230.

[228] See Yeo, Nicole. "China's Participation in Cobra Gold 2014: A Golden Opportunity for the United States?" China–US Focus. 11 March 2014. Web. http://www.chinausfocus.com/foreign-policy/chinas-participation-in-cobra-gold-2014-a-golden-opportunity-for-the-united-states/. Accessed 18 July 2014.

[229] Morring, Frank "NASA China meet on possible cooperation". Aviation Week & Space Technology. 16 January 2014. Web. http://aviationweek.com/space/nasa-china-meet-possible-cooperation. Accessed 21 July 2014.

[230] Also to be noted is that although the restriction remains in place in the current budget, the law includes a provision allowing discussions to go forward if NASA can certify that there is not a threat of revealing sensitive security information.

While the rest of this chapter will look into the partnership configurations that may be open to China under present circumstances, Chap. 7 will specifically focus on the posture of Europe and its influence on future pathways to manned lunar exploration.

6.4 Beyond the China–US Paradigms: Assessing Present and Future Cooperation Opportunities

6.4.1 Current Status of China's International Outreach

To appraise the potential cooperative schemes that could be utilised to achieve a lunar landing, we first examine the current status of China's international space cooperation.

Despite the isolation policies directly and indirectly pursued by the United States, China has succeeded in winning many new partners, as demonstrated by the impressive number of agreements it has signed since opening to the rest of the world in 1976, as well as by its participation in a wide range of space-related organisations. China has joined the International Telecommunication Union (ITU), the International Astronautical Federation (IAF), the United Nations Committee on the Peaceful Uses of Outer Space (UN COPUOS), the Committee on Space Research (COSPAR), the International Maritime Satellite Organisation (INMARSAT), and the International Organization for Standardization. China has also acceded to the main international outer space treaties, though not the Moon Treaty.

In terms of government-to-government cooperation, links have been built with over 40 countries,[231] and China appears to be moving towards what could be regarded as a "hub-role" for international space cooperation. While its international outreach is still not as broad as that of ESA, Roscosmos, or NASA, as an emerging space power, China has nonetheless been able to forge links with most of the world's major space powers, including Russia, ESA, and individual European nations, as well as with a number of other emerging space nations, within both multilateral and bilateral frameworks.

6.4.1.1 Cooperation with Established Space Powers

Besides a limited number of cooperative initiatives with Japan and the United States,[232] China has shaped fruitful cooperative relations with several major

[231] Harvey, Brian (2013). *China in Space. The Great Leap Forward*. Springer, New York: p. 100.

[232] Japan and the United States were, quite ironically, the first countries China had tried to establish space cooperation with after the launch of Deng Xiaoping's *Gaige Kaifang* in 1976. The process was initiated in 1977–1979 with a series of exchange visits and meetings, leading to some cooperative projects.

space powers, namely, Russia, ESA, and individual European nations. Europe's multilateral and bilateral cooperation with China will be assessed in greater detail in Chap. 7; for now suffice it to remark that, notwithstanding some ups and downs, relations appear to be moving towards a deeper form of engagement.

The most intensive cooperation has been with Russia, which has been central to the development of China's space programme. Cooperation was first initiated in the mid-1950s, when the USSR provided the Chinese with two R-1 missiles and technical assistance for the development of its missile programme. Although suspended in August 1960, as a result of the political tensions that culminated in the Sino-Soviet split, cooperation was restored immediately after the collapse of Soviet Union and has grown tremendously since then.[233]

Following the signature of the first 10-year intergovernmental agreement on space cooperation in December 1992, the Chinese were invited to study the Soyuz spacecraft, Russian ground and tracking facilities, and environmental control systems for manned spacecraft.[234] In March 1995 a new intergovernmental agreement was signed, specifying Russian assistance for China in human spaceflight and the sale of engines. The Chinese bought Russian RD-120 rocket engines, and later an entire life support system, a Kurs rendezvous system, a docking module, an entire Soyuz capsule—emptied of equipment and electronics—and a Sokol spacesuit.[235] The agreement also included the training of two Chinese astronauts in Star City, and the opportunity for 20–50 Chinese specialists to attend the training, which took place from 1996 to 1998.[236] Despite this strong assistance from Russia, relations in the field of human spaceflight have so far remained that of a "buyer–seller", with no active participation of either country in the other's human spaceflight programmes. [237]

In May 2000, cooperation was further institutionalised through the establishment of the Space Cooperation Subcommittee during the Russian and Chinese prime ministers' meeting. The subcommittee has held regular meetings ever since, with "two multiannual cooperation agreements... adopted, a first 5-year one from 2001 until 2006 and a second 10-year one running from 2007 until 2016".[238] The two

[233] Mathieu, Charlotte (2008). "Assessing Russia's space cooperation with China and India. Opportunities and Challenges for Europe". ESPI Report 12. European Space Policy Institute, Vienna: pp. 21–22.

[234] Harvey, Brian (2007). *The rebirth of the Russian Space Program. 50 Years after Sputnik, New Frontiers.* Springer—Praxis Publishing, Chichester, UK.

[235] The Russians, however, refused to sell RD-170 engines, a powerful LOX/Kerosene engine originally used for the first stage of Energya. See *Ibid.*

[236] Harvey, Brian (2004). *China's Space Program. From Conception to Manned Spaceflight.* Springer, New York.

[237] As noted by Charlotte Mathieu, it is remarkable that while the Chinese received training at Star City to become *taikonauts*, no Chinese ever flew with the Russians as "passenger" and vice versa. Mathieu, Charlotte (2008). "Assessing Russia's space cooperation with China and India. Opportunities and Challenges for Europe". ESPI Report 12. European Space Policy Institute, Vienna: p. 21.

[238] *Cit. Ibid.* p 21.

agreements have identified over 20 cooperation areas, including Earth observation, space science (e.g. ultraviolet space observatory, joint system of radio interferometers, Spektr UF, Radioastron, etc.), and deep-space exploration, including Mars exploration with the Russian Phobos-Grunt and the Chinese Yinghuo-1.[239] In contrast, satellite navigation is an area in which China and Russia have not come to terms. Although in the early 2000s China was interested in the development and use of the GLONASS system and Russia in China's financial contribution, no agreement was reached, eventually pushing Russia to reinforce cooperation with India.[240] However, on 30 June 2014 Moscow and Beijing signed a memorandum of understanding (MoU) on cooperation between GLONASS and China's BeiDou system. According to the MoU, each country will place three ground stations in the other country.[241]

In addition to these areas, Moscow and Beijing presented a joint diplomatic initiative to the Conference on Disarmament (CD), with the submission of a "Draft Treaty on the Prevention of the Placement of Weapons in Outer Space" in February 2008,[242] and the CNSA opened an office in Moscow in December 2008, after Roscosmos had opened a representation in Beijing in April 2008.[243]

Given the deepening political engagement between the two countries, it can be expected that the soon-to-expire 10-year cooperation agreement (2007–2016) will be renewed in 2015 and possibly extended to more active cooperation in the field of human spaceflight and space exploration.

6.4.1.2 Cooperation with Emerging Space Nations

According to Nicolas Peter, "the evolution of space technology development in a country can usually be divided into four stages. The first stage consists of purchasing satellites systems from other countries, the second stage consists of developing such systems in cooperation, and the third stage consists of developing the satellite system independently and the fourth of disseminating knowledge of satellite

[239] The mission, launched in November 2011, from Baikonur Cosmodrome was a failure. Harvey, Brian (2007). *The rebirth of the Russian Space Program. 50 Years after Sputnik, New Frontiers.* Springer—Praxis Publishing, Chichester, UK.

[240] In January 2006, Sergei Ivanov, "declared India–Russia's only cooperation partner in GLONASS". Rathgeber, Wolfgang (2007). "China Posture in Space. Implications for Europe". ESPI Report 3. European Space Policy Institute, Vienna: p. 48.

[241] It should be recalled that this agreement followed Russia's decision to disable American GPS ground stations on its territory in May 2014. See "China and Russia Continue to Deepen Space Cooperation". LaRouche PAC. 1 July 2014. Web. http://larouchepac.com/node/31183. Accessed 28 July 2014.

[242] The draft treaty can be read as a way for both actors to oppose alleged US space dominance, but for Russia it is also an instrument to prevent a dangerous arms race, whose costs could not be sustained by Russia.

[243] See Kondapalli, Srikanth (2010). "China's space programme and Asia". In Kai-Uwe Schrogl, Wolfgang Rathgeber, Blandina Banares, Christophe Venet (eds). *ESPI Yearbook on Space Policy* 2008/2009*: Setting New Trends*. Springer, Vienna: pp. 286–299.

development to other countries".[244] This is in a sense also true for China, which, after reaching a state-of-the-art technological level, has since the early 1990s increasingly extended cooperation to many emerging space nations. More practically, however, such cooperation was pursued also because the sanctions following the 1989 Tiananmen events had made it extremely difficult to cooperate with Western nations. In order to strengthen its nascent aerospace industry, China was obliged to look for non-Western partners, and that search started in the Asia-Pacific.

In the Asian context, China has signed a number of bilateral agreements (e.g. with Pakistan, Bangladesh, Thailand, Mongolia), although most of its efforts have been at multilateral level. Following the success of the Asia-Pacific Multilateral Cooperation in Space Technology and Applications (AP-MCSTA)—a workshop convened in 1992 to pursue regional scientific and technological exchanges between China and other developing states[245]—China took the lead in promoting a new and more solid cooperative scheme that culminated with the establishment of APSCO (see Chap. 2, Box 1).

Although APSCO's organisational structure and convention are similar to those of ESA,[246] APSCO is not a simple equivalent of its European regional counterpart, mainly because of the role China plays in it. There is an asymmetry in space capabilities between China and the other APSCO members. In a sense, China acts as a *primus inter pares*, providing the direction and decision-making process of the organisation, while it also sustains the major part of the financial cost.[247]

Many developing countries have, however, positively welcomed the initiative, as APSCO programmes have provided them with an important way to enhance their space capabilities. [248] At the same time, these programmes have both paved the way for Chinese leadership in the region and also "increased the demand for launching satellites through Chinese LM rockets, whose launch opportunity is severely limited by the US ITAR regime". [249]

---

[244] *Cit.* Peter, Nicolas "The changing geopolitics of space activities" Space Policy 22 (2): 102.

[245] For more on AP-MCSTA and its transformation into APSCO, see the organisation website: http://www.apsco.int/default.asp.

[246] The Secretariat of APSCO currently consists of four departments, namely, the Department of External Relations and Legal Affairs, the Department of Strategic Planning and Program Management, the Department of Education and Training and Database Management, and the Department of Administration and Finance.

[247] To avoid hegemony of the organisation by a particular state, Article 18 of the APSCO Convention states that financial contributions of any one state shall not exceed 18 % of the approved budget of the organisation. This provision, however, does not prevent China from leading the organisation in terms of decision-making. *Cit.* Aliberti, Marco (2013). "Regionalisation of Space Activities in Asia?". ESPI Perspectives 66. European Space Policy Institute, Vienna.

[248] The fields of cooperation identified by Article 6 of the Convention are space technology and applications, Earth observation, space science research, education and training, space law, policy, and regulations. Together with a development plan, concrete projects have been approved by council meetings in each of these fields. *Ibid.*

[249] *Cit.* Rathgeber, Wolfgang (2007). "China Posture in Space. Implications for Europe". ESPI Report 3. European Space Policy Institute, Vienna: p. 54.

Over the years, China has also been successful in expanding its space diplomacy to Latin American and African countries. In the former case, cooperative space relations are enjoyed with Brazil,[250] Peru (which is also a member of APSCO), Venezuela, and more recently Bolivia.[251] In the latter case, cooperation agreements or commercial contracts have been reached with Nigeria, Angola, and Kenya. The rationales for cooperating are commercial and strategic.[252] Indeed, for many analysts it is no accident that China is particularly keen on expanding space cooperation with resource-rich countries, using space technology as a bargaining chip to secure raw materials. As SES former Chief Executive Romain Bausch has stressed, "these nations do not pay cash but in raw materials, so in fact the satellite has no cost", a feature that could potentially disrupt the entire commercial market for space technology.[253]

While this might be an important motivation, the target is nevertheless more broadly that of using space to emerge as an alternative to the United States and claim the political leadership of developing countries. As such, space cooperation with the "South" could be part of China's veiled effort to replace the *Washington Consensus* with a *Beijing Consensus* within the African and Asian context.

As confirmation of this strategy, it is worth recalling the recently announced intention (June 2013) to open up the upcoming Chinese Space Station (CSS) to *all* interested countries, particularly developing ones. This initiative, which seems to emulate the Japanese-led KIBO-ABC programme (see Sect. 6.2.1), is however much bolder, as it is politically intended to establish a "more democratic" space environment and differentiate China's benign role in space from the nationalistic approach of the United States. What is also noteworthy is the prominent role assigned to the UN in promoting international cooperation on the CSS. As announced at the 55th plenary session of UNCOPUOS, the Human Space Technology Initiative (HSTI), launched by the UNOOSA in 2010, will work with the China Manned Space Agency to review possible collaboration in utilising the CSS.[254]

---

[250] China and Brazil, through the mechanism of the Space Cooperation Subcommittee of the Sino-Brazilian High-level Coordination Commission, have worked out a comprehensive bilateral space cooperation plan, actively promoted the research and development of the China–Brazil Earth resources satellites, continued to maintain data consistency of their Earth resources satellites, and expanded the application of their data into regional and global contexts.

[251] In 2013 China launched the first Bolivian telecommunications satellite and provided training for the (recently established) satellite operator. See Tiezzi, Shannon. "China's space diplomacy". The Diplomat. 24 December 2013. Web. http://thediplomat.com/tag/china-aerospace-science-and-technology-corporation/. Accessed 30 July 2014.

[252] For an analysis of Sino-Latin America cooperation, see Delgado-Lopez, Laura M (2012). "Sino-Latin American space cooperation: A smart move". Space Policy Vol. 28 (1): 7–14.

[253] De Selding, Peter. "Chinese Hardware, Financing Changing Satcom Landscape". Space News. 18 January 2013. Web. http://www.spacenews.com/article/satellite-telecom/33244chinese-hardware-financing-changing-satcom-landscape. Accessed 30 July 2014.

[254] See United Nations Office for Outer Space Affairs. Report on the United Nations/China Workshop on Human Space Technology. Beijing, China. 16–20 September 2013. Web. http://www.oosa.unvienna.org/pdf/sap/hsti/China2013/A.AC.1050_HSTI_WS.China.pdf. Accessed 31 July 2014.

In short, the message is both clear and powerful: unlike other "emerged nations", China is the only major power that is really complying with the "Declaration on International Cooperation in the Exploration and Use of Outer Space for the Benefit and in the Interest of All States, *Taking into Particular Account the Needs of Developing Countries*" (emphasis added).

6.4.2 Potential Cooperative Pathways to Lunar Exploration

6.4.2.1 Partnership SWOT

When considering the pathways to cooperation in lunar exploration, different forms and mechanisms must be explored. Cooperation can occur at various levels of a space programme (mission planning, mission design, mission operations, etc.)[255]; it can deliberately insert programmatic dependence into the architecture of the system or be limited to the provision of redundant elements (i.e. partners are kept out of the system's "critical path"); it can be implemented as a government-to-government endeavour (as with the ISS) or be based on industry-to-industry interactions (as in the case of the US Atlas V launcher) regulated by the respective governments, etc.[256]

For our purposes, cooperation is a "generic term denoting international governmental participation in a project".[257] By leaving aside for now the specific level of involvement of a partner and by utilising the ideal case of "true mutualism" (i.e. a cooperative scheme where partners benefit in the same way at the same time from cooperation),[258] China's potential partnership configurations can be determined on the basis of two different variables:

- Whether to engage with spacefaring nations or emerging space countries or both
- Whether to pursue cooperation on a bilateral or a multilateral basis

  With regard to the first variable, developing country participation should be expected to be largely at the behest of Beijing, which would thus exercise undisputed leadership in regard to mission planning, design, and implementation. Conversely, cooperation with established space powers implies a less prominent role for China. In the light of Beijing's long-standing strides to "come to the table" (*yi xi zhi di*) of the great powerhouses, it is, however, clear that for China cooperation with other major space powers should be as an equal or not at all!

[255] Correll, Randall, and Nicolas Peter (2005). "Odyssey: Principles for Enduring Space Exploration" Space Policy 21 (4): 251–258.

[256] Broniatovski, D.A., et al. (2006). "The Case for Managed International Cooperation in Space Exploration". Center for Strategic and International Studies. Washington DC.

[257] *Cit.* Peter, Nicolas "The changing geopolitics of space activities" Space Policy 22 (2): 100.

[258] *Ibid.*

Table 6.3 Matrix of potential cooperative pathways to the Moon

Cooperation potentials	Spacefaring nations	Emerging space nations
Bilateral	**Russia** =	*India* –
	U.S +	
	Europe =	
Multilateral context	*Europe–U.S–Russia* +	**APSCO framework** =
	ISECG countries +	

In *bold*, existing opportunities; in *italic*, potential opportunities
+, preferred configuration for China; =, good choice; –, residual partnership

As for the second parameter, China seems able to move comfortably in multilateral contexts, and there are subtle indications of its preference for such schemes. But the country remains open to potential bilateral frameworks. On such a basis, however, cooperation can be only pursued with an established space power—Russia, the United States, or Europe as a whole, but not with India nor Japan. These two regional space powers remain well behind China in the area of human spaceflight, although both have signalled an interest in developing an autonomous human spaceflight programme and in many respects Japan possesses far more advanced technological capabilities than China. The main reason why pure bilateral cooperation between Japan and China cannot take place is that Sino-Japanese political relations are neither stable nor friendly. In addition, potential Japanese involvement would require US endorsement: without the United States, no Japan. Finally, China does not seem particularly eager to explore such an option. Although recognising Japan as a crucial factor in its steady (and peaceful) development, as well as an important pillar for the regional integration process, Beijing does not appear to find much political benefit in bilateral cooperation with Japan in this domain. Any such cooperation could only be accepted in a multilateral context involving other established space powers. As for India, it would not be comfortable in a bilateral cooperative relationship under the guidance of the Chinese juggernaut, while China sees little reason to share the techno-nationalist benefits stemming from human spaceflight with the government of New Delhi.

In short, on a bilateral basis, only cooperation with the United States, Russia, and Europe would provide China with valuable political and technological benefits. After all, these powerhouses represent the most "structural" actors within the international system, as well as within the space arena.

By combining the two sets of variables, it is possible to construct a matrix containing alternative cooperation schemes (Table 6.3).

6.4.2.2 The APSCO Option Versus Major Power Cooperation

At first glance, a multilateral cooperation scheme with emerging space nations can be regarded as an ideal scenario for China, as this pathway would provide the Central Kingdom with the numerous benefits discussed in the previous sections.

The APSCO framework, in particular, appears an interesting and promising platform for Beijing. It is a fully fledged international organisation exclusively made up of developing countries that would guarantee China undisputed leadership. In addition, the organisation presents a certain flexibility in terms of membership. This flexibility is particularly valuable, as it would allow intercontinental participation stretching from Turkey to Peru, passing through Mongolia and Indonesia. In addition, China's elastic and pragmatic approach to international relations could eventually even open up participation in APSCO to African countries as well!

Notwithstanding these valuable benefits, there are structural impediments that make this scenario less interesting in the eyes of Beijing policymakers. It is clear that developing countries can only provide a limited financial and technological contribution. A manned lunar endeavour, however, presents demanding economic and technological efforts that can be best tackled through cooperation with the existing major powers.

As already noted, the *soft-landing* scenario projected for Chinese economic development (see Sect. 5.1) invites Beijing's policymakers to be more pragmatic in implementing highly ambitious programmes, and cooperation with existing spacefaring nations clearly moves in this direction. The large expenditure that will be required to launch and maintain in orbit the forthcoming CSS, combined with ever-increasing involvement in the whole spectrum of space activities, may also undermine China's capacity to devote large and parallel investments to a lunar endeavour. In terms of financial commitment, cooperation with existing spacefaring nations thus becomes preferable.

Furthermore, cooperating with established international partners offers the possibility of minimising the risk of failure. As already discussed (Chap. 2), Chinese political leaders have a low tolerance of failure: they are fully aware that if space successes are formidable instruments for boosting their political legitimacy, so too are space failures disastrous for their political standing. International cooperation with existing space powers is thus key to enhancing the robustness and performance of the programme and reducing the risks of failure. More broadly, this form of cooperation would substantially increase the possibility of hastening China's breakthrough in the key human spaceflight technology areas; improving its research capability, as well as the level of professionalism and project management in the field; providing access either directly or indirectly to new technologies; and enriching the pool of scientific and technical capabilities.

Although the participation of emerging space nations can be expected in the forthcoming CSS, it is clear that for the successful implementation of a lunar endeavour, China will prefer to cooperate with the more established space actors. The APSCO cooperative scheme, though not a priori excludable, can at best be regarded as a backup option for China.

6.4.2.3 The Russian Option

Compared to the APSCO framework, Sino-Russian cooperation looks highly relevant to the implementation of Chinese lunar plans. Russia is refocusing its space exploration programme to the Moon and has technological capabilities that would be extremely useful for the successful implementation of this endeavour. For its part, China could bring the required resources to make Russian space industry more innovative, competitive, and commercially self-sustainable.

Although Moscow appears particularly keen on protecting its technological gap with respect to China—as clearly demonstrated by the 2007 Moscow court prosecution of Igor Reshetin for transferring classified space-related information to the country[259]—it is also aware that its industry is in danger of losing its strategic edge and that industrial reforms are imperative. As Peter Hulsroj argues in the 2012 *ESPI Yearbook on space policy*, "Russia's industrial difficulties make Russia an ideal partner for China.... China is, of course, a master of industrial production, albeit not yet of the kind of advanced manufacturing required for space, and China would be likely to be keen to get access to Russian space technology for domestic production, something which seems to have happened quite a bit in China's current human spaceflight programme. Russia certainly has a very realistic view of China, and visa-versa, both are keen to leverage space, and neither is clearly inferior to the other in the overall scheme of things. From a partnership perspective an auspicious configuration!"[260]

Furthermore, given the current situation of the United States (prohibited from cooperating with China or with Russia, following broad sanctions over the latter's "imperial annexation" of the Crimean peninsula), cooperation between the two countries has in a sense become a "forced choice". Remarkably, during a roundtable discussion held at the first Russia–China Expo in Harbin on 30 June 2014, Russian Deputy Prime Minister Dmitry Rogozin stressed: "If we talk about manned space flights and exploration of outer space, as well as joint exploration of the Solar System—primarily it is the Moon and Mars—we are ready to go forth with our Chinese friends, hand in hand".[261]

---

[259] In December 2007, the Moscow court sentenced Igor Reshetin, the chief executive of TsNIIMAsh-Export, a producer of rockets and missiles working closely with the Russian Space Agency, to 11.5 years in prison for passing dual-purpose technology to China. "The other three defendants in the criminal case were sentenced to five to 11 years. Investigators said Reshetin and his co-accused had transferred know how that could be used to design nuclear missiles to China Precision Machinery Import-Export Corporation, causing losses to Russia of 110 million roubles". *Cit.* "Reshetin sentenced for 11.5 years for passing technology to China". RIA Novosti. 3 December 2007. Web. http://en.ria.ru/russia/20071203/90747889.html. Accessed 2 August 2014.

[260] *Cit.* Hulsroj, Peter (2014). "The Psychology and Reality of the Financial Crisis in Terms of Space Cooperation". In Al-Ekabi, Cenan, Blandina Banares, Peter Hulsroj, Arne Lahcen (eds). *ESPI Yearbook on Space Policy 2011/2012. Space in Times of Financial Crisis*. Springer, Vienna: pp. 159–168.

[261] Quoted from "China and Russia Continue to Deepen Space Cooperation". LaRouche PAC. 1 July 2014. Web. http://larouchepac.com/node/31183. Accessed 2 August 2014. See also "Russia, China Ready to Cooperate in Space, Explore Mars". Space Travel. 1 July 2014. Web. http://www.space-travel.com/reports/Russia_China_Ready_to_Cooperate_in_Space_Explore_Mars_999.html. Accessed 2 August 2014.

Notwithstanding this clear statement of interest and the diplomatic niceties in Dmitry Rogozin's speech, Russia still clearly visualises a potential partnership with China under its leadership. Imperial mind-sets are after all hard to let go of. More broadly, when looking at the Russian approach to cooperation with its neighbour, a key caveat is their mutual general mistrust.

The Chinese–Russian relationship is in fact "bedevilled by pervasive mistrust, rooted in historical grievances, geopolitical competition and structural factors".[262] Chinese officials are well aware of the so-called China-threat theory that is deeply rooted among both the general public and Russia's elites and harks all the way back to the Tsarist Empire. In addition, Russia appears extremely wary of Beijing's rising economic and political influence within the international system.[263] China, in turn, has long seen Russia as its voracious neighbour and witness Russia's part in China's century of shame and Mao's statements on Chinese readiness to let hundreds of millions die if necessary to defend the country against Soviet aggression.

Admittedly, at present Sino-Russian relations look to be the most stable bilateral relationship maintained by China.[264] The two countries are engaged in an all-dimensional, multi-tiered, and wide-ranging cooperation that also extends to military-to-military cooperation and is set out in the Shanghai Cooperation Organisation (SCO).[265] While trade relations do not play an important role in the current Sino-Russian interplay—especially when compared to China's profound economic interdependence with the United States and Europe—Russia remains an indispensable source of energy supply (in particular oil and natural gas) for Beijing.

This does not however mean that Russia has more bargaining chips than China. Quite to the contrary, the partnership is inherently "asymmetric", and in China's favour. Stephen Kotkin, professor of history at Princeton University, has said: "China extracts considerable practical benefits in oil and weapons from Russia.

---

[262] *Cit.* Kotkin, Stephen (2009). "The Unbalanced Triangle". Foreign Affairs. Vol. 88 (5): 130–138.

[263] See Medeiros, Evan S (2009). *China's International Behaviour. Activism, Opportunism and Diversification.* RAND Corporation, Santa Monica, CA: p. 100.

[264] As documented by Evan Medeiros. Beginning with Gorbachev's normalisation of Sino-Soviet relations in 1989, China–Russia relations have since undergone a sea change. Beijing has made gradual and consistent efforts to upgrade relations, driven largely, but not exclusively, by mutual concerns about US power and the US democracy-promotion agenda. In 1994, China and Russia formed a "cooperative partnership", followed by a "strategic cooperative partnership" in 1996, and the signing of a full treaty on "Good Neighbourliness, Friendship, and Cooperation" in 2001. These agreements led to a series of sustained high-level interactions, which remain the "thickest" part of this bilateral relationship. Since 1996, Chinese and Russian leaders have held annual summit meetings. *Ibid.* pp. 101–102.

[265] The SCO is a six-member security group founded in 2001 by China, Russia, Kazakhstan, Kyrgyzstan, Russia, Tajikistan, and Uzbekistan with the objective, at least in Russia's view, of forging a quasi-military alliance that could counter NATO. China is, however, quite hesitant in this regard: see China's reaction in the 2008 Georgia crisis. Within the organisation, cooperation mainly includes joint military exercises, intelligence sharing, and counterterrorism, but economic and cultural cooperation is also covered.

In return, Beijing flatters Moscow with rhetoric about their 'strategic partnership' and coddles it by promoting the illusion of a multipolar world. In many ways, the Chinese-Russian relationship today resembles that which first emerged in the seventeenth century: a rivalry for influence in Central Asia alongside attempts to expand bilateral commercial ties, with China in the catbird seat".[266] Also Bobo Lo, a former Australian diplomat in Moscow, has remarkably labelled their strategic partnership an "axis of convenience" primarily pushed by the need to constrain US diplomatic and military power, rather than by any real will to establish a mutually complementary and cooperative relationship.[267]

These considerations do not automatically imply that such an axis cannot become enduring and lead to a joint Sino-Russian manned lunar exploration programme. Should both Beijing and Moscow continue to be isolated—and in a sense contained—by the American grand strategy, the two great powers will inevitably see no other choice than to become closer allies.

Spurred by the recent standoff over the Ukraine, on 21 May 2014, the two countries signed a 30-year, $400 billion gas supply deal that was clearly intended to demonstrate to the "West" alternative cooperative schemes.[268] In addition, their expanding ties in the space arena could further cement such a partnership and ultimately lead to a "polarisation" of the international system—including the space community—between two opposing blocs pursuing parallel pathways to space exploration.

Although clearly not an appetising scenario for Beijing in the light of its current foreign policy behaviour, such a scenario could prove to become inevitable in the case of policy immobility in the West.

6.4.2.4 The Indian Option: Triangular Cooperation China–Russia–India

In light of the close cooperative ties that Russia has also built with India, a triangular cooperation China–Russia–India could come into the picture, at least theoretically. The concept of a strategic triangle between the three Eurasian giants was put forward by former Russian Prime Minister Y. M. Primakov in 1998 and relaunched by Russia's Ministry of Foreign Affairs in 2007.[269] As Moscow

[266] *Cit.* Kotkin, Stephen (2009). "The Unbalanced Triangle". Foreign Affairs. Vol. 88 (5): 130–138.

[267] See Lo, Bobo (2008). *Axis of convenience: Moscow, Beijing and the new geopolitics.* Brookings Institution Press, Washington DC.

[268] For an analysis on the agreement, see Koch-Weser, Iacob, and Craig Murray (2014). "The China-Russia Gas Deal: Background and Implications for the Broader Relationship". US–China Economic and Security Review Commission Staff Research Backgrounder, Washington DC.

[269] See a previous ESPI study by Charlotte Mathieu on Russian cooperation with China and India, also providing an assessment on triangular cooperation Russia–India–China. See Mathieu, Charlotte (2008). "Assessing Russia's space cooperation with China and India. Opportunities and Challenges for Europe". ESPI Report 12. European Space Policy Institute, Vienna.

visualises it, in light of their combined political, economic and demographic weight, the three countries have enough power to promote a "more democratic" international order, i.e. to counterbalance the prevailing hegemony of the United States in the international system. However, neither China nor India has seemed particularly interested in the idea, given the potential negative rifts such an alliance could generate in their respective foreign policies, whose liberty of manoeuvre and cooperative interplay with Western countries would become compromised.

All in all, a triangular China–Russia–India configuration for cooperative manned lunar exploration remains unlikely, inter alia because it might not be in Russia's interest to promote cooperation between the other two countries in space. In addition, at the moment neither India nor China seems to have "particular needs nor real opportunities to work together on space projects".[270]

6.4.2.5 Future Possibilities

All in all, the various options considered so far present a number of critical factors that make them undesirable in the eyes of Beijing policymakers. In spite of finding continuously closed doors, cooperation with the "West", especially the United States, still appears to China the preferable pathway, at least for now. Given the current freeze in Sino-American space relations, Europe's posture could eventually become a key variable in solving Beijing's strategic equation on the space chessboard.

[270] *Cit. Ibid.* p. 25.

Chapter 7
Europe and China in Space: Constraints, Opportunities, and Options

The aim of this final chapter is to further elaborate on the opportunities and challenges China's possible lunar ambitions are raising for Europe and to provide an assessment of the different policy options available to European stakeholders in this regard.

Such an investigation poses a specific analytical problem related to the appropriateness of discussing Europe as a single, distinct player within the international landscape for space activities—in other words, Europe's *space actorness*.[1] Europe is obviously not a nation state in a Westphalian sense nor has a polity emerged from a single, authoritative source as in the other spacefaring nations discussed so far. The European Union (EU)—which is the result of a long-standing project of integration initiated more than half a century ago and thus usually regarded as the geopolitical entity representing Europe—is itself problematic by virtue of the fact that "it is something more than an intergovernmental organisation, but still less than a fully-fledged European state".[2] In line with the definition of William Wallace, the EU can be regarded as a "partial, multilayered polity", that is, a "political entity which lacks, however, many of the features that one might expect to find in a traditional state".[3] In addition, when looking at the issue from a space policy perspective, the EU is neither the only nor the most directly involved actor in the management of European space activities. That role belongs to ESA.

---

[1] The concept of a European *actorness* has been developed by scholars arguing that the EU possesses the necessary structural prerequisites for action in world affairs that make it an international actor. Sjostedt, G (1977). *The External Role of the European Community*. Saxon House, Farnborough.

[2] Hill Christopher and Michael Smith (2005). *International Relations and the European Union*. Oxford University Press, Oxford: p. 4.

[3] Wallace, William (2005). "Post-Sovereign governance: the EU as a partial polity". In: Wallace, Horace et al. (eds). *Policy Making in the European Union*. Oxford University Press, Oxford: pp: 483–503.

M. Aliberti, *When China Goes to the Moon...*, Studies in Space Policy 11,
DOI 10.1007/978-3-319-19473-8_7

In fact, if a European *space actorness* is to be identified, this can be said to result from the complex interplay of three distinct constituencies, which create a triangular governance structure. At one tip there is ESA, an intergovernmental organisation, which over the past 40 years has taken the lead in carrying out the major European space endeavours and strengthening European space identity, though it lacks a political clout. At a second tip there is the EU, which has only recently started to position itself as an additional and effective space player, demonstrating the willingness—and in a sense the legitimacy—to complement ESA's actions with political leadership. At the third tip there are the different member states of both organisations which, despite a common basis of 18 states, do not exactly coincide. Each player in this composition has its own specific competences and interests.

Notwithstanding that the institutional divergences among the various actors exist and are reflected in the formulation of distinct views and strategic goals, it still appears appropriate to empirically discuss Europe as a unified, though sui generis, space actor. For one thing, over the years the path towards coordinated (if not integrated) governance has gradually yet consistently deepened. This is demonstrated by the ongoing enlargement of ESA towards all the EU Member States not yet members of the Agency, as well as by the closer cooperation between ESA and the EU. The two organisations have realised that both parties have specific complementary and mutually reinforcing strengths and need each other to fulfil public policy objectives and to provide an appropriate political profile and a more coherent framework for space activities in Europe. Since the publication of the first White Paper in 2003, the EU and ESA have thus committed themselves to working together for the implementation of space projects that are beneficial for both and avoiding duplication of efforts, in order to optimise available resources. Particularly in human exploration, it is clear that only a pan-European approach will do, meaning that both ESA and the EU must join forces.

It is thus positive that a common strategy for international space relations is gradually emerging, characterised by increasing coherence, synergy, and complementarity among the constituencies. The development of this nascent "European space diplomacy" is mainly being led by the EU, which since the entry into force of the Lisbon Treaty (2009) has taken primary responsibility for defining and representing the external dimensions of the European space programme.[4] What is, perhaps, still missing is a higher degree of coordination with national space agencies, which conduct many international cooperation activities under their own steam and which are central if Europe is to leverage its strength in high-profile activities such as a possible flight of humans to the Moon.

In the light of these considerations, it seems not only appropriate to assess Europe's policy posture vis-à-vis China's lunar ambitions as that of a unified, though sui generis, internationally acting body but also that such posture should—at least in principle—be crafted by the EU in a close and synergistic cooperation with all the other constituencies.

---

[4] Peter, Nicolas (2007). "The EU's Emergent Space Diplomacy". Space Policy Vol. 23 (2): 97–107.

Given the inherent geopolitical dimension of space activities, the following section will first provide an account of the most recent developments in the broader political relationship between the EU and China. An analysis of the long-standing framework of cooperation in space activities with the different European institutions will subsequently be provided. The two sets of analysis will in turn be used as a basis to discuss the strengths, weaknesses, opportunities, and challenges of potential Sino–European cooperation with regard to human space exploration and to identify a set of policy options for Europe. Finally, an assessment of the options and a series of recommended actions will be provided.

7.1 Towards a New Axis in World Politics: Rhetoric or Reality?

Initiated in 1975 with the establishment of formal diplomatic relations and reinforced in 1978 by the signature of a trade agreement,[5] the EU's (then European Community) relations with China were dominated by commercial interests until the end of the Cold War. The political interplay was of secondary importance and in a sense "derivative" of each side's relationship with the two superpowers.[6] In fact, as many scholars have noted, the Cold War constraints did not allow the relationship to develop its own independent dynamic and, as a consequence, little attention was paid by the two actors to a relationship seen as "weak" and "far away".[7]

The demise of the Soviet Union in 1991 brought about new possibilities for pushing forward the bilateral relationship beyond the scope of the economy and trade. The expansion of EU–China institutional links and bilateral cooperation has evolved across the board since then. In July 1995 the European Commission (EC) issued its first Communication on China (*A Long-Term Policy for China-Europe Relations*), followed in 1998 by a second policy paper entitled *Building a Comprehensive relationship with China*. In the same year an annual EU–China summit meeting mechanism was established, with host locations rotating between Beijing and an EU venue. The summits fostered the creation of a dense political dialogue between the two actors and increasingly expanded the portfolio of EU–China policy to a number of issues, such as nuclear non-proliferation, energy policy, human rights, and climate change.

---

[5] The EC-China Trade and Cooperation Agreement, signed in 1985, would replace this first agreement and form the basis for EU–China cooperation.

[6] Casarini, Nicola (2009). *Remaking Global Order. The Evolution of Europe-China Relations and its Implications for East Asia and the United States*. Oxford University Press, New York.

[7] Kapur, Harish (1990). *Distant Neighbours: China and Europe*. Printer Publisher, London. See also Casarini, Nicola (2006). "The evolution of EU-China relationship: from constructive engagement to strategic partnership". Institute for Security Studies Occasional Paper no. 64.

7.1.1 The Engagement

Unlike the USA, which, since the end of the Cold War, has remained divided on its China policy, the EU strategy towards China has followed a path of "constructive engagement", likely to culminate in a comprehensive and multidimensional partnership. This firm, broad European engagement policy towards China was above all greatly helped by the fact that, unlike the USA and Japan, Europe did not view China as a major strategic threat nor had Europe immediate strategic (i.e. security-related) interests on the East Asian chessboard: after all, no European military forces were based in the region and no territorial disputes or military alliances saw Europe's direct involvement.[8] As number of scholars has argued, China and Europe were left "free to forge a relationship unencumbered by the two factors that still prove to greatly complicate US-China relations: Taiwan and the potential clash of strategic interests".[9] To be sure, the rising political and economic dynamism of new China has since the mid-1990s posed a number of questions for European policy makers as well as the underlying difference with Washington explained by the fact that the EC and the majority of European capitals are convinced that cooperation would be preferable to confrontation.[10]

The belief—or at least the hope—implied in the 1995 and 1998 policy papers was that an engagement policy from Europe would gradually lead China to become a more open and democratic society, based on the rule of law and respect for human rights, and to adopt a peaceful and cooperative foreign policy behaviour in world affairs: the idea, in short, was to transform China into a "responsible stakeholder", as Robert Zoellick would later put it.[11] Behind the engagement, there was in addition the intention to capitalise optimally from China's burgeoning growth by fully integrating the country in the globalised world economy and benefiting from the potential of its huge market for European business. More broadly, Europe was willing to use its relationship with China to become a more influential and autonomous actor in the international system.

On the Chinese side, there was a recognition that the relationship with Brussels carried several advantages over relations with other major powers such as the USA, Japan, or even Russia, given the "civil nature" of European foreign policy projection, the absence of fundamental conflicts of interest, and a certain complementarity of the two economies. While Europe could not satisfy China's increasing appetite

---

[8] Casarini, Nicola (2009). *Remaking Global Order. The Evolution of Europe-China Relations and its Implications for East Asia and the United States.* Oxford University Press, New York: p. 175.

[9] *Cit.* Shambaugh, David (2004). "China and Europe: The Emerging Axis". Current History no. 670: 243–248.

[10] Grant, Charles, and Katinka Barysch (2009). "Can Europe and China shape a new world order?" Centre for European Reform, London. Web. http://www.cer.org.uk/sites/default/files/publications/attachments/pdf/2011/p_837-611.pdf.

[11] Casarini, Nicola (2009). *Remaking Global Order. The Evolution of Europe-China Relations and its Implications for East Asia and the United States.* Oxford University Press, New York: p. 41.

for raw materials and energy supplies, it was nonetheless an important source of development assistance,[12] business investment, and technology transfers that were crucial to the renovation of key industrial sectors and thus to its economic surge. To the Europeans, China in turn offered a low-cost manufacturing base and a potentially huge market for export. Developing closer ties with the EU was thus for Beijing an opportunity to diversify its sources of economic growth, as well as a way of strengthening its international legitimacy and security. Unlike the USA, Europe was, in fact, not seeking to contain China's rise but to promote its stable development. For China, Europe offered an opportunity to bring about a more multipolar balance of power in the international system, which could help it balance the overwhelming US pre-eminence.[13]

Against such a background, by the late 1990s both Europe and China began to place increasing emphasis on the development of the relationship, and the importance they attached to the expansion of bilateral cooperation was substantiated in two policy papers, *A Maturing Partnership*: *Shared Interests and Challenges in EU-China Relations* issued by the EC in September 2003 and *China's EU Policy Paper* in October 2003. The latter is notable as being the first ever White Paper on relations with a foreign partner to be published by China.

The almost simultaneous publication of the two documents was not accidental; it was intended to serve as a basis for the establishment of a comprehensive strategic partnership, which was officially launched on the occasion of the 8th EU–China Summit on 30 October 2003. The meaning of the label used to designate the relationship was well explained by then Chinese Prime Minister Wen Jiabao in a speech in Brussels in May 2004. As he emphasised: "Comprehensive means that cooperation should be all-dimensional, wide-ranging and multi-layered. It covers economic, scientific, technological, political and cultural fields, contains both bilateral and multilateral levels and it is conducted by both governments and non-governmental groups. Strategic, means that cooperation should be long-term and stable, bearing on the larger picture of China-EU relations. Partnership means that cooperation should be equal-footed, mutually beneficial and win-win".[14]

The launch of the EU–China partnership has deepened and broadened cooperation by establishing an institutionally fixed framework of regular high-level

[12] The importance of the EU aid programme as a source of investment should not be dismissed. In its China Country Strategy Paper for 2002–2006, the commission allocated Euro 250 million in aid, while in the most recent China Country Strategy Paper for 2007–2013, it allocated Euro 225 million. While the previous aid programmes focused on infrastructure and rural development, the last two have a focus on the environment, sustainable development, and good governance. See Medeiros, Evan S (2009). *China's International Behaviour. Activism, Opportunism and Diversification*. RAND Corporation, Santa Monica, CA: p. 119.

[13] Scott, David (2007). "China and the EU: A Strategic Axis for the Twenty-First Century?" International Relations Vol. 21 (1): 23–45.

[14] Lecture of Chinese Prime Minister Wen Jiabao in Brussels on 12 May 2004. Wen, Jiabao, "Vigorously Promoting Comprehensive Strategic Partnership Between China and the European Union". China–EU Investment and Trade Forum. 12 May 2005. Web. http://www.chinamission.be/eng/zt/t101949.htm. Accessed 18 April 2014.

dialogues and the so-called sectorial dialogues organised around three pillars: "political dialogue", "economic and sectorial dialogue", and "people-to-people dialogue". A graphical representation of the EU–China Dialogue Architecture is presented in Appendix C. These fora of consultation and cooperation have expanded considerably over the years (there were more than 50 sectorial dialogues by 2013) to cover a wide range of areas, including science and technology cooperation, enterprise regulation and consumer protection, environmental issues, education, aerospace cooperation, energy, and social affairs.[15]

This boost in political relations, which complemented increasingly healthy economic relations (since 2004 the EU has become China's top trading partner and China is the EU's second largest partner), gave international currency to the idea that a new axis in world affairs was about to emerge, as demonstrated by the scholarly output of the period and, even more importantly, by the public statements of European and Chinese stakeholders themselves.[16] Remarkable, for instance, is the speech given by Romano Prodi, then president of the EC, at the EU–China Business Forum in May 2004, in which he stated: "If it is not a marriage, it is at least a very serious engagement".[17]

If the birth of this "very serious engagement" was facilitated by the absence of serious impediments and frictions, an equally important thrust resulted from a convergence of geopolitical views about the emerging international order and, in particular, about the USA and its global behaviour. Indeed, the "transatlantic divergence" that developed between Washington and several European capitals (in particular Paris and Berlin) as a result of the US-led Iraq war and more generally US unilateral attitudes in world affairs during the first years of the G. W. Bush Administration ultimately acted as a powerful catalyst for pushing Europe and China to work closer together. While not all European countries shared French and German anxieties over the "imperial attitudes" of the hyper power, the EU and China would see in each other a potential supporter.

As the 2003 EU Policy Paper on China stated, "The EU, as a global power on the international stage, *shares China's concerns* for a more balanced international order based on effective multilateralism".[18] Along the same lines, China's 2003 EU

---

[15] "An overview of EU-China sectoral dialogue". European Union External Action Service (2014). Web. http://www.eeas.europa.eu/china/sectoraldialogue_en.htm#Space_cooperation. Accessed 18 April 2014.

[16] See, for instance, Shambaugh, David (2004). "China and Europe: The Emerging Axis". Current History no. 670: 243–248. Grant, Charles, and Katinka Barysch (2009). "Can Europe and China shape a new world order?" Centre for European Reform, London. Web. http://www.cer.org.uk/sites/default/files/publications/attachments/pdf/2011/p_837-611.pdf.

[17] Prodi, Romano. "Relations between the EU and China: more than just business". EU-China Business Forum. 6 May 2004. Web. http://europa.eu/rapid/press-release_SPEECH-04-227_en.htm. Accessed 18 April 2014.

[18] *Cit.* European Commission. A maturing partnership - shared interests and challenges in EU-China relations. Commission Policy Paper COM(2003) 533 final. Brussels, European Union. 10 September 2003. (Emphasis added)

Policy Paper subtly alluded to the importance of the EU–China relationship in promoting global multilateralism and the democratisation of international relations (code words for opposition to the US hegemonic approach in world politics).[19]

Each side shared dissatisfaction with Washington's behaviour and found common cause in strengthening the role of international institutions (in particular the authority of the UN) and promoting the importance of multilateral (as opposed to unilateral) solutions to address the global and transnational challenges of the twenty-first century. A crucial role in the establishment of the strategic partnership was played by balance of power considerations, i.e. the desire for Europe and China to team up and reinforce their weight in the management of global strategic issues in order to counterbalance or potentially constrain and delegitimise US action on the international stage.[20]

The idea that the EU and China were jointly adopting a strategy of "soft balancing", as analysts would label it, was strongly reinforced by two additional moves that accompanied the establishment of the strategic partnership: the decision to allow China to participate in the development of Galileo—the European GNSS alternative to the American GPS (see next section)—and the promise to start discussions on lifting the arms embargo on China imposed by the EU.

The two initiatives were not, in fact, merely driven by industrial and economic interests but had a highly symbolic and political dimension as well. On the one hand, they could be seen as a clear political act of EU recognition of a new international status for China, a recognition that China could be treated as a "responsible great power". On the other, they were an attempt to "open up new avenues in world politics outside the hegemonic interests of the United States". [21] The message to the USA was that the EU and China were ready and capable to join forces to promote their technological and political interests in the international arena. The intention, at least in Brussels, was not to challenge Washington but to promote greater European autonomy (the logic of the EU's search for identity through external policy) as well as to increase Europe's global competitiveness in key high-tech and defence-related sectors.

The EU embargo on arms sales to China was one of the key issues affecting relations and limiting the possibility of deepening S&T cooperation, including cooperation on Galileo. Imposed soon after the Tiananmen events in June 1989, the embargo was seen by Beijing and many European capitals as increasingly counterproductive in the light of the desired new strategic partnership. To be sure, its removal would be more symbolic than real, since national control regimes would still apply to direct arms sales to China. Lifting the embargo was nonetheless

[19] See Ministry of Foreign Affairs of the People's Republic of China. China's EU Policy Paper. Beijing, China. 13 October 2003.

[20] Gill, Bates, and Melissa Murphy (2008). "China-Europe Relations. Implications and Policy Responses for the United States". Center for Strategic and International Studies, Washington DC: p 2.

[21] *Cit.* Casarini, Nicola (2009). *Remaking Global Order. The Evolution of Europe-China Relations and its Implications for East Asia and the United States.* Oxford University Press, New York: p. 70.

perceived to be a necessary "green light" from the EU that could open up to European industries the very promising Chinese defence market and procurement budget, thereby greatly deepening the scope of Sino–European cooperation.[22]

The initiative was doomed to fail, however, and by summer 2005 the proposal had been shelved. Because of the different positions taken by member states,[23] and the strong pressure—and even threat of retaliation—from Washington,[24] it was impossible for the EU to achieve any concrete results.

The failure to push forward the lifting of the embargo deserves attention as it underscored important lessons on the boundaries and the real nature of the EU–China axis that emerged in 2003. Despite the rhetoric of official documents, the boundaries largely endure today.

The first lesson is that "transatlantic ties remain deep and strong even during periods of ardent disagreement".[25] Although it is true that the EU–China relationship has become broader and deeper since the launch of the partnership, it has not risen to the same level as the transatlantic relationship. The EU's reconsideration of the proposal to lift the embargo in the face of US pressure demonstrated that Europe is not willing to let its relationship with Washington deteriorate in order to deepen that with Beijing and shows that Washington still has a strong influence (if not a "veto power") on many European foreign and security policies. The possibility of implementing cooperation in the military domain thus remains unlikely, although the Tiananmen logic no longer carries the argument.

An additional point is that the embargo debacle made very visible the limited mandate and authority held by the EU in defence-related matters, as well as the lack of a "one voice system" in Europe. More broadly, this fragmentation calls into question the credibility of the EU as a foreign and security policy actor and thereby its ability to manage a strategic partnership with China, a partnership which, despite good intentions, is still far from strategic!

---

[22] *Ibid.* p.89.

[23] Of the 25 EU member states, only 16 were in favour of lifting the embargo. At the forefront, there were Germany, France, Italy, and Spain, while Denmark, Sweden, Ireland, Portugal, and Poland tenaciously opposed such a proposal.

[24] In response to the EU proposal, the US House of Representative and the Senate passed two resolutions condemning Europe and claiming that a lifting of the embargo would "potentially adversely affect transatlantic defence cooperation, including future transfers of United States military technology, services, and equipment to European Union countries". See US Senate Committee on Foreign Affairs. The Lifting of the Arms Embargo on China. 31 March 2005.

[25] Medeiros, Evan S (2009). *China's International Behaviour. Activism, Opportunism and Diversification.* RAND Corporation, Santa Monica, CA: pp. 114–125.

7.1.2 A Maturing Relationship?

Europe's difficulties in forging an autonomous political profile (as further demonstrated in 2005 by the failure to secure the passage of the EU Constitution) and the US ability to sway EU policy eventually prompted China to rethink EU's reliability as a strategic partner. In Beijing's eyes, the more the USA affects EU decision-making and the more larger member states persist in developing their own strategy, the less interested the EU becomes as a partner in the quest for a multipolar world.[26] As a result of these lessons, Beijing eventually found it more convenient to alter its strategy towards Europe: after 2005 it returned to focusing more on managing bilateral relations with individual EU countries and adopted a strategy of "divide and rule", exploiting differences among individual countries to push forward its diplomatic agenda and interests.[27]

In the same vein, Europe was also gradually forced to reconsider its ties with Beijing and to pragmatically reorient its foreign policy more in accordance with Washington's preferences. Different motives, however, account for this change. First, the European policy readjustment was prompted by evolution in the European governmental landscape: in particular by the emergence of a new political leadership in "the big three" EU Member States (Angela Merkel in Germany, Nicolas Sarkozy in France, and Gordon Brown in the UK). These three leaders had "more sceptical views towards China than their predecessors".[28] A further change was the formation of a new (and pro-American) European Commission, headed by Manuel Barroso; and the 2007 accession to the EU of the more Atlanticist Eastern European countries.[29]

Second, the reconsideration of EU foreign policy behaviour vis-à-vis China was accelerated by the emergence of several economic and diplomatic irritants. Particularly cumbersome were those disputes related to China's undervalued currency, the increasing trade deficit, low Chinese standards of food and product safety, and issues of IPR infringements and market access obstacles (e.g. the so-called bra war, among others). These economic frictions ultimately combined with a number of political issues, such as the Tibetan and ethnic minorities questions, the persistent violations of human rights, and China's close relationship with the so-called rogue states such as Iran, North Korea, Libya, and Sudan.

[26] Gill, Bates, and Melissa Murphy (2008). "China-Europe Relations. Implications and Policy Responses for the United States". Center for Strategic and International Studies, Washington DC: p 2.

[27] Medeiros, Evan S (2009). *China's International Behaviour. Activism, Opportunism and Diversification*. RAND Corporation, Santa Monica, CA: p. 124.

[28] *Cit.* Casarini, Nicola (2009). *Remaking Global Order. The Evolution of Europe-China Relations and its Implications for East Asia and the United States*. Oxford University Press, New York: p. 177.

[29] *Ibid.* p. 177.

The combination of all these dynamic factors gradually led Europeans to change their mood towards China, and as a consequence, the EU foreign policy posture was also affected. This was already becoming visible in the *Guidelines on the EU's Foreign and Security Policy in East Asia* adopted by the European Council in December 2007.[30] The document emphasises: "the US commitment to Japan, the Republic of Korea and Taiwan and the associated presence of the US forces in the region give the US a distinctive perspective on the region's security challenges. It is important that the EU is sensitive to this. Given the great importance of transatlantic relations, the EU has a strong interest in partnership and cooperation with the US on the foreign and security policy challenges arising from East Asia". In short, as Nicola Casarini has argued, the document eliminated those elements of the EU–China relationship (such as the transfer of strategic and military items to the region) perceived to be detrimental to the role and responsibility of the USA in East Asia, thus recognising the need for Europe to operate in line with Washington.[31] The return to multilateral diplomacy signalled by the Obama Administration subsequently contributed to fortifying the previously weakened transatlantic ties.

But on the Chinese side too, the decision to cancel the 11th EU–China summit at the end of 2008 in response to the meeting between the then-EU President Nicholas Sarkozy and the Dalai Lama seemed to show that the "very serious engagement" was over and no marriage was forthcoming. In reality, however, instead of having been broken, the relationship seems to have evolved into something more mature and complex, characterised by a mixture of cooperation and controversy.

For one thing, the quality and quantity of interactions and cooperation have continued to grow, as demonstrated by the increasing number of sectorial dialogues established (from 25 in 2008 to more than 50 by 2013) and by the creation in 2010 of a new high-level mechanism of consultation (the High-Level Strategic Dialogue chaired by the EU High Representative for Foreign and Security Policy and China's State Councillor), which complements the High-Level Economic and Trade Dialogue. Overall, the amount of political and diplomatic resources that the EU has continued to invest into expanding relations with Beijing (and vice versa) is impressive and is confirmed by the increasing number of agreements and joint projects launched in the past few years.

While differences and disputes still persist, these are seen by both actors more as occasional irritants rather than as structural barriers that thwart the achievement of a truly strategic partnership. Of course, utilising the terms "strategic" or "axis" remains potentially misleading insofar as it implies the formation of a Sino–European power bloc directed against the USA, which is clearly not the

[30] Council of the European Union. Guidelines on the EU's Foreign and Security Policy in East Asia. 16468/07. Brussels, European Union. 20 December 2007. A revised edition was published in July 2012, which is quite on the contrary characterised by a return to the policy of constructive engagement vis-à-vis China.

[31] Casarini, Nicola (2009). *Remaking Global Order. The Evolution of Europe-China Relations and its Implications for East Asia and the United States*. Oxford University Press, New York: pp. 163–164.

case, especially from the European perspective. However, an ever-increasing convergence and cooperation on issues having global (and thus strategic) implications (e.g. non-proliferation, climate change, Africa, economic governance) are considered crucial by the two actors.

Remarkably, at the 15th Summit held in Brussels in September 2012, both the president of the European Commission, Manuel Barroso, and Chinese Premier Wen Jiaobao emphasised the desire to elevate the partnership to a new and deeper level by strengthening cooperation on these major issues.[32] Such upgrading would be achieved by seeking and finding common ground on major issues, while shelving the non-essential problems that may understandably emerge in an all-dimensional relationship.

Besides the rhetoric and the diplomatic niceties contained in the official statements, the declared will to revive the axis of cooperation between Europe and China responds to a crucial set of considerations by the two actors. On the Chinese side, there is an awareness that China's development is intrinsically related to the affluence of Europe, as the recent "debt crisis" clearly demonstrated.[33] There is in addition recognition that the ongoing dynamic on the East Asian chessboard (in particular the increasing tensions with Japan and the launch of the "Pivot to Asia strategy" by the USA) requires support from third parties. Russia is an option, but Europe's support is seen as even more desirable, as it also offers the opportunity to balance the relationship with its Russian neighbour. Ultimately, closer cooperation with Europe remains a necessity for China to balance its relations with all the major powers. Although China regrets the fact that European leaders are unwilling to oppose the USA on key strategic issues (unlike Russia), in the long term it believes that the EU is likely to become a more confident and independent actor that will promote interests distinct from those of the USA. With the entry into force of the Lisbon Treaty and the creation of dedicated institutions for the management of the Common Foreign and Security Policy (CFSP), the European political profile has been gradually bolstered, and thus the appeal of the EU as a partner is also re-emerging. As recently as April 2014, Beijing released its second EU policy paper (with the expressive title "Deepen the China-EU Comprehensive Strategic Partnership for Mutual Benefit and Win-Win Cooperation") to define China's EU policy objectives and draw a blueprint for EU–China cooperation in the next 5–10 years. The document emphasises that the need "to grow China-EU relations is a priority in China's foreign policy" and thus expresses a commitment to cooperation with the EU to further increase the global impact of the relationship.[34]

---

[32] The joint press communiqué released at the end of the 15th summit was titled "Towards a stronger EU-China Strategic Partnership".

[33] See Westad, Arne (2012). "China and Europe: Opportunities of Dangers?". Global Policy. Vol. 3 (s1): 96–100.

[34] Ministry of Foreign Affairs of the People's Republic of China. China's Policy Paper on the EU: Deepen the China-EU Comprehensive Strategic Partnership for Mutual Benefit and Win-Win Cooperation. Beijing, China. 2 April 2014.

For Europe, the desire to strengthen a mutually beneficial relationship with Beijing reflects a different set of geopolitical interests and concerns, in particular the need to avoid the formation of an exclusive Sino–Russian power bloc. Such a bloc is potentially dangerous—and, clearly, not only for Europe.

But European geopolitical calculations include a consideration of a completely different nature as well. There is a concern that the much-discussed emergence of a Sino–American G2 (the *Chimerica* of Niall Fergusson) could geopolitically marginalise European relevance on the international scene, as the EU would inevitably be distanced from governance of global strategic issues. By simply aligning or *bandwagoning*, with the USA in its position vis-à-vis China, the EU would enable the "Western camp" to gain relative power but would also eventually put its fate in the hands of the USA, finding it harder and harder to gain political weight. Developing a strong relationship with Beijing and acquiring a pivotal role in international politics has become an ever-growing necessity for Europe.

Thus, a re-emerging convergence of interest towards closer cooperation between Europe and China can be seen. Such convergence has also recently been reflected in the *EU-China 2020 Strategic Agenda for Cooperation*, a comprehensive document jointly adopted at the 16th Summit held in Beijing in November 2013. The agenda sets out China's and the EU's shared aims of promoting cooperation over the coming years (up to 2020) in the four macro-areas of peace and security, economy, sustained development, and people-to-people exchanges. The objective is to take forward the China–EU Comprehensive Strategic Partnership through broad and concrete engagement in the implementation of policies and projects of common interest.[35] A major cooperative undertaking in space could, arguably, cement the EU–China strategic partnership.

7.2 Sino–European Space Cooperation: The Background

International cooperation has traditionally been one of the most striking features of Europe's space policy. Cooperation is structurally engrained in the inner workings of both the EU and ESA, being the result of multilateral construction. It is perhaps for this reason that European stakeholders today maintain cooperative relations with almost all other space actors worldwide. This is equally true for China, where Europe can boast a solid and long-standing record of cooperation dating back more than 30 years.[36]

The first ESA delegation visited China in February 1979, led by ESA Director General Roy Gibson, and a cooperation agreement was signed in 1980.

[35] European Union External Action Service. EU-China 2020 Strategic Agenda for cooperation. 131123/01. Brussels, European Union. November 2013.

[36] For an assessment of ESA's cooperation with China, see Bergquist, Karl (2014). "Cooperation with China in Space Science". ESA Bulletin No. 158: pp. 20–25.

The document promoted the exchange of scientific and technical information on space programmes and projects of mutual interest to ESA and the Commission for Science and Technology of China's State Council. Based on this initial collaborative formative tie, in 1992, a cooperation agreement between ESA and the CAS was signed to initiate collaboration on the ESA-led Cluster mission, a scientific programme of four satellites to study the effects of the Sun on the Earth's magnetosphere. Included in the agreement was the setting up of the Chinese Cluster data centre in Beijing, one of the 10 data centres that form part of the Cluster Science Data System.

The opportunity to participate in this international scientific venture was positively received in China and would in turn lead Chinese space officials to propose their own complementary project and to invite ESA as a partner. The invitation was forwarded in 1997 and the agreement on this joint project—generally known as the Double Star Programme—was signed in 2001. Under the agreement, China would provide two spacecraft, their launches and roughly half of the scientific payloads. ESA's contribution to the mission included eight scientific instruments, of which seven were spares from the Cluster mission, and support for the ground segment for four hours each day via its satellite tracking station in Spain.[37] The Double Star spacecraft—launched in December 2003 and July 2004, respectively—complemented ESA's Cluster quartet by forming a constellation of six satellites in different orbits.[38]

Since the early 1990s, ESA has also been cooperating with the National Space Science Centre (NSSC) of the Ministry of Science and Technology on the development of Earth Observation applications. In 1994, thanks to ESA assistance, the China Remote Sensing Ground Station in Beijing was upgraded to receive ERS data, and in 1997 a cooperative project for increased operational use of ERS data in China was launched. Five pilot projects addressing flood monitoring, land use, rice cultivation, and oceanography were created.[39] This fruitful cooperation was subsequently reinforced by the launch of a dedicated 3-year programme on the exploitation of earth observation data. Named Dragon, the programme focused on science and applications development in Chinese–European cooperation through the utilisation of ERS and Envisat mission data.[40] After its completion in April 2008, cooperation was renewed with the launch of extensive Dragon-2 (2008–2012)

[37] "Double Star Overview". European Space Agency. 6 June 2013. Web. http://www.esa.int/Our_Activities/Space_Science/Double_Star_overview2. Accessed 10 April 2014.

[38] "The First Sino-European satellite completes its mission". European Space Agency. 17 October 2007. Web. http://sci.esa.int/double-star/41400-the-first-sino-european-satellite-completes-its-mission/. Accessed 10 April 2014.

[39] Desnos, Yves-Louis and Zengyuan, Li (2006). "EO Science and Applications development in China". In: *Dragon Programme Mid-term results*. Proceedings of the 2005 Dragon Symposium, Santorini, Greece, June 27- July 1, 2005.

[40] "ESA-MOST Dragon Cooperation Programme". European Space Agency. 21 August 2011. Web. http://earth.esa.int/dragon/. Accessed 11 April 2014.

and Dragon-3 (2012–2016) programmes.[41] In November 2005, a cooperation agreement was also signed by ESA with the formal international interface of China's space programme, the CNSA.

Over the years, Sino–European space cooperation has been further strengthened by government-to-government agreements between individual European states and China, as well as by the signature of a number of industrial contracts. France was the first country to sign an international space agreement with China, during the Sino–Soviet split.[42] Apart from France, Germany, Italy, and the UK have been the most active vis-à-vis China, with bilateral projects covering space science, satellite applications, TT&C, as well as the delivery of industrial products and services. European companies such as the British Surrey Satellite Technology Ltd., the French Alcatel (now part of the Thales Alenia Group), and EADS Astrium (now Airbus) have also been important suppliers and partners for the Chinese space programme. Alcatel, for instance, has cooperated since 1984, when it delivered subassemblies for the Chinasat-1 satellite. This was subsequently reinforced through several contracts for the design and manufacture of advanced communication satellites.[43]

By the early 2000s, with the gradual involvement of the EU in space and the launch of its so-called flagship programmes, cooperation with China had acquired a political profile as well. For the EU, in fact, developing cooperative relations with third countries such as China was specifically intended to serve the furtherance of EU policies across a broad spectrum. As the 2003 White Paper on space activities underlined: "international cooperation is not simply a matter of scientific collaboration on technologies and applications, but should also be in function of serving the widest possible spectrum of the EU's objectives".[44]

It is clear, then, that, in order to support and strengthen its overall policy of constructive engagement towards China, the EU has been willing to engage the country in cooperative space projects, an objective that at least in the early 2000s was visibly stronger than the fears of reinforcing a competitor and facilitating sensitive technology transfers.

7.2.1 *The Galileo Experiment*

Cooperation between China and the EU gained momentum in 2003: on the same day the strategic partnership was declared (30 October); the two parties signed a

[41] "Dragon-3 Objectives". European Space Agency. Web. https://dragon3.esa.int/web/dragon-3/objectives. Accessed 11 April 2014.

[42] Harvey, Brian (2013). *China in Space. The Great Leap Forward.* Springer, New York: p. 100.

[43] Rathgeber, Wolfgang (2007). "China Posture in Space. Implications for Europe". ESPI Report 3. European Space Policy Institute, Vienna.

[44] European Commission. White Paper Space: A New European Frontier for an Expanding Union. An Action Plan for Implementing the European Space Policy. COM(2003) 673. Brussels, European Union. 11 November 2003.

cooperation agreement on the Galileo programme, the EU-led Global Navigation Satellite System (GNSS). According to the official wording, the Galileo Satellite Navigation Cooperation Agreement provided "for co-operative activities on satellite navigation in a wide range of sectors, particularly science and technology, industrial manufacturing, service and market development, as well as standardisation, frequency and certification".[45]

In conjunction with the agreement, the China–Europe Global Navigation Satellite System Technical Training and Cooperation Centre (CENC) was inaugurated in Beijing to serve as a focal point for Galileo activities,[46] while the designated industrial partner for Galileo was the NRSCC (which in October 2004 became a member of the Galileo Joint Undertaking (GJU), the temporary body established to manage the programme).[47] Around 12 projects relating to the manufacture of Galileo infrastructure elements and the development of navigation applications were identified by the GJU and contracted to China.

The number of projects, combined with the total investment pledged by Chinese stakeholders (200 million €, of which 5 million € was the entrance fee), made Beijing the most important non-EU partner in Galileo. Such a large involvement was clearly not accidental: Chinese participation was intended as something more than mere technical and industrial cooperation, as explicitly recognised in 2003 by François Lamoreux, then head of the EU Directorate General of Energy and Transport—"Never before ha[ve] the European Union and China embarked on a cooperation project of the same magnitude as in Galileo. The project goes well beyond industrial or standardization issues. It entails a strong strategic component, which will have far-reaching consequences on future Sino-European political relations".[48]

Cooperation with China on Galileo was in fact intended by the EU as a means of significantly deepening the scope and impact of the strategic partnership, as well as a counterweight to US primacy, by freeing Europe from over-dependence on US technology and promoting greater technological and political autonomy. Such

---

[45] Quoted from: "The EU and China are set to collaborate on Galileo, the European Global System of Navigation Satellite". European Commission Press Releases. 18 September 2003. Web. http://europa.eu/rapid/press-release_IP-03-1266_en.htm. Accessed 20 April 2014.

[46] *Cit.* Peter, Nicolas (2007). "The EU's Emergent Space Diplomacy". Space Policy Vol. 23 (2): 104–105.

[47] More specifically, the GJU was assigned with two main tasks: (a) implementation of the development phase, to this end, the GJU, by agreement, entrusted to ESA the carrying out of the activities required during the development phase in the space and associated earth segment, and (b) preparation of the subsequent phases of the programme. The GJU ceased to exist on 31 December 2006. Its activities were consequently transferred to the GNSS Supervisory Authority (GSA). See "Galileo Joint Undertaking". European Union—Summary of EU Legislation. Web. http://europa.eu/legislation_summaries/other/l24098_en.htm. Accessed 28 April 2014.

[48] Speech by François Lamoreux at the Opening of EU-China negotiations on satellite navigation. Brussels, 16 May 2003. Quoted from: Casarini, Nicola (2009). *Remaking Global Order. The Evolution of Europe-China Relations and its Implications for East Asia and the United States.* Oxford University Press, New York: p. 102.

cooperation could be clearly interpreted as a "geo-technological balancer", in the words of Joan Johnson-Freese and Andrew Erickson, aimed at closing the technological gap between second-tier great powers and the USA and ultimately countering US hegemonic interests in the aerospace sector.[49]

In the following 2 years, the EU–China "techno-political linkage" was reinforced by the creation of a High-Level Steering Group on China–EU Space Cooperation,[50] and in September 2005 the Space Dialogue Mechanism was endorsed as part of the so-called EU–China sectorial dialogues. Since then a series of high-level meetings discussing Sino–European space cooperation has taken place. The impression, at least in Washington, was that Europe was about to outpace Russia as China's main provider of advanced aerospace technologies as well as its principal international partner in space.

The ominous prospect of the dissemination of sensitive European technologies that would inevitably benefit the PLA and aid China's military modernisation was of particular concern to US policy makers, who began to put increasing pressure on European countries to terminate Chinese participation in the Galileo programme. An agreement on the use of Galileo by third countries and on interoperability between the US GPS and Galileo was reached at the EU–US summit held in Dublin on 28 June 2004.[51] Europe guaranteed via the agreement that China would not have access to the encrypted signals.[52] For the Chinese this represented tangible evidence of the impossibility of becoming a true and equal partner in the Galileo programme, a recognition that would eventually contribute to them transforming their regional navigation system (BeiDou) into a global one. As US pressure on Europe continued and the question of the compatibility between the nascent BeiDou/Compass system and Galileo remained unresolved, European policy makers were gradually reconsidering and downsizing China's participation in the programme. Hence, in the summer of 2008, Europe opted to put an end to satellite navigation cooperation with China. In the public procurement tender information package for the second phase of the Galileo system issued by ESA in July 2008, China was thus prevented from participating.[53] Although EU official documents

---

[49] Johnson-Freese, Joan, and Andrew S. Erickson (2006). "The Emerging China-EU Space Partnership: A Geo-technological Balancer" Space Policy Vol. 22 (1): 12–22.

[50] The creation of this group was intended to support the development of long-term perspectives for cooperation in space. The group includes representatives of the government administrations, agencies, and manufacturers, as well as operators and service providers.

[51] Agreement on the Promotion, Provision and Use of Galileo and GPS- Satellite-Based Navigation Systems and Related Applications. Dublin, Ireland. 28 June 2004.

[52] EC officials would reiterate that a "security firewall" would have to be put in place so to ensure that non-European countries (included China) could not have access to sensitive or secret information.

[53] As explained by Nicola Casarini, the ESA document stated that the tender was limited to member states of the EU or the states that are signatories of the multilateral Agreement on Government Procurement (GPA) adopted in the framework of the WTO. Casarini, Nicola (2009). *Remaking Global Order. The Evolution of Europe-China Relations and its Implications for East Asia and the United States*. Oxford University Press, New York: p. 179.

continued to emphasise that cooperation on the joint development of Galileo remained open, in fact the EU seemed less and less willing to promote cooperation as such and more inclined to address interoperability and compatibility issues between the European and Chinese systems.

As argued by several scholars and analysts, several motives for this decision can be discerned: questions of technology transfer and IPR enforcement, the perceived challenge to Galileo coming from *BeiDou* and in particular the emergent hurdle of the use of the same frequencies, strong US pressure, and China's 2007 ASAT test.[54] These concerns can in general be attributed to the inability of the EU to reconcile its desire to cooperate with China with the strategic concerns expressed by its American ally.

7.2.2 Recent Developments

In spite of this serious setback and the gradual realignment of Europe's space policy to that of the USA, the engagement approach to China pursued by the different European stakeholders did not come to an end, although it lost much of its previous political dimension.

At ESA level, cooperation has continued undisturbed and, not by chance, has been endorsed in a number of other important areas, such as space science and space exploration (both robotic and human). With regard to China's lunar exploration programme, for instance, ESA has provided ground support services through its ESTRACK network for the three Chang'e probes launched by China in 2007, 2009, and 2013. Indeed, ESA's role in these missions has been essential, considering that on several occasions ESOC was the entity transmitting commands to the probes. TT&C assistance has also been provided for tracking China's space laboratory Tiangong-1, with support from the stations in Spain, French Guiana, and Kenya. A mutual cross support agreement with the China Launch and Tracking Centre (CLTC) was negotiated by ESA in 2010.

At the bilateral level, ESA Member States have continued to promote closer ties with China's space programme. The German Aerospace Centre (DLR), for instance, signed an agreement with CAS for the provision of a package of scientific and medical instruments (SIMBOX) launched on the Shenzhou-8 mission in October 2011, while CNES has recently strengthened cooperation in satellite-based oceanography and astrophysics by establishing a joint venture with CNSA for the development of an EO satellite, the Chinese-French Oceanographic satellite (CFOSAT),[55] and for the construction of the Space Variable Object Monitor

---

[54] *Ibid.* p. 180.

[55] De Selding, Peter. "France, China Set Sail on Joint Ocean-surface Satellite Project". Space News. 28 March 2014. Web. http://www.spacenews.com/article/civil-space/40020france-china-set-sail-on-joint-ocean-surface-satellite-project?utm_content=bufferc6e27&utm_medium=social&utm_source=linkedin.com&utm_campaign=buffer. Accessed 29 April 2014.

(SVOM), an astronomy mission to study gamma-ray bursts.[56] Among other initiatives, a memorandum of understanding was also signed in September 2013 by ASI and the CNSA for the delivery of the scientific payloads of the China seismo-electromagnetic satellite (CSES), scheduled for launch in 2016.[57]

Finally, at the EU level, the Space Dialogue Mechanism established in 2005 has continued to result in annual meetings and work towards the establishment of a road map identifying cooperation projects and actions of mutual interest. With the adoption of the aforementioned EU–China 2020 strategic agenda of cooperation,[58] space relations can be seen to have finally recovered from the negative Galileo experience, and a series of major objectives and key actions to undertake has been listed. The three main domains are:

- Enhancement of the level of information exchange in the fields of Earth Observation, geoscience, space science, and exploration
- Reinforcement of cooperation in the fields of space science and space applications (including satellite navigation) through the establishment of a consultation mechanism identifying common objectives and joint actions
- Continuation of efforts to deepen exchanges and cooperation in manned spaceflight

As far as the latter point is concerned, between 2012 and 2013, a series of exchange visits and high-level meetings took place. In July and September 2013 (in conjunction with the 64th IAC), the ESA director general and the CMSA director met to discuss options for more intensified cooperation in the field of human spaceflight.[59] Probably as a result of these meetings, a series of minor, indeed symbolic, cooperative undertakings was subsequently launched. On 6 February 2014, during the annual press conference at the European Astronaut Centre (EAC), ESA Director of Human Spaceflight and Operations, Thomas Reiter, announced three specific initiatives. The first is the setting up of two working groups on "astronaut operations" involving cooperation with China.[60] Each of these groups will spend 1 week at the Astronaut Centre of China (ACC) and 1 week at the EAC. The second is an exchange on the topic of human behaviour

---

[56] Delrieu, Alain and Juliet Watelet. "CNES and the China National Space Administration strengthen cooperative ties". CNES Press Release. 27 March 2014.

[57] "CSES- China Seismo-Electromagnetic Satellite". Istitituto Nazionale di Fisica Nucleare. Web. http://spaceweather.roma2.infn.it/cses.html. Accessed 29 April 2014.

[58] European Union External Action Service. EU-China 2020 Strategic Agenda for cooperation. 131123/01. Brussels, European Union. 23 November 2013: p. 10.

[59] See "Wang Zhaoyao Met ESA Director in Beijing". China Manned Space Engineering. 8 July 2013. Web. http://en.cmse.gov.cn/show.php?contentid=1337 and at the IAC http://en.cmse.gov.cn/show.php?contentid=1353. Accessed 29 April 2014.

[60] As explained by Thomas Reiter at the annual press conference, "astronaut operations" include how to prepare mission operations when astronauts are on board the ISS and what needs to be done to keep them fit in space, to keep them exercising during their mission, and to provide them with medical advice and doctors in case of need.

and performance. Here, China has been invited to take part in the next "cave campaign" in Sardinia designed to study human behaviour and enhance performance training. Finally, a Chinese specialist from the ACC was invited to watch and assist ESA astronaut Alexander Gerst's rehabilitation process on his return from his mission on the ISS (November 2014).

These are all "soft" initiatives and forms of cooperation in comparison with the groundbreaking collaboration initiated on the Galileo programme. However, their importance should not be dismissed. For one thing, they create the required foundation for intensified cooperation in future manned spaceflight endeavours. After all, a major cooperative undertaking cannot emerge out of nothing. Especially where a country like China is concerned, to be effective, cooperation requires a proven track record and a certain degree of continuity, which is seen as an indispensable element for building up a "reservoir of trust" between partners.

In addition, these initiatives can be seen as a functional move intended to favour the inclusion of China in the international space club: in short the first step towards welcoming a Chinese astronaut on the ISS. It could even be argued that the real target of these initiatives is not China itself, but the European partners in the ISS programme and the USA in particular. In the light of the recent opening up on the American side, Europe might want to be a trailblazer in any potential effort by the international community directed at bringing China into broader international ventures. But Europe might also be aiming for some sort of involvement in the forthcoming CSS and through this might try to open up further avenues for cooperation on human space exploration.

This interpretation finds support in the cooperation agreement signed by ESA's Director General, Jean-Jacques Dordain, and the China Manned Space Agency's Director, Wang Zhaoyao, on 11 December 2014. Undoubtedly, this agreement marks an important milestone in the direction of both strengthening bilateral cooperation in the field of human spaceflight and actively promoting China's participation in the ISS (Fig. 7.1).[61]

CMSA reported that three possible cooperation areas were identified under the agreement:

- Implementation of joint scientific experiments and studies in different fields (including space life and physical sciences, microgravity research, space biology and medicine, and technology research) by utilising in-orbit infrastructures (namely, the ISS and CSS) and ground facilities
- Astronaut selection, training, medical operations, and astronaut flights
- Space infrastructure cooperation in human exploration in LEO and beyond[62]

[61] "China Manned Space Agency Signed Cooperation Agreement with European Space Agency". China Manned Space Engineering. 16 December 2014. Web. http://en.cmse.gov.cn/show.php?contentid=1471. Accessed 20 December 2014.

[62] *Ibid.*

Fig. 7.1 ESA–CMSA Signing Cooperation Agreement (*Source*: CMSA)

It was also anticipated that a "Human Spaceflight Consultation Committee co-chaired by joint chairmen will be established to assess the status of the cooperation, examine proposals for conducting specific cooperative activities and make recommendations for implementation".[63] In this light, it can be expected that Europe's objective of sending a European astronaut aboard the CSS may in good time be explicitly endorsed.

7.3 The Way to the Moon: Europe's Opportunities and Challenges

As shown in the previous sections, Europe's path towards closer cooperation with China in both the political and the space arenas seems quite clear. Europe is very conscious of the emergence of China as one of the leading space powers of the world and is thus preparing the ground for what it believes is a promising important partner in future space endeavours. Certainly the enhancement of space cooperation with China is not—and should not be—intended as bloc building with this emerging superpower but as a way to best tackle the challenges and grasp the opportunities that the evolution of the international landscape for space activities is yielding (see Sect. 6.2).

For Europe, there is the chance—not to say the necessity—to ready itself for a future context in which it is not critically dependent on a single partner's capabilities but where relations are ideally characterised by broader and diversified

[63] *Ibid.*

partnerships. In fact, Ukraine on the one side and the political volatility in the goals that the USA is pursuing for human spaceflight on the other have started to make Europeans hesitant about relying too heavily on their two historical partners in space. Specifically with regard to the USA, the frequently demonstrated uncertainties over the continuation of the ISS programme,[64] together with the abrupt cancellation of the Constellation programme and the withdrawal of NASA from the *ExoMars* missions, have brutally demonstrated the need for Europe to rethink future cooperation approaches, either by looking for alternative valid partners or by developing a more autonomous profile in areas such as space exploration and human spaceflight. In the light of its impressive ascent, much thought now seems to be going into whether China could also be an auspicious partner for Europe.

It is clear, however, that, before assessing its future cooperation potential with China, Europe should first define what interests and activities it wants to pursue and in which form. Specifically, Europe needs more broadly to define what role it wants to play in the future space exploration context. Without a clear decision on whether to become a major player or to join as a junior partner, Europe could find Chinese plans imposing trade-offs that it will not be able to tackle. The risk for Europe is to completely lose the possibility of shaping the priorities and timing of the future international space agenda, as well as the ability to attract the best partners in order to capitalise materially—and politically—on such cooperation. Needless to say, inaction would inevitably lead Europe to become a follower.

7.3.1 What Direction for Europe in Future Human Spaceflight? An Ongoing Debate

As emphasised in a previous ESPI study, "Europe cannot avoid the necessity to have a long-term view of its ambitions and actions in space exploration".[65] To date, however, a comprehensive and long-term vision for Europe's role in the post-ISS period is a long way off. To be sure, ambitious plans for a European space exploration programme started as early as 2001, with the presentation at the ESA Ministerial Council in Edinburgh of the Aurora programme. The programme's objective was to formulate and then implement a long-term European plan for the robotic and human exploration of the solar system. The programme was intended as the European building block in a broader international effort for the robotic and human exploration of Mars, with the Moon as an important stepping stone.[66]

[64] The nominal ISS programme ended in 2015 and was subsequently extended to 2020 and now to 2024.

[65] *Cit.* Peter, Nicolas (2008). "*Space Exploration 2025: Global Perspectives and Options for Europe*". ESPI Report 14. European Space Policy Institute, Vienna: p. 64.

[66] Messina, Piero, et al. (2006). "The Aurora Programme: Europe's Framework for Space Exploration". ESA Bulletin No. 126: pp. 10–15.

This long-term plan was derived from the human spaceflight experience on the ISS and the development of robotic planetary exploration. While the former was to be continued and enhanced so that human spaceflight could be extended beyond LEO, the latter was to be pursued through the Aurora programme with the aim of extending capabilities towards larger spacecraft suitable for human exploration of the solar system. According to ESA, the intertwined development of capabilities in the two strands would eventually have resulted in Europe being able to play a key role in a future international human mission to Mars.[67] The proposed road map was ambitious and contemplated a series of milestones to be achieved in cooperation with international partners:

- 2009: The aforementioned ExoMars, an exobiology mission to send a rover to Mars in order to search for traces of life and characterise the nature of the surface environment
- 2011/2014: Mars sample return, a split mission to bring back to Earth the first samples of Martian material
- 2014: Human mission technologies demonstrator to validate technologies for orbital assembly and docking, life support, and human habitation
- 2018: A technology precursor mission to demonstrate aerobraking/aero-capture, solar electric propulsion, and soft landing
- 2024: A human mission to the Moon to demonstrate key life support and habitation technologies, as well as aspects of crew performance and adaptation and in situ resources utilisation technologies
- 2026: An automatic mission to Mars to test the main phases of a human mission to Mars
- 2030/2033: A split mission that would culminate in the first human landing on Mars[68]

From this ambitious plan, only the *ExoMars* mission was formally approved at the ESA Ministerial Council of December 2005. This mission has subsequently become much delayed, evolved into two missions, and is currently being implemented in collaboration with Roscosmos and slated to launch in 2016 and 2018. As for the human side of the programme, this was not sufficiently backed with high-level political commitment (or financial support) and did not move beyond the study phase of the Crew Space Transportation System (CSTS).[69]

[67] "Aurora Roadmap to Mars". European Space Agency. 19 December 2003. Web. http://www.esa.int/Our_Activities/Human_Spaceflight/Exploration/Aurora_s_roadmap_to_Mars. Accessed 30 April 2014.

[68] *Cit. Ibid.*

[69] The CSTS was specifically intended for use in human exploration missions to the Moon (both in orbit and on the surface) via LEO assembly, in addition to supporting missions to the ISS. Aware that not being involved in the next generation transportation systems would have meant to remain forever a second-class partner, ESA eventually succeeded in participating to the development of the NASA-led Orion Crew Vehicle, after the cancellation of the CSTS activity.

A new political impetus towards the elaboration of a long-term vision in space exploration was provided, thanks to the increased involvement of the EU, specifically since the publication of the first European Space Policy (ESP), adopted in 2007 by the 29 member states of the EU and/or ESA. Among its various objectives, the ESP clearly highlighted the need for a growing role for Europe in space exploration in order to ensure a higher degree of political visibility for space in Europe and for Europe on the global arena. Although the Union had by that time identified satellite systems and applications—specifically Galileo and Copernicus—as its priority policy areas, the expansion of its mandate over space matters eventually influenced the EC to consider a possible contribution to space exploration and to provide political backing for its development. In May 2009 the Space Advisory Group (SAG) of the EC formed a subcommittee on space exploration (SAG-SP) with the aim of providing expert advice to the commission on Europe's future role in a global space exploration strategy. The group recommended that the EU should take a central role to ensure the success of future European space exploration.[70]

An important milestone in this regard was achieved with the first EU–ESA Conference on Human Space Exploration, held in Prague in October 2009. On that occasion, Ministers expressed their support for a major financial investment in space exploration and agreed on the need for active EU involvement in this domain in order to ensure an appropriate political profile and financial framework. Although initiatives have so far failed to materialise—principally due to the strong budgetary pressures prompted by the financial crisis—there is an ongoing debate on the possibility of devoting the next EU–ESA flagship programme to space exploration, possibly human space exploration. Needless to say, the high financial investment required for a more active human spaceflight programme makes the EU contribution and political support indispensable.

In the meantime, a *Baseline European Roadmap* was proposed in two studies awarded by ESA in 2009 to industrial teams led by Thales Alenia Space, Italy, and Astrium, Germany.[71] The road map offered an outlook on the possible European contribution to future exploration scenarios and architectures and was intended as a starting point for promoting both political commitment and international discussion, in line with the orientations that had emerged at the second ESA–EU Conference on space exploration.[72] This second conference was held in Brussels in October 2010. Among its major conclusions, it identified the need for policy discussion at international level and thus called for the organisation of an initial

[70] Space Advisory Group of the European Commission. Space Exploration, a new European flagship Programme. European Commission Framework Programme 7—Space Theme. Brussels, European Union. 10 October 2010. See also Horneck, Gerda et al. (2010). "Towards a European vision for space exploration: Recommendations of the Space Advisory Group of the European Commission". Space Policy Vol. 26 (2): 109–112.

[71] Perino, Maria Antonietta (2013). "Outlook of possible European contributions to future exploration scenarios and architectures". Acta Astronautica Vol. 88 (1): 25–34.

[72] *Ibid.* p.25.

meeting of an international, high-level space exploration platform, in which future directions and cooperation schemes could be discussed.[73] Accordingly, the third Space Exploration Conference transformed the proceedings into a high-level global discussion involving 28 countries. To highlight this evolution, the conference was renamed the First High-Level International Exploration Platform. The first meeting took place in Lucca in November 2011, while the second was held in Washington in January 2014.

Building on the recommendations of the SAG-SP and the conclusions of the different conferences, the EC issued a working document entitled, "A Role for Europe within a Global Space Exploration Endeavour". The document, released on 20 August 2013, emphasises the importance of an integrated approach (at both European and international level) in the field of space exploration and proposes building the current European long-term scenario, consistent with international plans, in a three-step sequence:

- First step, 2015–2020: Utilisation of the ISS, robotic missions (including *ExoMars*), R&D for preparing the next step, and demonstration of human transportation capabilities
- Second step, 2020–2030: Continued robotic missions including Mars Sample Return, human missions beyond low Earth orbit, R&D for preparing the next step
- Third step, >2030: Sophisticated robotic missions in the Solar System, continued human exploration missions, possibly including human missions to Mars[74]

It appears clear therefore that for Europe any future space exploration scenario must be achieved through international cooperation, and thus a degree of international consensus regarding the priorities and direction of human spaceflight and exploration is a *conditio sine qua non* for moving forward. But such consensus remains difficult to establish, particularly because of the unilateralist direction being taken by NASA. Indeed, the current US policy of pursuing an Asteroid Retrieval Mission does not represent a very attractive scenario for ESA.[75] Although they might be interested in a robotic mission to an asteroid,[76] the majority of

[73] Ehrenfreund, Pascale, et al. (2012). "Toward a global space exploration program: A stepping stone approach". Advances in Space Research No. 49: 2–48.

[74] European Commission. A Role for Europe within a Global Space Exploration Endeavour. Commission Staff Working Document. SWD (2013) 301 final. Brussels, European Union. 20 August 2013.

[75] The ARM aims to capture a very small near-Earth asteroid of less than 10 m in diameter onto a Moon-like orbit to subsequently enable in situ human exploration. It is expected that such a mission could ideally take place in the early 2020s.

[76] Several projects are currently under study, the most important of which being the Asteroid Impact and Deflection Assessment (AIDA) mission, an ESA joint effort with NASA aiming at studying the effects of crashing a spacecraft into an asteroid and testing the ability to deflect an asteroid on collision course with the Earth. More information on ESA website: http://www.esa.int/Our_Activities/Technology/NEO/Asteroid_Impact_Deflection_Assessment_AIDA_study.

European stakeholders seem to lean towards manned lunar exploration, which is still considered the best stepping stone for enabling a future human mission to Mars and the best way to capitalise on European expertise.[77]

In addition to the mismatch between the respective goals in space exploration, the extent of Europe's possible contribution to US plans remains largely undetermined. To date, it is limited to the development of the service module for the Orion Multi-Purpose Crew Vehicle (MPCV) slated to launch on the maiden flight of the Space Launch System (SLS) in 2017/2018.[78] However, it should be noted that American acceptance of this European contribution is paradigm breaking in the sense that this contribution puts Europe on the critical path of all future exploration plans in the USA.

Nevertheless, even with Europe's involvement in the US Space Launch System, other options in the crafting of partnership configurations for future human spaceflight endeavours must be considered by the EU and ESA, particularly those that will not force them to choose sides. Given that ESA's interest in a lunar exploration might to some extent converge with Chinese plans, the temptation to engage in a cooperative venture with China is becoming stronger. Further, such convergence rests on a "psychological factor". Europe perceives cooperation with the USA in human spaceflight as "inherently difficult, because it can never result in a true partnership—such a partnership is economically impossible for the EU and politically unacceptable for Washington".[79] Therefore, it could be argued that where the USA, and to some extent Russia are concerned, Europe appears to suffer from the same "complex" as China in the past: the difficulty of coming to the table as an equal (*yi xi zhi di*). For two second-tier spacefaring nations like Europe and China, joining together in a groundbreaking cooperative venture could therefore be a promising strategy for developing their space programmes and positioning themselves as part of the top of the space pecking order.

7.3.2 Europe's Strengths, Weaknesses, Opportunities, and Threats

When assessing the potential for European cooperation with China on future space exploration, specifically on manned lunar exploration, the strengths and weaknesses on which such cooperation would rest and the opportunities and challenges it could

[77] See Horneck, Gerda et al. (2010). "Towards a European vision for space exploration: Recommendations of the Space Advisory Group of the European Commission". Space Policy Vol. 26 (2): 109–112.

[78] Perino, Maria Antonietta (2013). "Outlook of possible European contributions to future exploration scenarios and architectures". Acta Astronautica Vol. 88 (1): 25–34.

[79] *Cit.* Johnson-Freese, Joan, and Andrew S. Erickson (2006). "The Emerging China-EU Space Partnership: A Geo-technological Balancer" Space Policy Vol. 22 (1): 12–22.

potentially generate must be assessed. A SWOT analysis is hence an appropriate instrument in this regard and will be used as a basis for discussing a set of policy options for Europe.

As in the case of the SWOT analysis on China, the level of analysis is not limited to the space dimension but is also placed within the broader political context, which will ultimately be the chief element in determining the European posture vis-à-vis China. It is clear that strengths, weaknesses, opportunities, and challenges will be shaped by the broader political dynamics that surround the potential implementation of a cooperative venture in a highly political domain like human spaceflight.

7.3.2.1 Strengths

Among its major strengths is the fact that Europe is an established space actor in space exploration, boasting long-term experience and invaluable contributions to robotics missions and human spaceflight. Over the years, it has reached a state-of-the-art technological level and acquired a solid set of critical capabilities, often making it a so-called partner of choice for international cooperation. Furthermore, Europe has demonstrated the ability to provide essential elements to large robotic missions (e.g. Cassini-Huygens) and to the ISS infrastructure (e.g. the Columbus Orbital laboratory, the Automated Transfer Vehicles (ATVs) and other infrastructure elements such as the Multi-Purpose Logistics Module (MPLM), Node 2, and Node 3.[80] All these activities have built an excellent "heritage" for a future human mission beyond LEO.[81] ATV-derived technologies are, for instance, currently utilised for the development of the MPCV Service Module, and other derivatives are being studied as potential key elements in an international lunar exploration architecture (e.g. the Cis-Lunar Logistic Vehicle and the ESA Lunar Lander).[82] In short, Europe could very well leverage its experience and capabilities to build a partnership with China and would be in a position to provide crucial contributions to a joint manned lunar landing programme.

Europe's attractiveness also stems from the fact that, once approved, its space programmes and funding have demonstrated a very high degree of stability, which is ultimately an important guarantee for international partners. As Jean-Baptiste Thépaut has argued, European reliability as a partner can be explained by the different nature of the targets it has pursued in its space programmes: "unlike other leading space nations, which have made space exploration a national priority

[80] The Node 3, also known as Tranquillity, although not a European module, was built for NASA by ASI.

[81] Horneck, Gerda et al. (2010). "Towards a European vision for space exploration: Recommendations of the Space Advisory Group of the European Commission". Space Policy Vol. 26 (2): 109–112.

[82] For an analysis of the potential European contributions to future exploration, see Perino, Maria Antonietta (2013). "Outlook of possible European contributions to future exploration scenarios and architectures". Acta Astronautica Vol. 88 (1): 25–34.

to suit political and strategic agendas, the European space programme has traditionally been science based and technologically focused. This pragmatic approach has in turn protected public funds and space projects from the political volatility experienced in other countries".[83] In addition, the increasing involvement of the EU and the ongoing enlargement of ESA, possibly by up to 10 further member states in the medium term, are bound to enlarge the scope of the European space programme.

Another major asset for Europe is the wide and robust network of cooperative relations ESA has built up with all the other space actors worldwide, while the EU has increasingly promoted S&T cooperation as a tool of foreign policy to both reinforce international relations and make Europe an important centre of gravity in global S&T affairs. This puts the continent in a strong position to promote and harmonise a broad international programme by acting as a bridge-builder and matchmaker between today's isolated parties.

Specifically with regard to China, Europe has a long-standing record of cooperation, built at the various levels of European governance (national, ESA, and EU levels), which has in many cases proved to be of valuable importance for the development of the Chinese space programme. Although the level of Sino–European space cooperation is not comparable to the technological and operational assistance provided by Russia, Europe has often been the leading edge for Beijing. In cooperation on the Double Star programme or the Dragon programme, for instance, Chinese stakeholders were not designated by ESA as "recipients" but came to the table as equals, thus pooling resources and scientific and technological expertise. This European posture, coupled with the tangible and valuable benefits the Chinese have gained from their cooperation with Europe (e.g. the crucial TT&C support for its exploration missions), has conferred on ESA a "trust capital" and laid a solid basis for embarking on a major cooperative undertaking. In short, compared to other spacefaring nations, the European position appears robust.

The space-related assets that Europe could potentially bring into future cooperation with China are complemented by equally important strengths in the broader bilateral interaction. Unlike the USA, Japan, India, or even Russia, Europe and China have built their relationship free of disturbing elements that could potentially lead to a clash of strategic interests. In spite of the emergence of some serious economic and diplomatic irritants in recent years, the overall political interplay remains positive and appears to be heading towards maturity.

Thanks to sound and ever-increasing economic interdependence that now sees a trade volume exceeding $550 billion annually, Europe and China have promoted all-dimensional, multi-tiered, and wide-ranging cooperation. The Comprehensive Strategic Partnership established in 2003 has since its inception made important headway, and on both sides there is a will—and in a sense the need—to further deepen the scope and quality of the partnership. The institutionalised framework of

[83] *Cit.* Thépaut, Jean-Baptiste (2012). "Analysis of Cooperation Opportunities for Europe in Future Space Exploration Programmes (COFSEP)". Proceedings of the 63rd International Astronautical Congress, Naples, Italy, October 1–5, 2012. Paper: IAC-12-A3.1.3.

political dialogue and cooperation, including the EU–China 2020 Strategic Agenda, creates a strong political basis for enhancing possible cooperation in the area of space exploration.

7.3.2.2 Weaknesses

Europe's numerous and valuable strengths are to some extent offset by a number of critical issues and pitfalls that could potentially inhibit or even undermine future cooperation efforts. These weaknesses are as much inherent in the European space programme as in the broader bilateral relationship with China.

Although it possesses unique capabilities relevant to the architecture of future manned lunar exploration, Europe has not mastered the skills for an autonomous human spaceflight programme; it also lacks critical technologies required for non-dependence (e.g. life support and protection, Earth re-entry capsule, radioisotope-based power systems, advanced propulsion). This may constitute a major issue negatively affecting European attractiveness as an international partner. But it is also clear that the budgetary situation, coupled with the large ISS-related expenditures that are bound to continue over the next 10 years, prevents any short- and midterm large investment in human spaceflight for Europe unless there is a political impetus that makes such an endeavour a priority issue.[84] Even if an EU contribution comes into play, the difficulties encountered in the funding of Copernicus and the significant outlays on Galileo will inevitably continue to negatively affect the major financial investment required for human spaceflight.

The science-based and technologically focused approach of the European space exploration programme highlighted above as a source of strength has a mirror-image downside, namely, the lack of a strong political dimension within the current governance of space activities in Europe. This political dimension is an indispensable element for implementing a long-term vision in space exploration and becoming involved in a domain like human spaceflight, which will inevitably entail a large degree of cooperation with non-European countries.

A closely related set of weaknesses and potential obstacles is cumbersome European governance. In spite of long-standing efforts to create an integrated pan-European space programme, the current governance of space activities remains characterised by a degree of fragmentation. Whereas the critical mass required to deal on a peer-to-peer level with the major space powers has been achieved through the creation of "centralised" multilateral bodies, political authority over space matters has remained de facto largely in the hands of the various national entities. As a result, for the implementation of more ambitious space projects, both the EU

[84] The total European contribution to the ISS programme amounts to approximately 8 billion € spread over the whole programme. See "International Space Station: How Much Does It Cost?". European Space Agency. 13 May 2013. Web. http://www.esa.int/Our_Activities/Human_Spaceflight/International_Space_Station/How_much_does_it_cost. Accessed 30 April 2014.

and ESA are ultimately dependent on a "green light" from their member states—an aspect that could, in turn, slow down (or even undermine) the development of a major European initiative vis-à-vis China.

In addition, much work remains to be done in the definition of the respective roles of ESA and the EU and coordination with powerful national agencies. At present, no clear structure of the relationship between the EU and ESA has been achieved. This may cause of a number of uncertainties and drawbacks in the decision-making processes that could ultimately make Europe unable to develop a clear and coherent strategy for future space exploration. This governance issue would also count as a weakness in the implementation of an agreed programme, as China will no doubt remember the inability to deliver on the lifting of the arms embargo.

It should be noted, however, that governance issues affect cooperation opportunities as much on the Chinese side as on the European one. In the former case, the "byzantine maze" of organisational structures set up to manage space activities, coupled with the involvement of the PLA in the running of the human spaceflight programme, is a hindrance to cooperation in this domain.

A further and much more tangible constraint is the technology transfer issues regarding China that bring into play well-known intellectual property rights concerns and the International Traffic in Arms Regulations (ITAR) of the USA. The ITAR restrictions, which apply to US technology incorporated in foreign equipment, including space systems, have de facto created an international embargo based on US export licences.[85] Given that a number of European space systems include critical US components, it could be argued that the USA essentially has a veto power over possible Sino–European space cooperation. Of course, several European countries and companies have since the late 1990s made increasing efforts to develop an ITAR-free business model for the manufacture of commercial satellites,[86] and in recent years ESA, together with the EU and the European Defence Agency (EDA), has more broadly started to invest resources in order to achieve strategic non-dependence in key technologies. The underlying issue, however, remains broader than just loosening this dependency or circumventing legal obstacles. The existence of a European arms embargo, which has also become a serious hindrance for the further development of Europe–China space cooperation, indicates that it is difficult to unbundle space cooperation from military concerns. The underlying difficulty is thus whether the attraction of cooperating with China might be stronger than the military concerns and whether US concerns can be placated.

The analysis of the broader political dynamic provided in the first section of this chapter has clearly shown how sensitive Europeans are about US preferences. This "sensitivity" ultimately points to the fundamental structural weakness in

[85] Mineiro, Michael C. (2011). "An inconvenient regulatory truth: Divergence in US and EU satellite export control policies on China". Space Policy Vol. 27 (4): 213–221.

[86] Yun Zhao, Yongmin Bian (2011). "Export control regime for space items in China: Opportunities and challenges in the new era". Space Policy Vol. 27 (2): 107–112.

Europe–China relations, namely, the current European inability to manage a strategic partnership with Beijing when it directly affects US interests and security concerns.

From the provision of military equipment to collaboration on the development of advanced space technology, Europeans have been unable to handle any major initiative vis-à-vis China. When it comes to a highly sensitive domain like human spaceflight, it can be anticipated that this inability might eventually play a crucial role, unless significant efforts of coordination are put in place. At the moment, however, it is safe to agree with Joan Johnson-Freese and conclude that Europe is left with the "dilemma of either not expanding cooperation with China, a restriction that it wants to avoid, or of risking the wrath of the USA, which it neither wants, nor most probably can afford, to do".[87]

7.3.2.3 Opportunities

It is widely recognised that countries approaching space exploration as a mutually beneficial endeavour are rewarded with significant paybacks. These paybacks have been well documented in research: among others they include improving and complementing each partner's capabilities, ensuring robustness and redundancy of the programme's systems, generating higher programmatic and political stability, and increasing the total level of available resources.[88]

Importantly in a period of financial constraints, cooperation will also eliminate useless duplication of effort and thus offset economic costs. Additional overhead costs increase the overall cost of any international cooperative endeavour as well, but these costs are spread among partners.[89] Participating in a large cooperative undertaking with China would enable Europe to expand the scope of its space programme much beyond its individual capabilities by tapping into the vast and valuable resources China would contribute. The expansion of resources that would accrue through cooperation would not just be financial but also scientific and technological.[90] Europe would certainly be empowered to develop new core capabilities (e.g. human-rated lunar ascent stage, lunar cargo lander, versatile mobility

[87] *Cit.* Johnson-Freese, Joan, and Andrew S. Erickson (2006). "The Emerging China-EU Space Partnership: A Geo-technological Balancer" Space Policy Vol. 22 (1): 12–22.

[88] Cooperation enhances the domestic legitimacy of space programmes and gives them international credibility and it consequently makes them less vulnerable to cancellation due to domestic political or financial problems. Correll, Randall, and Nicolas Peter (2005). "Odyssey: Principles for Enduring Space Exploration" Space Policy 21 (4): 251–258.

[89] See Broniatovski, D.A., et al. (2006). "The Case for Managed International Cooperation in Space Exploration". Center for Strategic and International Studies. Washington DC. For an interesting case study in this regard, see Lahcen, Arne (2013). "EUMETSAT-NOAA Collaboration in Meteorology from Space. Review of a Longstanding Trans-Atlantic Partnership". ESPI Report 46. European Space Policy Institute, Vienna.

[90] Peter, Nicolas "The changing geopolitics of space activities" Space Policy 22 (2): 100–109.

platform, advanced shielding, and inflatable structures[91]), which could in turn contribute to expanding Europe's overall space ambitions and would provide an important steppingstone for future human Mars exploration.

While all these benefits could accrue to Europe from any "partnership configuration" with at least one of the major space powers, others would specifically be the result of a cooperative undertaking with China.

Associating with China would allow Europe to gain access to Chinese space-related infrastructure (e.g. launching sites, TT&C stations) and, more importantly, would be likely to open up alternative flight opportunities for European astronauts, this time on board the forthcoming CSS. ESA would appear to have considerable interest in being able to use Chinese ground and space infrastructures in the future[92]; Sino–European cooperation would consolidate the likelihood of achieving these objectives.

More broadly, cooperation with China would allow a diversification of the portfolio of European partnerships. As already mentioned, such diversification is particularly important, as it would promote greater political autonomy for the continent (thus also enhancing its "soft power") and specifically reduce European critical dependency on the USA and Russia, while at the same time providing more back-up opportunities for the implementation of programmes. Potential back-up opportunities are important, as they might ultimately prevent the collapse of a mission in case of withdrawal by another partner (cf. the *ExoMars* mission).

Equally important is the possibility for European industries to increase their presence in the Chinese market for the provision of space-related hardware and space-based services and through this perhaps also to enhance overall high-tech exports. While such opportunity would not be a direct outcome of a cooperative venture with China on a manned lunar mission, the emergence of some form of *guanxi* relations between China and European aerospace companies would seem likely.[93] The Chinese demand for space-related products and services is expected to increase dramatically over the coming years in several domains, thus creating unprecedented opportunities that could be best seized by European stakeholders if closer cooperation was pursued.

An additional potential benefit of expanded cooperation with China could be the avoidance of the isolation of current Chinese space technology efforts. As highlighted in a US Congressional Research Service report in 2007, collaborating with China—instead of isolating it—could ultimately lead the country to rely on Western technology rather than forcing it to develop technologies and programmes

[91] Perino Maria Antonietta., "Outlook of possible European contributions to future exploration scenarios and architectures". Acta Astronautica, Volume 88, July–August 2013. pp. 25–34.

[92] As mentioned, a Mutual Cross Support Agreement with China's Launch and Tracking Centre (CLTC) was already negotiated by ESA in 2010.

[93] It could even be expected that China would accord some forms of privileged access for European companies in the Chinese market.

alone.[94] Paradigmatic in this regard is the case of the CSS programme, which by public admission would have been abandoned if China had been allowed to join the ISS. Great geopolitical benefits would have been the result.

Besides giving important leverage in other areas of the relationship, offsetting the need for China's unilateral development would help Europe maintain its competitive advantage and strengthen the positioning of the European space sector in the global market for space technologies and services. As well as contributing to the dissemination of European technologies, it could also embed the use of related standards, which is an aspect of significance in a major cooperative venture. Having the ability to define the standard interfaces between the space exploration systems of the countries involved in the architecture of a programme is one of the most efficient ways of maintaining a prominent role in the management of the venture: once these interface specifications are in place, any new participant is required to adhere to the standards.[95]

Besides these potential benefits, cooperating with China can also advance diplomatic goals: above all, it would allow Europe to emphasise and actively promote its core values within the Chinese space community with regard to the sustainability and governance of space activities.

Cooperation could be conducive to more active Chinese engagement in joining the EU-originated International Code of Conduct for Outer Space Activities, implementing transparency and confidence-building measures (TCBMs), as well as more appropriate policy measures with regard to—among others—planetary protection (specifically protection of the lunar environment).[96] Indeed, cooperation can be considered an important TCBM in itself, as it leads to a higher degree of mutual trust and understanding between partners, reducing possible tensions. Specifically with regard to China, it would also allow partners to learn more about Chinese space activities and would help to make China's intentions more transparent to the international community.

The skilful crafting of a major cooperative space endeavour could also advance the broader political objectives of the EU. Independent of the specific partnership configuration, European involvement in a major space exploration programme like a manned lunar endeavour would be of paramount importance in promoting Europe's status in space, as well on Earth, and would fulfil its policy objectives. To echo the EC's Space Advisory Group, space exploration is a field where Europe can assert itself globally and where the EU institutions can bolster their image in the

[94] Logan, Jeffrey (2007). "China's Space Program: Options for US-China Cooperation". Congressional Research Service Report for Congress, Washington DC.

[95] Broniatovski, D.A., et al. (2006). "The Case for Managed International Cooperation in Space Exploration". Center for Strategic and International Studies. Washington DC.

[96] Planetary protection programme activities linked to the establishment of a permanent base on the Moon require the elaboration of more solid international frameworks. ESA has initiated valuable efforts in this regard.

eyes of their citizens and of the world.[97] Space exploration can be a political driver for the EU on the international scene. A major European space exploration programme would also contribute to reinforcing a European identity. A manned lunar landing programme with European participation would have a much more dramatic impact than Copernicus and Galileo on the long-standing efforts to promote a "meta-national" European perspective. Such an achievement could serve as one of the icons in this respect! Associating with China thus offers Europe (the EU in particular) a promising means of fulfilling long-term policy objectives.

China's growing international relevance and the strong Sino–European economic interdependence clearly invites Europe to deepen its cooperative relations with Beijing. Space cooperation could become an integral part of a broader initiative aimed at enhancing EU–China economic integration, as well as furthering the global impact of the partners on the international system. This opportunity should not be dismissed lightly, as it will be key to the broader advancement of Europe's international positioning and thus an instrument enabling the continent to avoid the much-feared geopolitical marginalisation in the future governance of world affairs.

Finally, greater cooperation with China could also contribute to inhibiting the formation of a stronger strategic partnership between Russia and China. In both the space and political arenas, a strong partnership would clearly not be in the interest of Europe and is indeed one of the key drivers that should guide the grand strategy of Brussels in the coming years.

From a space perspective, if Russia seizes the opportunity of partnering bilaterally with China—as recent events seem to be suggesting—a European or a broader multilateral contribution could risk becoming less relevant for China. Such a prospect would in addition reinforce—albeit inadvertently—competitive dynamics in the field of human space exploration and might ultimately become cause for the re-emergence of a "space race mentality". In all likelihood the USA would then be compelled to embark on a more ambitious endeavour, possibly stepping up the human exploration of Mars as an attempt to eclipse a Sino–Russian lunar endeavour.

From a geopolitical perspective, Sino–Russian cooperation in space exploration would become an additional piece reinforcing the ongoing convergence between Russia with China. Although the Sino–Russian strategic partnership currently still appears to be little more than an "axis of convenience",[98] it is clear that with their current status as "isolated powers" an ambitious joint project would be tempting. However, it would stand in the way of an emerging broader world order of cooperation. To put it in the blunt geopolitical terms of the influential "neo-

[97] Horneck, Gerda et al. (2010). "Towards a European vision for space exploration: Recommendations of the Space Advisory Group of the European Commission". Space Policy Vol. 26 (2): 109–112.

[98] Lo, Bobo (2008). *Axis of convenience: Moscow, Beijing and the new geopolitics*. Brookings Institution Press, Washington DC.

comms" scholar Yan Xuetong,[99] if China makes a formal alliance with Russia, it could change the current unipolar structure of the international system and establish a new bipolar world order. The much-discussed projections made in 2007 by Robert Kagan about the emergence of an *axis of autocracies* (composed of China, Russia, and possibly Iran) versus an *axis of democracies* (composed of the USA, Europe, Japan, and perhaps India) could eventually become reality.[100]

By necessity, European foreign policy action should be directed to preventing the emergence of this ominous scenario, and cooperation in a highly symbolical domain like manned lunar exploration could become one of the key actions helping to transform the potential trend of power politics competition into more cooperative and win–win approaches. From an American point of view, it is much better to have Europe as a partner of China than Russia.

7.3.2.4 Challenges

Together with opportunities, a number of challenges and potential threats would inevitably accompany the crafting of a cooperative undertaking with China. It goes without saying that European policy makers should be mindful also of all the potential (space-related and political) pitfalls that might undermine the venture.

First, cooperation typically adds layers of complexity to the management of a programme and creates a number of challenges related to the control of the "critical path of systems".[101] Not having developed an independent manned transportation system and other critical technologies required for independent human spaceflight, Europe could run a serious risk of being kept out of the "critical path" and ultimately losing influence on programme implementation, also in terms of decision-making. For political and prestige-related reasons, China might be eager to ultimately keep Europe in the backseat, while maximising its contribution. In addition, given that cooperation inserts an element of programmatic dependence into the architecture of a system, Europe could be "held hostage" by the policy, schedule, or technological difficulty of its partner. At the same time, cooperation with China may require more significant technology transfers, assistance, and financial investment than European stakeholders would be expecting or ready to agree to. The US experience with spiralling economic costs in the ISS programme is still paradigmatic in this regard.

The possibility of inadvertent technology transfer is another major issue of concern, particularly in light of China's record of technology acquisition by ques-

[99] See Xuetong, Yan (2012). "The weakening of the unipolar configuration". In: Leonard, Mark (ed). *China 3.0*. European Council on Foreign Relations, London: pp. 113.

[100] Kagan, Robert (2007). "End of dreams, return of history". Policy Review, Hoover Institution.

[101] Correll, Randall, and Nicolas Peter (2005). "Odyssey: Principles for Enduring Space Exploration" Space Policy 21 (4): 251–258.

tionable means and its current policies towards technology development.[102] Besides the concrete risk of defying the requirements of the ITAR regulatory framework, a serious threat for Europe would be losing its high-tech competitive edge and unintentionally reinforcing the technological capabilities of a potential future competitor. China might even use the (directly or indirectly) acquired technologies to gain market share in foreign institutional markets at the expense of European industrial stakeholders.

Furthermore, given the active involvement of China's General Armaments Department in the management of the human spaceflight programme, there is the much more worrisome prospect that technologies directly or indirectly acquired by China through cooperation with Europe would benefit the technological and operational capacities of China's armed forces. Although GAD's involvement is—as explained in Chap. 2—not directly used to advance the military component of the space programme but as a catalyst for broader innovation throughout the PLA, it is evident that any major cooperative undertaking involving dual-use technologies would likely contribute to the speeding up of China's military modernisation. It can be anticipated that security-related concerns, coupled with a sort of "moral compromise", will probably engender strong pushbacks from some European constituencies (e.g. sceptical political leaders, part of the military establishment bureaucracies), making the likelihood of reaching a broad and firm European consensus doubtful.

It is also safe to say that, under present circumstances, if Europe pursues cooperation with China on human spaceflight, it will run the risk of encountering opposition from third countries as well, first of all the USA. As the Galileo and arms embargo initiatives have demonstrated, any unilateral venture vis-à-vis China that does not see the participation of—or at least consultation and coordination with—the American ally is inevitably bound to generate wrath. On those occasions, the USA even threatened retaliatory measures, including repercussions on the transatlantic defence cooperation that encompassed transfer of US military technology, services, and equipment to EU countries. Similar hostile reactions can be expected from Washington if Europe ultimately opts for cooperation: one concrete reprisal measure that might be adopted by the USA in an extreme case would be the suspension of NASA's cooperation with ESA.

Independent of the specific measures the USA could implement, it is clear that Sino–European space cooperation might more broadly have detrimental effects on transatlantic solidarity. Similarly, Europe's relations with other key strategic partners such as Japan could also be seriously affected, especially if the dynamics of regional power politics lead to increasingly adversarial behaviour between Beijing and Tokyo. Even though there appears to be a gradual opening—at least on the US

---

[102] For a more detailed assessment of China's record of technology acquisition, see Seedhouse, Erich (2010). *The New Space Race. China vs. the United States*, Springer—Praxis Publishing, Chichester, UK: p. 213–14. For an analysis of China's policy towards technology development, see Sect. 5.3.

side—towards the possibility of cooperating with China, such an overture currently remains much more virtual than concrete and does not extend to the possibility of Europe freely embarking upon a bilateral strategic undertaking with China without arousing Washington's irritation. The flipside of this, however, is that Europe could become a bridge-builder between the two alienated countries, at least if the engagement faction in the USA becomes dominant.

Compromising its current positive relationship with two key strategic partners would be an ominous prospect for Europe, which would become even more substantial when seen in the light of another potential downside implied in space cooperation, which of political ineffectiveness. For European policy makers, there is no certainty that diplomacy in space will yield tangible, sizeable benefits on Earth. As the recent standoff over Ukraine shows, the extensive US–Russian cooperation initiated with the 1975 "handshake in space" has not prevented the rise of serious tensions on Earth. Quite to the contrary, the "space interdependence" created by this cooperation has become a source of mutual vulnerability, to be used as a political weapon.[103] One lesson from this development is that the prospective political benefits of a collaborative Euro–Chinese human spaceflight programme (e.g. improving bilateral relations, enhancing mutual trust, making China space programme more transparent) should not be overestimated.

An ultimate and closely related set of potential threats springs from possible developments in China's international behaviour. Although Europeans have so far taken a rather positive view of China's development—especially compared to the American ally—there is still a risk that China could evolve in a way that makes Europeans unable to reconcile such evolution with strategic cooperation on a manned lunar exploration programme.[104] A major fear is the prospect that with its rising economic capacity and military potential, China might develop an "imperial temptation" that would translate into a more aggressive foreign policy seeking to achieve regional or even global hegemony and replacing the current world order with a different one: the ancient tributary system. Europeans are increasingly aware of the forces shaping the internal debates about Chinese foreign policy and are starting to fear that the strongly nationalist component might eventually get the upper hand.[105] The increasing assertiveness shown by China in East Asian, including in its territorial disputes (most notably the dispute over the Senkaku/

[103] In response to US sanctions following the annexation of Crimea, the Russian authorities have announced the intention to retire from the ISS programme in 2020, instead of 2024. In addition, they have announced to cut off supplies of rocket motors critical to US military and civil launches and threatened to deny access to 11 ground stations on Russian territory which support the GPS constellation. See Butler, Amy. "The Empire Strikes Back". Aviation Week & Space Technology. 19 May 2014: 8–9.

[104] For this interpretation on the evolution of China's foreign policy, see Luttwak, Edward (2012). *The rise of China* vs. *the logic of strategy*. Harvard University Press, Cambridge, MA.

[105] See Sect. 5.2. See also a dedicated collection of essays published by the European Council on Foreign Relations: Leonard, Mark (ed) (2012). *China 3.0*. European Council on Foreign Relations, London.

Diaoyu Islands with Japan and over the Paracels with Vietnam[106]), provides Europeans with evidence that contradicts the self-proclaimed peaceful nature of China's rise.

Clearly, the more aggressively China asserts its global presence, the more Europe will find itself in an uncomfortable situation if it decides to pursue cooperation with Beijing in space. If tension increases in the future and China's foreign policy behaviour dramatically changes trajectory, how will Europe be able to explain, to its citizens and the world, close cooperation—let alone a strategic partnership—with China? In the end the trade-off is between the benefits of trying to anchor China in a peaceable world and the danger that such an endeavour will fail, with Europe being seen as having chosen the wrong side of history.

7.3.2.5 Summary

All in all, there appears to be a complex set of strengths and weaknesses in the current state of play in Europe–China relations and an even more delicate balance between promising opportunities and menacing challenges arising from a cooperative Sino–European undertaking in lunar exploration (see Table 7.1). In light of this complexity, it is therefore hard to predict a fixed, "natural" course of action for Europe. Quite to the contrary, the relative weight assigned to each of the elements in this composition could support different policy postures by Europe, some of which might be facing in diametrically opposed directions.

7.4 Europe's Policy Options

In order to assess what posture Europe should adopt China's lunar ambitions to maximise opportunities and mitigate possible threats, a set of policy options can be elaborated on the basis of the SWOT analysis. Specifically, the interaction between the four macro-dimensions comprised in a SWOT analysis enables the identification of four different policy options. Each of these options would ideally instrumentalise one of the two internal dimensions in current Sino–European relations (either strengths or weaknesses) in order to gain future opportunities or avoid the potential threats.

The four policy options, as represented in Table 7.2 and described below, provide general directions for how Europe could move forward relative to China's manned lunar exploration programme.

---

[106] For the most recent development in the dispute over the Paracel Islands, see Panda, Ankit. "Why Did China Set Up an Oil Rig in the Vietnamese Waters". The Diplomat. 13 May 2014. Web. http://thediplomat.com/2014/05/why-did-china-set-up-an-oil-rig-within-vietnamese-waters/. Accessed 20 May 2014.

Table 7.1 SWOT analysis for Sino–European cooperation in manned lunar exploration

	Helpful	Harmful
	Strengths	Weaknesses
Current conditions	At space level	At space level
	• Europe is an established space actor with long-term experience • Critical capabilities for future exploration endeavour. Attractive partner/state-of-art technological level • High stability of programmes and budget once decided • "Trust capital", in particular with ESA • Cooperative relations with all other space programmes worldwide	• Europe has not yet defined its long-term priorities nor does it have a clear vision of its future role • Lack of an independent human spaceflight programme • Technology transfer and ITAR issues • Budget constraints • Governance fragmentation and high-level duality • Lack of transparency and PLA involvement in China's manned spaceflight
	At political level	At political level
	• Overall positive relations • Strong and increasing economic interdependence • Wide-ranging cooperation • Institutionalised framework of political dialogue and cooperation • Lack of strategic conflict of interests	• Lack of a political dimension in the European space exploration programme • Solidarity with the US partner • Lack of coherence/shared vision and strategies among EU countries • Difficulty of managing a strategic partnership
Future prospects	Opportunities	Threats
	At space level	At space level
	• Diversify the portfolio of European partnerships • Offset economic costs • Improve transparency and enhance mutual trust • Maintain a key profile in the space hierarchy • Avoid global perception of "cultural decline" • Strengthen the international framework of cooperation • Provide a stepping stone for future Mars exploration • Provide industrial partners of choice	• Lack of control of the programme • Inadvertent transfer of sensitive technologies and ITAR infringement • Europe in the backseat • Spiralling financial investment • Lose high-tech competitive edge • Possible repercussions on cooperation with NASA
	At political level	At political level
	• Increase Europe's profile and strengthen its identity • Boost the overall relationship • Gain autonomy from the USA and Russia; avoid geopolitical marginalisation • Foster a multipolar world • Prevent the formation of Russia–China axis	• Stronger position of China • Direct benefits for the PLA and Chinese military modernisation • Deterioration of relations with the USA and Japan • Cooperation in space brings no real benefits on Earth • Uncertain geopolitical future/trajectory of China

Table 7.2 Europe's policy options

	Opportunities	Threats
Strengths	Europe as a strategic partner	Europe as a bridge-builder
Weaknesses	Europe as a limited partner	Europe as a competitor

The question is how to reap the best balance between opportunities and threats, by maximising strengths and minimising weaknesses. A limitation of the SWOT analysis is that it does not prioritise the issues identified within the four areas and, as a consequence, does not provide any qualitative guidance on which policy alternative should be given preference.[107]

For this reason, in the following the SWOT will be complemented by a qualitative assessment of the policy options. Such appraisal is performed by weighing up the major factors that should be taken into account by European decision makers when choosing between various policy alternatives. Four parameters, or evaluative criteria, have been specifically identified as relevant:

- The first evaluative criterion is political feasibility, which is a measure of how well the considered option would generate convergence of interests and a broad consensus among European stakeholders. For a policy alternative to be enacted and implemented, it must be politically acceptable or desirable. A feasible policy alternative is thus one that has a high probability of receiving sufficient political push and support to be implemented.
- The second criterion, affordability, refers to the projected costs associated with the option. While it is extremely difficult to make a financial projection for the European space programme over the next 15 years, this criterion looks at the short-term ability of Europe to implement the option considered. It appears clear that any political decision related to the future role of Europe in space exploration should be made in the near future.
- The third criterion used to evaluate the options refers to their likely effectiveness. From a "space perspective", an option is considered effective if it brings valuable benefits to the European space programme; specifically if it allows Europe to: (a) engage in a large, complex, and costly exploration programme that is beyond its current capabilities; (b) enrich its pool of scientific and technological expertise while sharing costs and avoiding duplication of efforts; (c) gain access to non-European ground and space systems (e.g. TT&C stations, launching sites, new spacecraft, alternative transportation systems, etc.); (d) maintain or advance its current position in the international "space hierarchy"; and (e) diversify the portfolio of its potential partnerships.
- Considering that the only path to the realisation of an ambitious programme such as human lunar exploration is for it to also serve political objectives, a final parameter looks at the broader political benefits that would accrue from the

[107] The information gathered might also be oversimplified and crucial data overlooked. Some information may fit into more than one category or be slotted into an inappropriate category.

implementation of the considered policy alternative. From a political perspective, an effective option must (a) contribute to the implementation of EU policies, (b) fulfil the EU's foreign policy goals, (c) strengthen Europe's profile and promote its core values in the international arena, and (d) avoid negative rifts in international relations.

Each of the policy options is evaluated using these four criteria in order to provide a general appraisal of the most appropriate choice for Europe.

7.4.1 Europe as Strategic Partner

A first policy option for Europe would be the creation of a long-term strategic partnership aimed at achieving the ambitious goal of human lunar exploration with China. Pursuing this option would reflect the recognition that China's ascendance in space and its ambitious plans for space exploration offer Europe unprecedented opportunities to promote its policy objectives in both the space and political arenas (and advance the overall positioning of its space exploration programme). Such a groundbreaking cooperative undertaking could be crafted by leveraging Europe's "trust capital", as well as the numerous and valuable strengths Europe possesses, and directing them to best reap all the above-listed opportunities.

The landing of European and Chinese astronauts on the Moon would be the most visible manifestation of the Sino–European strategic partnership. However, in order to maximise the benefits associated with such a strategic partnership, it is also reasonable to expect that cooperation would not materialise solely through the implementation of a lunar endeavour but would entail the pursuit of widespread engagement in various areas of space and probably beyond. Cooperation could be extended to the upcoming CSS and be complemented by joint development of brand new manned transportation systems to enable future joint steps in space exploration and to secure the achievement of the ultimate destination of human space exploration: a manned Mars landing.

From a strictly utilitarian perspective, a strategic partnership could be a win–win situation for Europe and China, as it would enable the two actors to ideally complement their respective strengths: Europe's valuable technological expertise and innovation capabilities and fast-rising China's financial and engineering resources. As such, the costs associated with this option could be expected to remain within Europe's financial capabilities, barring the emergence of serious economic downturns. Likewise, the space effectiveness of this option should be high, as it would enable Europe to reap all the above-listed benefits and ultimately reach a top position in the international space pecking order.

In terms of geopolitical effectiveness, the option would yield a number of positive pay-offs for Europe: it would allow the continent to improve its relationship and political profile in China tremendously, to fulfil important policy objectives, and more broadly to gain a political emancipation that would allow it to play an active and autonomous role in world affairs. However, implementing a bilateral strategic partnership with China would also involve serious drawbacks for Europe, the most notable being the probable rupture of its current political relations with other partners.

The USA, as well as Japan and perhaps India, would become alarmed by the prospect of Sino–European space cooperation of this kind and would accuse the EU of lacking political integrity by promoting the material and political standing of an authoritarian regime and triggering a negative shift in the East Asian balance of power. Thus, embarking on a highly ambitious endeavour with China could have far-reaching consequences, including disruptive effects on the current geopolitical configuration.

The political feasibility of this option would thus be very low: it is in fact hard to imagine all European stakeholders accepting—let alone supporting—its implementation. At the national level in particular, the more Atlanticist of the EU Member States would resolutely oppose the prospect of a stronger bilateral partnership with China, especially if the participation of, or at least a "green light" from, the USA was not first secured. For the EU and ESA therefore exploring alternative routes seems preferable, as it may, indeed, for China.

7.4.2 Europe as Competitor

At the opposite end of the spectrum of policy alternatives, Europe could choose the path of not pursuing cooperation at all. The decision to take this option could stem from a recognition that the structural weaknesses associated with a Sino–European cooperative undertaking are too large to be tackled and the threats emerging from the prospect of Sino–European cooperation too overwhelming. In particular, if the aforementioned prospect of destroying positive relations with a number of key partners is combined with the unpredictable evolution of China's foreign policy behaviour and the hard issues related to dual-use technology development, Europe could conclude that competition is more desirable than cooperation.

An independent human spaceflight programme can be considered desirable as it would enable Europe to catch up with the current space superpowers and achieve relevant technological and political objectives. It would, however, require enormous investments and an even stronger political will. The Hermes programme, which was eventually terminated in 1992 for lack of solid financial and political backing, offers a highly relevant cautionary precedent in this regard. In addition, considering the current climate of strong budgetary pressures, it is clear that the goal of an independent human spaceflight programme could be achieved only at the expense of other major undertakings. The political feasibility and affordability of this sub-option is thus very low. Promoting such a policy would not be in line with the orientations emerging over the past few years among all European stakeholders.

If cooperation with China is not pursued, the alternative path is to associate with other partners, something that might practically be limited to the remaining two countries with autonomous capabilities in the area of human spaceflight: Russia and the USA. The former is currently refocusing its national space programme to lunar exploration with a long-term view of setting up a permanent lunar base.[108] However, cooperating with Russia would entail similar political pitfalls to those associated with a Sino–European partnership. Indeed, recent geopolitical dynamics are making the prospect of Euro–Russian cooperation even less appetising than the Chinese option.

This leaves Europe with the practical option of associating itself with the USA. Europe could either opt to join in US plans for an Asteroid Retrieval Mission (ARM) or alternatively try to influence the direction of the NASA towards a more desired lunar scenario. The latter could have some chance of success given that there is still no US national consensus over the asteroid mission. In addition, Europe could advantageously use the prospect of a "Red Moon" as a bargaining chip to channel US plans towards a lunar scenario and be enabled to play a greater role in the architecture of the programme, both in terms of decision-making and concrete involvement in mission implementation. This would allow Europe to demonstrate the willingness and the ability to take responsibility, while in the meantime capitalising politically as well as materially from the valuable resources the USA would provide.

The underlying issue is, however, pursuing such an option would contribute to stimulating—albeit inadvertently—a competitive dynamic in space exploration and could ultimately revive a "space race mentality". This prospect is politically undesirable for Europe, as it could weaken its political relations with China and create threatening political rifts in world affairs.

Admittedly, compared to the "strategic partnership with China" option, this alternative has a higher degree of political feasibility, especially because it would be a natural continuation of the traditional European space policy of close transatlantic cooperation. But the appetite for it would probably remain modest. In fact, if it is eventually pursued, European stakeholders are aware that they are likely to remain a second-class partner in space exploration and be prevented from achieving much-needed diversification in their partnership portfolio and in the fulfilment of broader policy objectives vis-à-vis China. In both the space and political arena, Europe would almost inevitably end up putting its fate in the hands of the USA.

As a result, the political effectiveness of this option would doubtless remain low.

[108] "Russian Space-Based Activities' Development Strategy until 2030 and Beyond" Aviation Explorer. 27 April 2012. (Russian language source). Web. http://www.aex.ru/docs/8/2012/4/27/1561/. Accessed 18 May 2014.

7.4.3 *Europe as a Limited Partner*

A third policy option for Europe is based on the assumption that maintaining a degree of cooperation with China would be in the continent's best interests: China is steadily emerging as an indispensable pillar in future space exploration; for all the major spacefaring nations experiencing budgetary pressures, it makes sense to capitalise on China's openness to cooperation and to tap into the resources this country can contribute to the pursuit of common interests.

While, as mentioned, there are structural weaknesses that undermine the prospect of a strategic partnership, these could be minimised and valuable opportunities reaped by adopting a cautious and pragmatic policy approach. This could assume the form of a limited partnership, whose enactment and implementation would primarily be guided by ESA, with the limited involvement of the EU, so as to reduce the political dimensions related to such cooperation and prevent possible resistance from European stakeholders or international partners.

In such a configuration, the contribution provided by Europe to China's lunar exploration would be limited to the delivery of some smaller elements at different levels: module, subsystem, equipment, or even operational levels. It could include the provision of scientific instruments for conducting experiments, selected advanced robotic tools, and advanced shielding structures or TT&C services and astronaut training.

By focusing on the autonomous development of a few transferrable technology niches (i.e. that are ITAR-free), Europe could also try to prudently extend cooperation to the provision of a few major elements of the architecture of China's exploration programme. Examining the outlook for possible European contributions to future space exploration contained in the aforementioned European Baseline Roadmap, for instance, suggests that the delivery of a lunar cargo lander could be envisaged. This would be able to deliver some key assets to the lunar surface (i.e. logistics and spares) or alternatively to deliver systems related to mobility surface activities, life support, or rendezvous-and-docking operations. Although not explicitly intended to enable active participation in a manned lunar landing, these elements would be directed to providing valuable pay-offs to European industries or barter capabilities for other endeavours and might also secure the chance to fly European astronauts onboard the CSS.

If this option were to be pursued, Europe would more broadly continue to cooperate with China when mutual interests happen to converge and external factors open windows of opportunity; in this way it would avoid compromising cooperation schemes with other partners.

A limited partnership could yield several benefits for the European space programme, as it would secure a sustainable role for European industry in an increasingly competitive environment, and allow Europe to develop new capabilities and continue to maintain its commitment in certain human spaceflight activities. It would also enable Europe to diversify its cooperation portfolio for the implementation of future endeavours and—from a broader political level—

markedly improve Europe's cooperative relationship with Beijing and the scope of their strategic partnership.

In addition, compared to the strategic partnership option, the costs associated with the implementation of this alternative are much lower and likely to be within the current financial capabilities of European stakeholders. Indeed, although Europe would provide some elements to the architecture of China's lunar endeavour and specialise in some technologies that could serve as building blocks for an eventual long-term mission to Mars, in all other respects such an approach would largely allow the European focus to remain on LEO for the next 10 years or so.

The political feasibility of this option would be relatively high, as European stakeholders would probably be willing to support an option with some interesting potential and moderate cost for Europe. However, the ambitions of such an option are rather limited and imply an inadequate role for ESA in the upcoming global space exploration context. This option would provide few possibilities for Europe to project influence in the global arena; it would also perpetuate Europe's role as a junior partner of the big players and feed the impression of being only a follower in future space exploration. Although demonstrating willingness to take responsibility as a reliable partner, pursuing this option could thus also preclude the enhancement of European competitiveness and attractiveness as a partner in the long run. As a result, the policy's space effectiveness would not be particularly high: while it would bring some tangible benefits to the European space programme, Europe's overall position would appear weakened in the light of the much more visible achievements of other spacefaring nations, China above all. The mere provision of some elements of the architecture of China's lunar exploration programme would not deeply affect European public perceptions. This would only happen if a European astronaut were to go to the Moon. Being a witness to China's achievement, and a limited partner, would not be likely to avert the profound ontological shock which China's landing of a taikonaut on the Moon would cause, not only in Europe but even more so in the USA. This might make the political effectiveness of this option low.

7.4.4 Europe as Bridge-Builder

A final policy path for avoiding the most menacing drawbacks of either a bilateral strategic partnership with China or a lack of cooperation could be to become a bridge-builder between currently opposing forces, primarily China and the USA.

Within this option, Europe could leverage its long-standing partnership with Washington and its maturing relations with Beijing to become an interlocutor between currently isolated players. Europe could become a trailblazer for a broader international endeavour in space exploration. The basic partnership configuration for such an endeavour would be Europe, China, and the USA but could possibly also extend to Russia and other spacefaring nations, such as Japan and India.

The rationale for taking this option would be that by assuring the simultaneous participation of the USA and China in a joint cooperative undertaking, Europe would also secure its active involvement without facing the dilemma of choosing between mutually exclusive partnership configurations or being limited to providing minor contributions to multiple partners' endeavours. Being a bridge-builder would project European geopolitical skills and European values.

Needless to say, under this option Europe's role would not solely be that of a matchmaker between the interests of the two juggernauts but also that of an active participant in the implementation of a unified road map that would be shared multilaterally—and in turn conducive to an overall cooperative approach to exploration.

Active involvement is natural and necessary for Europe if it wants to avoid being pushed aside once a Sino–American *detente* has been reached. Europe must thus work to secure the chance to provide major cornerstones to the architecture of the programme, while also tapping into the technological and financial resources of its partners. Possible European contributions to space exploration identified in the European Baseline Roadmap include some ATV derivatives (e.g. the service module that ESA is currently developing for the Orion), or the so-called Cis-Lunar Logistic Vehicle, intended to deliver pressurised and unpressurised payloads to cis-lunar space, the Lunar Cargo Lander, or the Human Rated Lunar Ascent Stage.

By pursuing this option, the European exploration strategy would be more broadly embedded in a multilateral context, thus exploiting the synergies between the programmes of different players and reinforcing the European exploration programme by taking full advantage of international collaboration mechanisms that are currently embedded within the GES and the ISECG.

As some parties in the USA begin to acknowledge the need to cooperate with China and as the latter continues to emphasise its interest in cooperating with the USA, this strategy appears to have some degree of viability. To be sure, Europe would have to work hard in order to ensure a convergence between the two juggernauts and coordinate the implementation of a multilateral endeavour, but a basis for such efforts seems to exist.

First, it would be essential for Europe to encourage full inclusion of China within the international space community (e.g. symbolic participation in the ISS programme and ISS partner participation in the CSS programme). To effect such a "paradigm shift" vis-à-vis China in the USA would require accomplishing the twofold goal of shaping US intentions and posture over China in a direction more in tune with Europe's policy of constructive engagement, while at the same time inducing China to make its strategic objectives in space (as well on Earth) more open to the scrutiny of the international community.

For a multilateral and complex endeavour to become viable, it would also be necessary to enhance the existing fora and mechanisms of consultation into a fully fledged governance body. Ideally, an upgrading and transformation of the International Space Exploration *Coordination* Group into an International Space Exploration *Cooperation* Group is required. The benefits accruing from the successful implementation of this option would far exceed its cost and clearly not just for Europe.

Besides avoiding useless duplication of effort and promoting cooperative relations on Earth, in the long run the joining of international forces to implement human lunar exploration could also enable the realisation of an idea on which the International Lunar Exploration Working Group (ILEWG) has been working for more than a decade: the building of a Manned International Lunar Base to conduct scientific research based on the model of the Antarctica bases.[109] The skilful implementation of this idea could in time become a crucial stepping stone for a unified, cooperative approach to manned Mars exploration, which has been globally acknowledged as the "final destination" of twenty-first century exploration, but whose cost remains prohibitive without a multinational (global) cooperation scheme.

In terms of both space and political effectiveness, the option is likely to be positive, as it would enable Europe to engage in a large, complex, and costly exploration programme that is beyond its current capabilities; to reinforce its space exploration programme by taking full advantage of international cooperation mechanisms (e.g. the GER, the ISECG, the ILEWG); and to enrich its partnership portfolio together with its pool of scientific and technological capabilities. It could secure access to non-European ground and space systems. While avoiding the drawbacks emerging from the prospect of closer cooperation with China, other important opportunities could also accrue: playing the role of matchmaker might enable Europe to influence and coordinate additional and future space exploration programmes.

From a political perspective, this option would foster the consolidation of European identity around an objective of great appeal but also reinforce the role of Europe as a centre of gravity in international affairs and promote cooperative international relations, while undermining the emergence of confrontational stances in the emerging global world order.

Relative to the "strategic partnership" and the "competitor" options, the costs associated with this policy alternative would remain affordable, as they would ideally spread across a number of partners. However, if Europe wants to play the role of active participant, its contribution should not be limited solely to the provision of redundant systems directed to ensure programmatic robustness. A large capital investment devoted to the development of core products and unique capabilities would logically be required; all European stakeholders should be able to unite behind this goal given the consistency of this option with the policy

[109] ILEWG is an international public forum established in 1994 and designed to support cooperation in the exploration of the Moon. The working areas of ILEWG's task groups include science of, on, and from the Moon, utilisation of lunar resources, and infrastructures for lunar bases. ILEWG has developed a road map, the final phase of which foresees the transition from short missions to permanent human presence at international bases. See Ehrenfreund, Pascale, et al. (2012). "Toward a global space exploration program: A stepping stone approach". Advances in Space Research No. 49: 26–27.

orientations that emerged at the 2010 Brussels summit and were consolidated at the Lucca (2011) and Washington (2014) Space Exploration Forums.[110]

On the other hand, major difficulties will probably be encountered in the process of securing a green light for fully fledged development of the option from all the international partners involved (particularly the USA), as well as for its actual implementation. Because of the groundbreaking nature of this strategy, many steps are required to reach a unified space exploration governance structure and to transform the current, informal mechanisms of coordination and cooperation into proper tools of governance.

7.4.5 *Summary*

In order to summarise the findings of this assessment of the policy options, each policy was given a score on each of the four criteria (political feasibility, affordability, space effectiveness, and political effectiveness) on a scale of 1–5 (1 representing the lowest score and 5 the highest). These were subsequently compared so as to have a general appraisal. The overall performance of the four options is presented in Table 7.3.

As can be seen from the table, the "bridge-building" option obtained the highest score. Relative to the other options, this policy alternative would allow Europe to leverage and maximise its strengths, while also overcoming its weaknesses and preventing the emergence of serious threats.

A comprehensive and enduring international consensus might, however, prove difficult to reach, and pursuing this strategy will inevitably be slow. There are potentially cumbersome political hurdles as well as a number of undetermined

Table 7.3 Assessment of Europe's options

	Political feasibility	Affordability	Space effectiveness	Political effectiveness	Total
Europe as strategic partner	1	2	4	2	10
Europe as limited partner	4	4	2	1	11
Europe as competitor (US option)	3	3	2	1	9
Europe as bridge-builder	4	3	4	5	16

Key to numbers: 1, very low; 2, low; 3, moderate; 4, high; 5, very high

[110] See Horneck, Gerda et al. (2010). "Towards a European vision for space exploration: Recommendations of the Space Advisory Group of the European Commission". Space Policy Vol. 26 (2): 109–112.

operational issues that need to be addressed. Accordingly, the following section will elaborate on the concrete policy actions Europe needs to undertake in order to ensure the viability and effectiveness of the "bridge-building option".

7.5 Proposed Way Forward

The successful crafting and implementation of a groundbreaking multilateral endeavour in space exploration will require Europe to undertake a series of enabling policy actions to give effect to its strategy. The actions recommended below are divided into three different categories: actions to be undertaken within the European context, vis-à-vis China, and with respect to the international space community:

- As far as the first set of actions is concerned, it should go without saying that all European constituencies must coordinate to elaborate and then adopt a coherent policy based on "one voice system". As the arms embargo imbroglio has made clear, no concrete policy measures should be publicly announced—let alone enacted—until a solid and enduring intra-European consensus has been reached.

  Since ESA cannot be regarded as a real political actor able to conduct fully fledged foreign policy and since the EU is not a programme-implementing body with regard to space activities, it is essential that the action of the two main European constituencies not only be coordinated but also be mutually reinforcing.[111] A specific step in ensuring more coherence in Europe's China space policy would be offered by the creation of an ad hoc coordinating mechanism within the framework of the EU–ESA Space Council. Such a mechanism should be dedicated to elaborating, together with ESA–EU Member States, the policy actions Europe intends to implement with regard to potential Sino–European cooperation in space exploration. At the same time, ESA involvement in the ongoing EU–China Dialogue Architecture (and EU–China 2020 Agenda for Cooperation) should be better defined and strengthened.

  Based on these premises, it is recommended that the elaboration of Europe's policy action vis-à-vis China in space should consistently be defined within the framework of the broader China policy of the European External Action Service (EEAS). Without a clear European definition on whether to become a major player or join as a junior partner, Chinese space ambitions are likely to entail unmanageable trade-offs.[112] Achieving concerted policy actions among all European constituencies might prove challenging and slow, but a lack of consensus on these issues will be counterproductive or even self-defeating.

[111] It is clear that, to be fully effective, a coherent European China policy cannot be pursued separately from the broader problem of ESA–EU relations.

[112] See a more detailed description in Sect. 7.1.

- With respect to China's expanding space ambitions, Europe will need to gradually deepen the scope of *soft* cooperation with Chinese stakeholders, in all the three main areas of data and information exchange, policy dialogue, and joint activities. Continuity in cooperation is an important enabling element to further cement relations and accumulate "trust capital", the essential precondition for Chinese stakeholders who are unwilling to work on the basis of one-offs. Specifically with regard to human spaceflight and space exploration, the recently announced initiatives concerning information exchange on astronaut training methods should be gradually broadened to include joint training programmes and even more ambitious projects (e.g. the joint development of life-support systems). The midterm objective of these actions should be to make Chinese participation in the ISS programme—and European participation in the CSS programme—possible.

  Because the promotion of closer ties between Europe and China is vital, the possibility of opening an ESA representation office in Beijing should be considered.[113] At the same time, a promising strategy would be to deepen links with the Chinese Academy of Sciences. Besides the scientific benefits to be achieved from such ties, there are political pay-offs at stake. More than CNSA, the CAS has proven to be a crucial stakeholder in the overall set-up and decision-making processes of China's space programme. CAS not only comprises the intelligentsia behind the programme, but it is also one of the most supportive stakeholders in terms of both promoting cooperation with international partners and increasing the transparency of China's space activities. In addition, closer ties with CAS could indirectly contribute to making the civilian stakeholders even more influential and prominent within the Chinese space programme.

  Additional cooperation-enabling mechanisms need to be defined and implemented:

  (a) The participation of Chinese stakeholders in the EU Framework programmes should be further stimulated, so as to get the scientific communities to work more closely together. EU Framework programmes are important instruments for reaching out to potential partners, and their prudent utilisation can serve to increase the feasibility and effectiveness of the "bridge-building option".[114]

---

[113] The pay-offs to be harvested by such a move should not be overlooked when performing a cost-benefit analysis. An ESA Beijing office would in fact help to (a) build mutual trust and have a better understanding of the internal forces shaping the Chinese space programme; (b) increase the opportunities for Europeans to liaise with the Chinese stakeholders and identify potential areas of cooperation; (c) promote ESA activities and image in China; (d) signal to Beijing that it is regarded as an equal, valuable partner of Europe; and (e) influence the Chinese space programme in a direction that is more in tune with Europe's interests.

[114] More than 300 Chinese organisations have already participated in the EU FP5, FP6, and FP7, but the commission should increase its efforts to encourage researchers from China to participate in the traditional "core" of the Framework Programmes on similar terms to other participants. As noted by several analysts, the commission "should also establish a mutually agreed *EUChinaLink*

(b) The possibility of setting up ad hoc ESA–EU Trainee programmes for Chinese students should be explored to create future network possibilities. The pay-off would doubtless exceed the potentially cumbersome legal and operational issues involved, and the move would promote Europe's soft power in space and create a good contact basis for further interaction once the trainees return home. The so-called third pillar of the EU–China Dialogue Architecture (people-to-people exchange) already offers an effective institutional platform in this regard.

(c) The so-called Track II study groups should be established within the framework of EU–China Space Technology Cooperation (see Appendix C) to explore possible areas of cooperation and each partner's specific contributions to a joint human exploration programme. These informal groups may also serve as useful instruments for articulating and defining the broader conditions under which cooperation can take place. Through a platform of regular, informal dialogue, European stakeholders would be able to impress upon Chinese counterparts the need to have a space programme open to the scrutiny of the international community.

- In terms of actions to be undertaken with respect to the international space community, European stakeholders should become promoters of an international "paradigm shift" vis-à-vis China. Such a shift would be intended to foster a process of real and enduring inclusion of China within the international space community, an inclusion that would, in time, be conducive to an overall cooperative approach to space exploration and space activities in general.

  In order to achieve this radical paradigm shift in the posture of the leading space powers, cooperation-facilitating mechanisms should be defined and implemented with respect to the USA. The EU, ESA, and their member states should invest consistent and coordinated diplomatic resources to spur a political overture by the USA.

  Finally, European stakeholders should continue to make coordinated efforts to reinforce the current mechanisms of dialogue and coordination in the governance of space exploration activities. Despite their few concrete achievements, the GES, ISECG, and ISEF have all been important platforms in making a compelling case for taking the next step in space exploration cooperatively. Supporting their action (and raising their political profile) is crucial for Europe.

  European stakeholders' long-term objective must be to move away from the overwhelming *coordination* paradigm to one of real *cooperation*. In this regard, a key issue is finding the most appropriate framework for the implementation of a

programme with a ring-fenced budget", which could potentially include also space cooperation among its components. For an interesting evaluation of Chinese participation in the EU Framework Programmes and a set of recommendations aimed at stimulating more active Chinese participation, see Arnold, Erik, et al. (2009). "Evaluation of Chinese participation in the EU Framework programmes". EPEC. Web. http://www.eurosfaire.prd.fr/7pc/doc/1237308017_china_fps_final_2009_03_07.pdf. Accessed 25 September 2014.

multilateral endeavour in space exploration. Needless to say, Europe should also take a leading role in suggesting and defining the optimal cooperation model to be deployed.

There are, of course, several models that could be explored, and many scholars and space policy analysts have started to suggest new paradigms for future space exploration.[115] Using the examples of large, non-space R&T cooperative ventures such as the International Thermonuclear Experimental Reactor (ITER), as suggested by Alain Dupas, could provide an interesting framework for large-scale future space exploration programmes, since many characteristics of the ITER model are applicable to human space exploration as well:

(a) It is the largest R&T effort apart from the ISS, with a total cost in the order of $10 billion.
(b) Its participants are the major current (European Union, Japan, Russia, the USA) and future (China, India, Korea) economic and political powers.
(c) It works according to the principle of the non-exchange of funds, with each partner managing the development and procurement of its own part of the system.
(d) It has an integrated managerial structure.[116]

In addition, Dupas advises looking at the ESA model, particularly at its optional programme framework, which "was and is very successful in enabling different nations to pool resources for major space undertakings", although he also recognises that the "ESA system was adapted to the economically and politically quite integrated [Western] European context and [it] might thus be hard to make it work on the world scene".[117] Indeed, the ongoing process of ESA enlargement to include the quite dissimilar economies of Central and Eastern Europe is already posing challenges in this regard.[118]

[115] See, among many, Correll, Randall and Nicolas Peter (2005). "Odyssey: Principles for Enduring Space Exploration" Space Policy 21 (4): 251–258. The two authors have elaborated in particular on the concept of open-systems architectures and meta-principles to organise cooperative undertaking in space exploration. See also Ehrenfreund, Pascale, et al. (2012). "Toward a global space exploration program: A stepping stone approach". Advances in Space Research No. 49: 2–48; Broniatovski, D.A., et al. (2006). "The Case for Managed International Cooperation in Space Exploration". Center for Strategic and International Studies. Washington DC; Finarelli Peggy, and Ian Pryke (2006). "Implementing International Cooperation in Space Exploration". Space Policy Vol. 22 (1): pp. 23–28.

[116] Dupas, Alain (2009). "International Cooperation in space exploration: Lessons from the past and perspective for the future". In: Schrogl, Kai-Uwe, Charlotte Mathieu, Nicolas Peter (eds) (2009). *ESPI Yearbook on Space Policy 2007/2008*: *From Policies to Programmes*. Springer, Vienna: p. 185.

[117] *Ibid*. p. 186.

[118] For additional considerations on the ESA enlargement process, see Klock, Erich and Marco Aliberti. "ESA Enlargement. What Interested Countries can do to prepare themselves for Ultimate Access". ESPI Report 47. February 2014.

Specific models notwithstanding, it is clear that only by being proactive can Europe continue to play a crucial role in the governance of space activities and claim a place in the driver's cabin. As a multilateral, multilayered construction, with international cooperation engrained in its identity, Europe has the *intellectual* and *material* resources to play the role of a "thought leader" in the promotion of a broad and beneficial international partnership in future space exploration.

Chapter 8
Epilogue: Enabling a New World Order

Driven by its sense of centrality, the Central Kingdom has rediscovered its greatness. In what is truly a transformative process, China has joined the USA as a "careful" *producer* of global governance and *shaper* of the international system, marking the end of the unipolar moment. According to the prevailing wisdom, the future global order will be ultimately centred on the systemic and strategic interaction between these two powers.

Yet, the assumption of the emergence of a G-2 world—be it cooperative or competitive—is off target. China and the USA certainly are two protagonists and indispensable pillars in the contemporary quest for a new global order, but there are other "poles", India in the long run and Europe even in the short run.

Although Europe remains an *unfinished project*, suspended—as Henry Kissinger has put it—between a past it has not overcome and a future it has not yet defined,[1] Europe cannot be simply dismissed as a political midget. Indeed, Europe possesses unique *ideational* and *material* strengths that make it a most prominent variable in the world affairs equation.

Europe has been, and continues to be, a driving force behind global developments. After all, it is the entity that has designed and spread across the globe the structures of the international system and many of its current rules of engagement, such as diplomacy, international law, and multilateral organisations. It has played a leading role in the creation of regimes and institutions to govern the international economy as well as in the definition of the economic rules that have enabled globalisation to flourish. What is more, Europe offers a strong and alternative model to that of China and the USA for addressing the challenge of global governance; it is breaking new ground in terms of regional integration, and its ideas and values continue to permeate the inner workings of the international system.

---

[1] Kissinger, Henry (2014). *World Order*. Penguin Press, New York: p. 93.

M. Aliberti, *When China Goes to the Moon...*, Studies in Space Policy 11,
DOI 10.1007/978-3-319-19473-8_8

In addition to its intellectual capital, Europe is the only power with the critical mass to stand on an equal footing with the Chinese and American giants. Its aggregated GDP is the largest in the world, representing over 20 % of the world's total, and its population of more than 500 million is well above that of Japan, Russia, and the USA, though much less than China's. Taken together, European countries represent the world's largest trading block and exporter of capital. The common currency, the Euro, has emerged as one of the world's major reserve currencies, despite its prolonged woes, and with a new wave of enlargement, the economic and demographic weight of the continent is likely to be further strengthened. Finally, in terms of technological innovation, scientific output, and industrial and financial assets, Europe remains a true protagonist in the world arena.

The failure to see Europe as a collective geopolitical whole stems from the fact that it often lacks a "one voice system" and that its achievements in the field of economic integration are unmatched in matters of collective defence, adding to the impression that it can do little more than project soft power. In addition, the recent financial crisis has made Europe increasingly inward looking and curtailed its global assertiveness. And as Pope Francis has perceptively mentioned in his 2014 address to the European Parliament, "despite a larger and stronger Union, Europe seems to give the impression of being somewhat elderly and haggard, feeling less and less a protagonist in a world which frequently regards it with aloofness, mistrust and even, at times, suspicion".[2]

However, even if the time when American scholars were warning about a "rising Europe"[3] as the primary counterbalance to the US power in the world has been overshadowed by the more striking ascendancy of China, this does not mean that Europe can no longer be a producer of global governance and architect of the global order as much as China and the USA. Europe remains a "structural power" and it must be conscious of its great potential.

To be sure, such consciousness should not breed complacency but should lead the continent to play an ever more proactive role on the international stage. Inaction will inevitably condemn Europe to the loss of its international power of creation and ultimately to geopolitical irrelevance. For Europe, the admonition of Tancredi in Lampedusa's *The Leopard* rings true: *If we want things to stay as they are, things will have to change.*

Europe must thus continuously reinvent itself in order to promote its interests and continue to exercise that power to shape global forces which it has possessed for much of modern history. The European integration process has gone in the right direction: Europe has come a long way through almost constant reforms aimed at integrating the valuable resources of its constituent states. But a more ambitious

[2] Pope Francis: Address to the European Parliament. Official Vatican Network. 25 November 2014. Web. http://www.news.va/en/news/pope-francis-address-to-european-parliament. Accessed 25 November 2014.

[3] Kupchan, Charles A (2002). *The end of the American Era. US Foreign Policy and the Geopolitics of the Twenty-first Century.* Vintage Book, New York.

political evolution is proving even more indispensable if Europe is still to count in the international system. At the same time, however, it is also clear that Europe would be in danger of "cutting itself off from the contemporary quest for world order by identifying its internal construction with its ultimate geopolitical purpose".[4]

While compelled to create *more* Europe—at pain of having *no* Europe at all—Europeans must also rediscover international assertiveness, an assertiveness commensurate with their place and directed to ensure Europe's active participation in the construction of the international order they want to see.

There are many playing fields where Europe can strive for a higher profile or even a pivotal role. Space is one of those fields, perhaps the most symbolic, as it possesses unique transversal qualities. Since the onset of the space age, space endeavours, and human spaceflight in particular, have proved to be a domain of "high politics", with a marked geopolitical value and the potential to accompany great societal changes and deliver a myriad of benefits in terms of innovation, inspiration, and economic growth. A more proactive European approach in this field must be understood as a fundamental strategic necessity.

For Europe, exercising assertiveness in space does not necessarily mean embarking upon a solo space exploration programme that is intended to eclipse the past and future achievements of other countries. Pursuing this path would be nonsensical or even counterproductive. Assertiveness rather means that Europe must promote itself as a centre of gravity in international space affairs and provide new solutions to the challenges facing global space endeavours; it means seizing the opportunities that are emerging from the changing space and geopolitical context and taking a lead in the transformation of the current governance of space activities in a direction that is more supportive of its geopolitical interests.

China's arrival on the international space stage and its possible ambitions to send a taikonaut Moon-wards offer unprecedented opportunities in this regard, and, as has been shown, it is in the best interest of Europe to find ways of engaging Beijing in joint space endeavours, while continuing to cooperate with its traditional partners, above all the USA, Japan, and Russia. For this scenario to become effective, an appropriate strategy is needed to promote an international "paradigm shift" vis-à-vis China. Such a shift should be aimed at integrating China's rising space ambitions within the global space exploration context. Specifically Europe should propose itself as an interlocutor between currently isolated players—namely, the USA and Japan on the one side and China on the other—and subsequently be a trailblazer of a broader international endeavour in space exploration.

The successful crafting of a broad and enduring partnership for space exploration would, in fact, not merely avoid confrontational stances in space; it would also prevent their projection onto terrestrial politics and hence undermine an unpromising polarisation of international relations between two competing blocs. Should China's quest to have a "seat at the table" continue to be snubbed by the West, it

[4] Kissinger, Henry (2014). *World Order*. Penguin Press, New York: p. 93.

will be certainly be tempted to reach the Moon on its own, or with alternative partners such as Russia. And if this scenario materialises, the new battle lines will be inescapably drawn.

This would be an unfortunate outcome and it behoves Europe not to let it happen. A polarisation between two competing blocs would leave the continent with little room for geopolitical manoeuvring and would *in fine* force it to take a black or white stance vis-à-vis the two poles. To echo the admonition of Parag Khanna and Mark Leonard, such a polarisation would all but guarantee a repeat of history rather than a break from it. It would be steering the world towards a century as unstable as the last.[5] Therefore, if Europe wants a twenty-first century of progressive governance and a degree of global harmony rather than another war, cold or otherwise, leveraging space to build bridges among nations could be a good way to start. Space possesses unique potential in this respect and achieving cooperation in a highly symbolic domain like human space exploration could thus become one of the key actions for Europe in transforming potential power politics competition into more cooperative win-win approaches.

Real political will is the first enabling condition for having a more assertive Europe and an overall cooperative approach to space exploration. While this objective might have great political appeal for all European constituencies, which are thus likely to unite behind the common goal, major difficulties will doubtless be encountered in the negotiations with the USA, as discussed previously.

To overcome this, a strong political commitment at the highest level will be required. Every great achievement—be it in space or elsewhere—dictates such commitment. Should cooperation between Europe, the USA, and China eventually fail, it will do so because the political leaders have eventually proved incapable of overcoming the obstacles associated with this path.

To be sure, cooperation needs ultimately to be achieved at a technical and operational level, but it first requires statespeople to forge a grand vision and commit to it. When in 1991 the deputy director of China's Aerospace Ministry proposed the State Council to revitalise the manned space programme—the programme guiding China's present efforts—he made the point very clearly: "Whether or not we go ahead with a human spaceflight programme is a political decision, not purely a technical question, not something scientific and technical people can decide by themselves".

This should be understood as an explicit call for the newly appointed leaders of European institutions to provide solid and enduring political backing. There is much at stake and it would be an uplifting scenario to see the President of the European Commission, with prominent European heads of government, meet with the leaders of China and the USA, indeed with the leaders of *all* interested nations, to promote a joint initiative in human space exploration. Would not such a move confer upon Europe the much-needed assertiveness and enable it to play a visible

[5] Khanna, Parag, and Mark Leonard. "Why China wants a G3 world". European Council on Foreign Relations. 12 September 2011.

and pivotal role in the years to come (provided, of course, that such an initiative not only consisted of words but was backed by the availability of credible resources)?

The second key factor Europe must necessarily confront is time. It is evident that the more Europe postpones concrete policy actions, the more Chinese space ambitions will take their own path. The risk for Europe is to completely lose the possibility of shaping the priorities of the future international space agenda, as well as the capability to attract partners in order to materially—and politically—capitalise from such cooperation. It would be extremely hard for Europe to harvest valuable paybacks in a policy reality it had not contributed to shaping. China shows a willingness to cooperate with the West but its space ambitions will be achieved with or without a European contribution.

It should also be acknowledged that windows of opportunity do not last forever and must be seized before alternative, undesirable outcomes take root. In the light of the recent reorientation of Russian foreign policy towards closer cooperation with China and of the expected renewal (and likely upgrade) of their 10-year cooperation agreement by 2016 at the latest, first steps should be taken by Europe in the very near future, before Russia seizes the opportunity of bilaterally engaging with China to reach the Moon in a partisan move.

As we know, such a prospect would destroy the possibility of achieving an overall cooperative approach to exploration—and to international relations in general. On the contrary, it would be a clear invitation to a new space race. It is highly likely that Sino-Russian cooperation would compel the USA to embark on a more ambitious endeavour, possibly stepping up the human exploration of Mars as an attempt to eclipse a Sino-Russian lunar endeavour. In that scenario, Europe may have no choice but to bandwagon with the USA or stand aside altogether. As happened during the space race, the fragmentation of the international community will once again be demonstrated, and in this case also prompted, by space.

To prevent such a gloomy outcome, it is thus vital for Europe to endorse a coherent and proactive grand strategy aimed at including—rather than confronting—China's space ascendancy and to take real steps towards moving away from the *talking shop model* that, ISS apart, has so far been overwhelming in the governance of human spaceflight endeavours.

The most symbolic and beneficial measure in this regard would be clearly offered by the eventual participation of China in the ISS programme. Such participation would in itself constitute an invaluable stepping stone in laying the foundation for a broader and bolder cooperative undertaking in space exploration. Equally importantly "democratising" the ISS would provide the West with a move of "smart diplomacy", forcing China to show its hand on its announced willingness to cooperate. It would in a sense compel China to *really* open its upcoming CSS to the world. There are, of course, technological and political hurdles that need to be solved to enable a taikonaut to reach the ISS and an astronaut the CSS, but these are not insurmountable if there is political will. The only critical constraint is time: here too windows of opportunity are inexorably closing and steps must be taken before the ominous path of "two separate highways for exploration" congeals and a new Cold War scenario emerges.

The world needs a new departure, not a reiteration of history. Achieving such a departure will certainly entail formidable challenges. Many hurdles will stand in the way and many steps will have to be taken in order to overcome them. But only by seeking and finding common ground on this grand vision, while putting the political issues of the day to one side, can cooperation be achieved. The rewards will prove invaluable.

When in 1971 the USA and the USSR agreed on the Apollo-Soyuz Test Project, it was because they were able to raise their sights beyond their political rivalry. Great geopolitical benefits have been the result. Their symbolic "handshake in space" emerged as a political act of peace between two foes, marking the end of the space race and easing dangerous tensions on Earth.

The ball is now in Europe's court to bring the world to a similar historical handshake, this time on a joint flight to the Moon. Europe can enable a new world order!

Erratum to: When China Goes to the Moon...

Marco Aliberti

M. Aliberti, *When China Goes to the Moon...*, Studies in Space Policy 11,
DOI 10.1007/978-3-319-19473-8

The original version of this book was inadvertently published with an outdated version of the Editorial Advisory Board.

The online version of the original book can be found under
DOI 10.1007/978-3-319-19473-8

M. Aliberti, *When China Goes to the Moon...*, Studies in Space Policy 11,
DOI 10.1007/978-3-319-19473-8_9

Appendix A: China at a Glance

Key dates	
October 1949	Proclamation of the People's Republic of China
1958–1960	Great Leap Forward
November 1960	Sino–Soviet Split
1966–1969	Cultural Revolution
November 1971	Replacement of the Republic of China in the UN Security Council
February 1972	President Nixon's visit to China
September 1976	Death of Mao Zedong
December 1978	Launch of the *gaige kaifang* and Four Modernisations
December 1982	Adoption of the new Constitution
June 1989	Tiananmen Square protests
July 1997	Return of Hong Kong to China
December 1999	Return of Macau to China
December 2001	Access to the WTO
November 2002	Election of Hu Jintao (President) and Wen Jiabao (Premier)
August 2008	Beijing Olympics
December 2010	Beijing outstrips Japan as the second world economy
November 2012	Election of Xi Jinping (President) and Li Keqiang (Premier)
October 2014	Beijing outstrips the United States as the first world economy in PPP terms

Geography and population	
Total land area	9,327,489 km^2
Rural	55.6 % of land area
Forest	22.6 % of land area
Administrative division	22 provinces, 5 autonomous regions, and 4 municipalities
Total population	1357 billion (2013)
Population growth	+0.5 % (annual)

(continued)

M. Aliberti, *When China Goes to the Moon...*, Studies in Space Policy 11,
DOI 10.1007/978-3-319-19473-8

Geography and population	
Life expectancy at birth	75.2 (2012)
Workforce population	787,632 million (2013)
Literacy rate	95 % of the adult population (2010)
Population age (0–14) (% of total)	18 % (2013)
Population age (15–64) (% of total)	73 % (2013)
Population age—(over 65) (% of total)	9 % (2013)
Access to electricity	99.8 % of the population (2011)

Politics	
Type of government	Communist state
Leadership	China's Communist Party (CCP) and Paramount Leader Head of the State: President Xi Jinping (November 2012) Head of the Government: Premier Li Keqiang (November 2012)
Executive branch	State Council Premier, 4 vice-premiers, and 25 ministries
Legislative branch	Unicameral National People's Congress 29,857 seats 5-year terms Elected by municipal, regional, and provincial People's Congresses Last elections held in 2012
Judicial branch	Supreme People's Court and Supreme People's Procuratorate 340 judges 5-year term Courts of the SPC: criminal, civil, economic, administrative, ad hoc Judiciary subordinates the written law to the "three Supremes": 1. "Supremacy of the business of the CCP" 2. "Supremacy of the interests of the people" 3. "Supremacy of the Constitution and the laws"
Military branch	People's Liberation Army Controlled by the Central Military Commission of the CCP Four main departments: General Armaments Department General Logistics Department General Political Department General Staff Department

Economy[a]		
GDP (current USD)	$9240 trillion (2013) (second rank)	IMF
GDP per capita	$6807 (2013)	IMF
GDP growth rate	7.7 % (2013)	WB
GNI at PPP (current USD)	16,080,584,813,552 (2013) (second rank) > see table below	WB
GNI (per capita)	$6560 (2013)	WB
Income level	Upper middle income	WB

(continued)

Economy[a]		
Public debt (% of the GDP)	26 % (2013)[b]	WB
Total reserves (current USD)	3,880,368,275,099 (2013)	WB
Gross savings % of GDP	49.71 % (2013)	IMF
Gross investment % of GDP	47.78 % (2013)	IMF
Current account balance (current USD)	+193,139,152,768 (2012) +182.807 billion (2013)	IMF
Inflation rate	2.6 % (2013)	IMF
Unemployment rate	4.5 %	WB
Poverty headcount ratio	4.5 %	WB
Military expenditures	2.1 % of GDP	WB
R&D expenditures	1.98 % of GDP	WB

According to the updated data on the world economy released in early October 2014 by the International Monetary Fund (IMF), China's economy ranks as the world's biggest in purchasing power parity terms, followed by the United States and India. As amply documented by historians, the new element in this configuration is represented by the United States, not China and India, whose economies are merely regaining a title they have held for much of recorded history

[a]Source: WB indicators. http://data.worldbank.org/indicator#topic-20

[b]http://www.tradingeconomics.com/china/government-debt-to-gdp

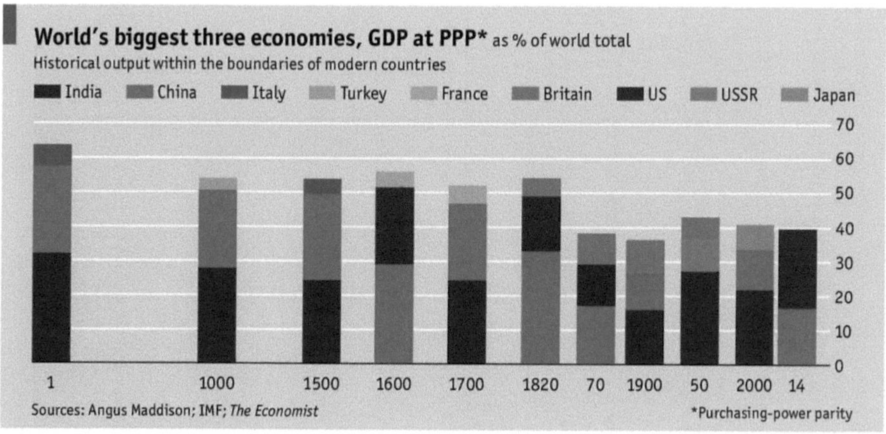

Appendix B: China Long March into Space. Past Achievements and Future Prospects

Date	Event
8 October 1956	Founding date of China's space programme
24 April 1970	Launch of China first Earth satellite
5 October 1970	First squad of *yuhangyuan* selected
26 November 1975	First Earth observation satellite
29 January 1984	First launch of the LM-3
7 September 1988	First experimental meteorological satellite in polar orbit
7 April 1990	Launch of the first commercial communications satellite
10 June 1997	First Geostationary Meteorological Satellite
19 November 1999	First flight of the *Shenzhou*-1
31 October 2000	First navigation satellite
15 October 2003	Flight of Yang Liwei on-board *Shenzou*-5
2005	Three-crew flight
24 October 2007	*Chang-e* 1 mission, lunar probe
25 September 2008	First extravehicular activity
2009	Second lunar probe
29 September 2011	Launch of the first space laboratory, Tiangong-1
16 June 2012	Launch of the first space woman and of the first crew of Tiangong-1
June 2013	*Shenzhou*-10 mission
3 December 2013	*Chang'e*-3 mission, lunar lander
2015	*Tiangong-2*
2016	*Shenzhou-11*
2017	*Chang'e-4 sample return mission*
2018	*Launch of the CSS Core module*
2019	*Chang'e-5 sample return mission*
2022	*Completion of the CSS*
2023	*First crewed visit to the CSS*
2025	*Shenzhou circumlunar mission*[a]

(continued)

M. Aliberti, *When China Goes to the Moon...*, Studies in Space Policy 11,
DOI 10.1007/978-3-319-19473-8

Date	Event
2030	*Manned lunar landing*[a]
2033	*Return of first samples from Mars*[a]
2040	*Construction of a lunar outpost*[a]
2050	*China's first manned Mars landing*[a]

[a]Missions not yet part of any official plan (*Source*: CAS)

Appendix C: CASC Organisation (Author's Visualisation)

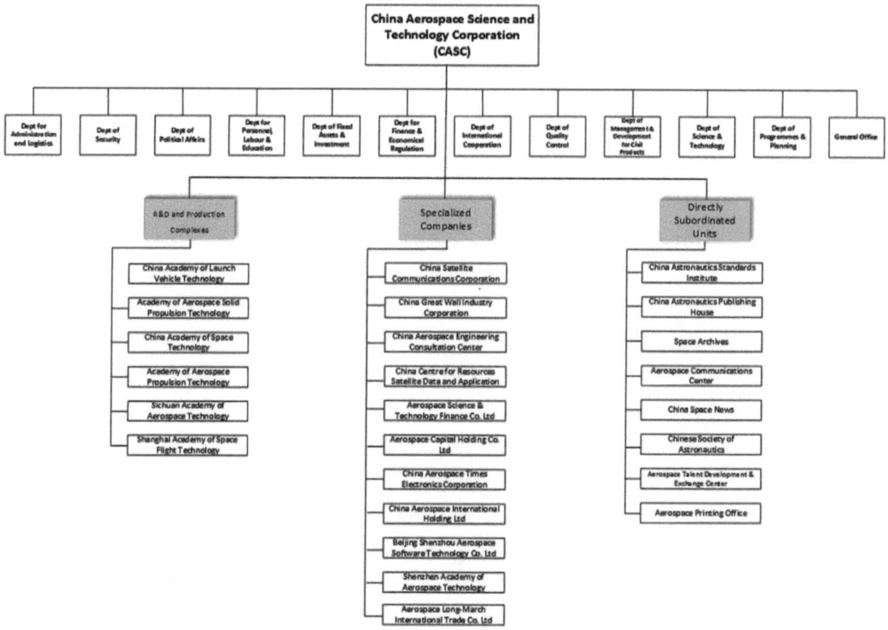

M. Aliberti, *When China Goes to the Moon...*, Studies in Space Policy 11,
DOI 10.1007/978-3-319-19473-8

Appendix D: EU–China Dialogue Architecture

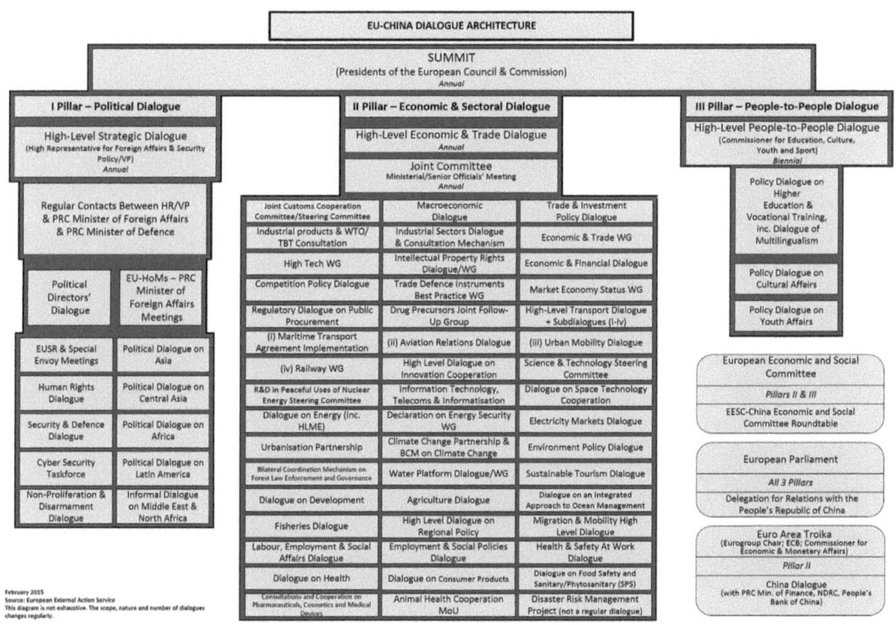

M. Aliberti, *When China Goes to the Moon...*, Studies in Space Policy 11,
DOI 10.1007/978-3-319-19473-8

Appendix E: Space Exploration Programmes of National Space Agencies

This appendix summarises the space exploration programmes of selected countries. These are the United States, Russia, Europe, Japan, and India, which are the most active actors in this domain. Space exploration comprises both robotic and human spaceflight missions.

E.1 United States/NASA

Key data	
National Space Agency	National Aeronautics and Space Administration—NASA
Civil budget (2013)	$19,770 million
Budget for robotic exploration (2013)	$3126 million (16 % of the budget)
Budget for manned spaceflight (2013)	$7809 million (39 %)
International membership	ISS, IAF, ISEF, GER, ISECG, IMEWG, ILEWG
Priorities in space exploration	Robotic Exploration of the Solar System
	Human spaceflight LEO
	Human spaceflight beyond LEO

Thanks to the Apollo legacy, NASA has undoubtedly become the most well-known brand when dealing with space exploration and human spaceflight. Following the collapse of the Soviet Union, NASA's space exploration programme further consolidated its leadership position in terms of both budget and capabilities and programmes. To date, NASA remains the biggest civilian space agency in the world, boasting the broadest set of space activities.

As stated in the ISECG 2013 Annual Report, "in 2013 NASA continued to make significant progress in space exploration that includes critical scientific and technical achievements from research on the International Space Station (ISS), the debut

M. Aliberti, *When China Goes to the Moon...*, Studies in Space Policy 11,
DOI 10.1007/978-3-319-19473-8

of a new NASA mission to capture and redirect a small asteroid, the successful launch of a second commercial resupply service provider to ISS, and the launch of new science missions".

NASA is the leading partner of the ISS programme and at the ISEF held in Washington in January 2014 proposed extending its utilisation up to 2024. In anticipation of future human missions beyond low Earth orbit, NASA continues to utilise ISS for human research and to test new technologies. This includes studies to better understand the effects of prolonged spaceflight on human health and performance (a year-long mission involving a US astronaut and a Russian cosmonaut is planned for 2015), the testing of an innovative inflatable module on ISS in 2015, and the utilisation of a 3D printer to investigate the concept of in-orbit fabrication of replacement parts in 2015.[1]

Following the decommissioning of the space shuttle and the cancellation of the Constellation programme, NASA has focused its efforts on the development of the Space Launch System (SLS), a space shuttle-derived heavy launch vehicle, and of the Orion Multi-Purpose Crew Vehicle (MPCV), a manned spacecraft designed for mission beyond LEO. On 16 January 2013, NASA and ESA held a joint press conference to announce their cooperation on the service module of the MPCV.[2] The Obama Administration has proposed using the future SLS to launch a crewed Orion MPCV for an Asteroid Redirect Mission (ARM) that will robotically capture a small near-Earth asteroid, redirect it into a Moonlike orbit, and explore it with astronauts sometime between 2021 and 2023.[3] NASA, however, has not formally committed to this Asteroid Redirect Mission (ARM), and there appears to be little enthusiasm, let alone a firm national consensus, in this regard. Besides feasibility considerations, NASA's international partners have given a lukewarm response to the ARM proposal. The goal of an ARM is, in fact, not aligned with that of US partners, which still seem to lean in favour of lunar exploration.

In terms of robotic exploration, NASA has been the most active space agency in the exploration of the solar system, extending its presence to visit to different celestial bodies, including the Moon, Mercury, Mars, Jupiter, and Saturn. The Moon and Mars, however, continue to stimulate the main interest of NASA. In addition to the missions that are already underway—the Lunar Reconnaissance Orbiter launched in October 2009, the Gravity Recovery And Interior Laboratory (GRAIL), the Mars Odyssey mission, the Mars Reconnaissance Orbiter (MRO), and the Mars Science Laboratory (MSL) rover named Curiosity, NASA launched new science missions in 2013, including the Lunar Atmosphere and Dust

[1] International Space Exploration Coordination Group (2014). Annual Report 2013. ISECG Secretariat. Web. http://www.globalspaceexploration.org/wordpress/

[2] ESA, in particular, will be in charge of supplying NASA with an ATV-derived service module in order to provide the Orion spacecraft with propulsion, power, and thermal control, as well as water and gas for the Orion's crew.

[3] International Space Exploration Coordination Group (2014). Annual Report 2013. ISECG Secretariat: p. 32. Web. http://www.globalspaceexploration.org/wordpress/

Environment Explorer (LADEE) and the Mars Atmosphere and Volatile EvolutioN Mission (MAVEN).[4]

LADEE, launched on 6 September 2013, will provide unprecedented information about the environment around the Moon and give scientists a better understanding of other planetary bodies in our solar system and beyond. [5] MAVEN is the second mission for NASA's Mars Scout Programme, launched on 18 November 2013. It is intended to obtain critical measurements of the Martian upper atmosphere to help understand how the climate has changed over the Red Planet's history.[6]

The United States has also announced a plan to create an International Lunar Network (ILN) involving robotic landers, orbiters, and instrumentations by March 2018 and to send an unmanned probe to investigate the internal structure and crustal deformation of the Red Planet in March 2026.[7]

E.2 Russia/Roscosmos

Key data	
National Space Agency	Roscosmos
Civil budget (2013)	203,304 million ($6414 million)
Budget for robotic exploration (2013)	14,323 million (7 % of the budget)
Budget for manned spaceflight (2013)	40,923 million (20 % of the budget)
International membership	ISS, IAF, ISEF, GER, ISECG, IMEWG, ILEWG
Priorities in space exploration	Robotic lunar exploration
	Robotic Mars exploration
	Human spaceflight LEO
	Human spaceflight beyond LEO (planned)

Thanks to the considerable investment made during the Cold War, together with the United States, Russia remains an undisputed master in human spaceflight and space exploration. The country boasts a number of landmark achievements in space, including the first artificial satellite (Sputnik), the first human in space (Yuri Gagarin), the first unmanned lunar landing (Luna-1 in January 1959), and the first space station (the Salyut).

[4] Al-Ekabi, Cenan, Blandina Banares, Peter Hulsroj, Arne Lahcen (eds) (2015). *ESPI Yearbook on Space Policy 2012/2013. Space in a Changing World.* Springer, Vienna: p. 122

[5] International Space Exploration Coordination Group (2014). *Annual Report 2013.* ISECG Secretariat: p. 32. Web. http://www.globalspaceexploration.org/wordpress/

[6] *Ibid.* p. 32.

[7] Al-Ekabi, Cenan, Blandina Banares, Peter Hulsroj, Arne Lahcen (eds) (2015). *ESPI Yearbook on Space Policy 2012/2013. Space in a Changing World.* Springer, Vienna: p. 121

Although the "capabilities acquired by Russia through its numerous programmes during the Soviet era might have been lost or deactivated given the low volume of exploration missions conducted" between 1990s and the early 2000s,[8] since the approval of the long-term Federal Space Programme in 2006, Russian space power is gradually resurging. Russia has shown the intention to regain the status of a leading space power. [9]

With regard to its human spaceflight activities, Roscosmos continues to fully utilise the ISS in preparation for future human spaceflight missions. Using Soyuz and Progress, Roscosmos provides regular crew transportation to the ISS and ISS cargo resupply services with Europe's ATV and Japan's HTV providing auxiliary support. Between 2012 and 2013, Russia launched eight Soyuz spacecraft and eight launches of the Progress cargo transfer vehicles. It should also be noted that, following the decommissioning of NASA's Space Shuttle, Roscosmos has remained the sole launch provider able to regularly transport astronauts to the ISS. In April 2013, an agreement was signed between Roscosmos and NASA to continue using the Soyuz spacecraft for sending American astronauts to the ISS until June 2017. In the same year, Roscosmos and NASA also announced a year-long mission aboard the ISS involving Russian cosmonaut Mikhail Kornienko and American astronaut Scott Kelly for 2015. The mission, which is twice as long as a typical astronaut mission to the ISS, will study the effects of prolonged spaceflight on human health and prepare the ground to embark upon missions beyond the LEO framework. Roscosmos is currently also envisaging a manned lunar mission sometime between 2025 and 2030.

In terms of robotic exploration, Russia is focusing its efforts on the Moon and Mars. Roscosmos' Lunar Exploration Programme encompasses two main series of missions: Luna-Glob and Luna-Grunt. Due to financial constraints, the former series is now scheduled to launch its first Luna-Glob 1 (a Moon orbiter and landing probe) in 2016. The follow-up Luna-Glob 2 or Luna-Resurs 1, an orbiter–rover mission initially planned in combination with the Indian Chandrayaan-2 mission, has been postponed to 2018,[10] while the Luna-Resurs 2—a multielement mission comprising a lander, a rover, and retransmitting satellite—to 2019.[11] Following the

[8] Thépaut, Jean-Baptiste (2012). "Analysis of Cooperation Opportunities for Europe in Future Space Exploration Programmes (COFSEP)". Proceedings of the 63rd International Astronautical Congress, Naples, Italy, October 1-5, 2012. Paper: IAC-12-A3.1.3

[9] Harvey, Brian (2007). *The rebirth of the Russian Space Program. 50 Years after Sputnik, New Frontiers.* Springer - Praxis Publishing, Chichester, UK

[10] Due to the loss of the Phobos-Grunt mission, which was intended to test the landing system of Luna-Resurs 1, Roscosmos has found itself unable to provide the lander within the proposed timeframe (end of 2015), and as a consequence India has decided to develop its Chandrayaan-2 lunar mission independently.

[11] Al-Ekabi, Cenan, Blandina Banares, Peter Hulsroj, Arne Lahcen (eds) (2015). *ESPI Yearbook on Space Policy 2012/2013. Space in a Changing World.* Springer, Vienna: p. 122

Luna-Glob series, the Luna-Grunt series will comprise two missions to be launched in the early 2020s. The first will comprise a lunar orbiter and a lander carrying a rover capable of in situ soil analysis. The second will be a sample return mission, featuring a rover and an ascent stage that will return up to 1 kg of rock samples.

As for Mars Exploration, in November 2011 the Phobos-Grunt mission was launched in cooperation with China to the Martian Moon Phobos, but was lost due to an engine firing failure. Roscosmos now plans on launching another probe Phobos-Grunt 2 to Mars's moon by 2022.[12] Roscosmos is also cooperating with ESA in the development of the joint ExoMars programme: in particular, it will provide the launches of the 2016 and 2018 missions, in addition to the scientific instruments and the descent module and surface platform of the 2018 mission. By the mid 2020s, Russia also intends launching Venera-D (a Venus orbiter/lander) and Mercury-P, the first lander on Mercury.

E.3 Europe/ESA

Space exploration in Europe	
Principal Space Agency	European Space Agency—ESA
ESA budget (2013)	3449 € ($4550 million)
Budget for robotic exploration (2013)	696 € (28 % of the budget)
Budget for manned spaceflight (2013)	305 € (11 % of the budget)
International membership	ISS, IAF, ISEF, GER, ISECG, IMEWG, ILEWG
Priorities in space exploration	Robotic lunar exploration
	Robotic Mars exploration
	Human spaceflight in LEO

Europe, defined as the sui generis space actor denoting the interplay of ESA, the EU, and their Member States, can boast a rich and long-standing tradition of space exploration, be it with robotic or manned missions. European capabilities, which are mainly shared between Germany, France, Italy, and the United Kingdom, have been traditionally pooled through the efforts of ESA in an integrated programmatic approach.

With regard to robotic exploration activities, ESA has undertaken a number of successful programmes on its own or in partnership with other space agencies (SMART-1, Mars Express, Venus-Express, Cassini-Huygens, Rosetta, etc.). ESA has in addition assisted both India and China in the implementation of their lunar exploration programmes, contributing with three instruments to India's

[12] *Ibid.* p. 126.

Chandrayaan-1 lunar mission and with ground station control and data assistance for China's *Chang'e* missions.[13]

However, due to financial constraints, the ESA's Lunar Lander mission that was announced in 2010 with the target of an autonomous landing on the Moon's South Pole in 2018 was not funded at the ESA Council Meeting at Ministerial Level in November 2012.[14] The decision to put the programme on hold was taken to favour the Launcher development and better tackle the emerging uncertainties with regard to the Mars Exploration Programme. In fact, the 2012 withdrawal of NASA from the joint *ExoMars* programme forced ESA to rethink the cooperation scheme with the programme eventually becoming an ESA–Roscosmos joint cooperation. The reform was finalised in March 2013, with the signature of a cooperation agreement between ESA and Roscosmos for the joint development of *ExoMars*. The programme now consists of a split mission to be launched in January 2016 and May 2018, respectively.[15] The 2016 "mission will carry the ESA-provided Trace Gas Orbiter (TGO) and the Entry Descent and Landing Demonstrator Module (EDM) and will be launched by a Russian Proton launcher. The 2018 mission will consist of an ESA-provided Carrier Module, bringing the Russian Descent Module and Surface Platform and the ESA Rover to Mars".[16] The Rover will investigate Mars' surface, searching for past and present signs of life. The scientific instruments of the two missions will be provided by both partners. A number of potential missions for the post-ExoMars period have already been identified by ESA.

Europe's human spaceflight programme is currently focused on LEO activities. ESA and a number of its Member States (France, Germany, Italy, the United Kingdom, the Netherlands Belgium, Sweden, Norway, Denmark, and Switzerland) are partners in the ISS programme and have provided significant contributions to this programme: suffice it to mention the Automated Transfer Vehicles (ATVs), the Columbus Orbital Laboratory, and other ISS infrastructures such as Node 2, Node 3, and the Cupola.

However, not having developed independent manned access to space, Europe currently relies on Russia's Soyuz for launching its astronauts to the ISS. The newest ESA squad of astronauts, who graduated from the European Astronaut Centre (EAC) in November 2010, is composed of Luca Parmitano, Alexander Gerst, Samantha Cristoforetti, Timothy Peake, Andreas Mogensen, and Thomas Pesquet.[17] Luca Parmitano and Alexander Gerst have already

---

[13] Ehrenfreund, Pascale, et al (2012). "Toward a global space exploration program: A stepping stone approach". Advances in Space Research No. 49: pp. 26–27

[14] Al-Ekabi, Cenan, Blandina Banares, Peter Hulsroj, Arne Lahcen (eds) (2015). *ESPI Yearbook on Space Policy 2012/2013. Space in a Changing World*. Springer, Vienna: p. 120

[15] International Space Exploration Coordination Group (2014). Annual Report 2013. ISECG Secretariat: p. 26. Web. http://www.globalspaceexploration.org/wordpress/

[16] *Ibid*. p. 26.

[17] "ESA—Human Spaceflight and Exploration—Astronauts—Graduation of Europe's new astronauts". European Space Agency. 25 August 2011. Web. http://www.esa.int/esaHS/SEMRFLIRPGG_astronauts_0.html

flown to the station, respectively, in May 2013 and May 2014, while the others have been assigned to missions scheduled to fly between the end of 2014 and 2017.

ESA utilisation of the ISS comprises a number of experiment series in medical, biomedical, and physiological sciences that are also intended as preparatory activities to pave the way for future human spaceflight beyond LEO. In this regard, ESA is making progress in the study and development of the European eXPErimental Re-entry Testbed (EXPERT) and of the Micro-Ecological Life Support System Alternative (MELiSSA).[18] ESA has in addition agreed with NASA to develop and supply NASA with the ATV-derived service module of the Orion MPCV, which will provide the spacecraft with propulsion, power, and thermal control.

A long-term vision for ESA's role in human space exploration beyond LEO is, however, still missing, and possible EU involvement in this domain is increasingly deemed necessary. Although the Union initially identified satellite systems and applications—specifically Galileo and Copernicus—as its policy priority areas, the expansion of its mandate over space matters has eventually influenced the Commission to consider a possible contribution to space exploration and to provide political backing for its development.

Starting with the European Space Policy adopted in 2007, the EU has increasingly acknowledged the political dimension of space exploration and thus the necessity to become more actively involved in this domain. Looking at its most recent initiatives in this regard, on 20 August 2013, the EC issued a working document entitled, "A Role for Europe within a Global Space Exploration Endeavour". The document emphasises the importance of an integrated approach in the field of space exploration and proposes building the current European long-term scenario, consistent with international plans, on a three-step sequence:

- First step, 2015–2020: utilisation of the ISS, robotic missions (including ExoMars), and R&D for preparing the next step and demonstration of human transportation capabilities
- Second step, 2020–2030: continued robotic missions including Mars Sample Return, human missions beyond low Earth orbit, and R&D for preparing the next step
- Third step, >2030: sophisticated robotic missions in the solar system, continued human exploration missions, including possibly human missions to Mars[19]

[18] International Space Exploration Coordination Group (2014). Annual Report 2013. ISECG Secretariat: p. 26. Web. http://www.globalspaceexploration.org/wordpress/

[19] European Commission. A Role for Europe within a Global Space Exploration Endeavour. Commission Staff Working Document. SWD (2013) 301 final. Brussels, European Union. 20 August 2013

E.4 Japan/JAXA

Key data	
National Space Agency	Japan Aerospace Exploration Agency—JAXA
Civil space expenditures (2013)	¥190 billion ($1957 million)
Budget for robotic exploration (2013)	¥25 billion (13 % of the budget)
Budget for manned spaceflight (2013)	¥34 billion (18 % of the budget)
International memberships	ISS, ISEF, GER, ISECG, IMEWG, ILEWG, IAF
Priorities in space exploration	Robotic Lunar Exploration
	Robotic Exploration of NEAs
	ISS utilisation

Japan is a partner in the ISS programme and has invaluable mature expertise in robotic exploration. Already in 1990, thanks to the successful launch of its *Hiten* lunar probe, Japan succeeded in breaking the monopoly on missions to the Moon held by the United States and the USSR.[20] Robotic exploration of the Moon continued with the launch of SELENE-1 in September 2007. The probe, also known as *Kaguya*-1, impacted on the Moon after the successful completion of its operation in June 2009. Japan was also the first country to successfully implement an asteroid sample return mission. The *Hayabusa* (Peregrine Falcon) mission, launched in May 2003, explored the near-Earth asteroid Itokawa and successfully returned some samples to Earth in June 2010. Less fruitful has been Japan's Mars Exploration Programme. The *Nozomi* (Wish) Mars orbiter launched in June 1998 was in fact unable to reach Mars orbit due to an electrical failure. Operation was terminated in December 2003 and the Mars Exploration Programme put on hold.

JAXA is currently preparing the sample return mission named *Hayabusa-2* to be launched in 2014/2015 with an expected arrival at the target asteroid in 2018 and an expected return to the earth in 2020.[21] JAXA is also working on the lunar landing explorer SELENE-2. The probe, which is currently scheduled for launch in 2018, will land on the Moon and deploy a robotic rover to investigate the surrounding areas. In addition to SELENE-2, a Moon sample return mission (SELENE-3) is also planned.

With regard to human spaceflight activities, Japan's participation in the ISS programme has two important cornerstones: exploitation of the Japanese experiment module KIBO (Hope) and the H-II Transfer Vehicle (HTV), the automated cargo spacecraft used to resupply the KIBO and the ISS. Japan's HTV, also known as *Kōnotori* (Oriental Stork), performed its maiden flight to the ISS in 2009, in total

[20] Hiten, was Japan's first lunar probe, the first robotic lunar probe since the Soviet Union's Luna 24 in 1976 and the first lunar probe launched by a state other than the Soviet Union or the United States.

[21] International Space Exploration Coordination Group (2014). Annual Report 2013. ISECG Secretariat: p. 29. Web. http://www.globalspaceexploration.org/wordpress/

providing four successful deliveries to the station. Three additional HTV flights are scheduled, respectively, in February 2015, 2016, and 2017. In November 2013, Japanese astronaut Koichi Wakata started his third stay on the ISS, becoming the first Commander of the ISS during the 39th expedition period. JAXA is currently investigating the possibility of a human mission beyond the LEO framework and is studying the concept of a new launch system (H-X) and of an Orbital Transfer Vehicle to contribute to future space exploration.

While a decision on the development of independent spaceflight capabilities is expected before the end of the decade, financial commitment to the human spaceflight programme has continued to sensibly reduce in recent years. Indeed, with the release of the updated "Basic Plan for Space Policy" in January 2013, the priority has shifted to favour the areas of satellite navigation and satellite reconnaissance systems.[22]

E.5 India/ISRO

India	
National Space Agency	Indian Space Research Organisation—ISRO
Budget (2013)	67,779 million ($1173 million)
Budget for robotic exploration (2013)	5213 (8 % of the budget)
Budget for manned spaceflight (2013)	270 (0.3 % of the budget)
International memberships	ISEF, GER, ISECG, IMEWG, ILEWG, IAF
Priorities in space exploration	Robotic lunar exploration
	Robotic mars exploration
	Human spaceflight (planned)

Space exploration and human spaceflight have been rather marginal activities at ISRO, which has up to recent years decided to give priority to the development of space applications (in particular telecommunications and earth observation). The situation, however, is rapidly changing, with ISRO embarking on new space endeavours that include lunar and Mars exploration.

The successful launch and realisation of *Chandrayaan-1* (Moon-Vehicle), India's first planetary mission in 2008, was a landmark achievement for the Indian Space Programme.[23] Within the 12th Five-Year Plan (2012–2017), ISRO is working on the launch of *Chandrayaan-2*, the follow-up Lunar Lander mission. The mission was initially planned as a joint venture with Roscosmos, featuring an ISRO-supplied orbiter and rover and a Roscosmos-supplied lander, scheduled for

[22] Al-Ekabi, Cenan, Blandina Banares, Peter Hulsroj, Arne Lahcen (eds) (2015). *ESPI Yearbook on Space Policy 2012/2013. Space in a Changing World.* Springer, Vienna: p. 118

[23] Planning Commission, Government of India. Twelfth five-year plan (2012/2017). Faster, More Inclusive and Sustainable Growth. New Delhi, India. 2013: pp. 264–268

launch in 2015. However, due to technical and financial hurdles, Roscosmos has said it is unable to provide the lander within the proposed timeframe and as a consequence India has decided to develop its *Chandrayaan-2* lunar mission independently. The launch is currently planned for the end of 2016 or beginning of 2017.

In addition to lunar exploration, a key focus of the Indian space exploration programme during the 12th Five-Year Plan is represented by Mars exploration. ISRO's Mars Orbiter Mission (MOM), also known as *Mangalyaan* (Mars-vehicle), was launched on 5 November 2013, successfully reaching Mars orbit on 24 September 2014. The probe, which carries five Indian-built scientific payloads, will investigate the Red Planet's surface, atmosphere, and exosphere intending to gain comprehensive understanding of its geologic and possible biogenic evolution.

With regard to human spaceflight activities, India has so far had only one astronaut flight in April 1984 under the Soviet Intercosmos programme.[24] However, since the mid-2000s ISRO has conducted technological studies on human spaceflight scenarios, which have led to a proposal to the Indian government for a first manned mission in the 2016 timeframe and an ambitious programme to follow. The government has not yet accepted this proposal.

While the government has allocated funds in the 12th Five-Year Plan to conduct pre-project studies and develop critical technologies associated with the proposed mission, no human spaceflight mission is expected to take place before the end of the decade. For one thing, the financial commitment to the development of human spaceflight capabilities remains all in all too modest. In addition, while ISRO has already validated its re-entry technology, it still lacks an operational GSLV-Mk III that is needed to launch a two-member crew to LEO and have them return safely to Earth.[25] Due to a number of launch failures of the GSLV-Mark I and II, the GSLV-Mark III remains under development.

[24] Makish Rakesh Sharma flew under the Soviet Intercosmos programme in April 1984 for a seven-day mission.

[25] "India Not to Undertake Human Space Flight Before 2017: ISRO". 17 Sept. 2012. The Economic Times. Web. http://articles.economictimes.indiatimes.com/2012-09-17/news/33902713_1_cryogenic-engine-gslv-mk-iii-radhakrishnan-today. 30 May 2014

List of Acronyms

Acronym	Explanation
A	
ACC	Astronaut Centre of China
APRSAF	Asia-Pacific Regional Space Agency Forum
AP-MCSTA	Asia-Pacific Multilateral Cooperation in Space Technology and Applications
APSCO	Asia-Pacific Space Cooperation Organisation
ARM	Asteroid Retrieval Mission
ASAT	Anti-Satellite Test
ASEAN	Association of Southeast Asian Nations
ASI	Agenzia Spaziale Italiana (Italian Space Agency)
C	
CALT	China Academy of Launch Vehicle Technology (under CASC)
CAS	Chinese Academy of Sciences
CASC	China Aerospace Corporation
CASIC	China Aerospace Science and Industry Corporation
CAST	China Academy of Space Technology (under CASC)
CCP	China's Communist Party
CSSTEAP	Centre for Space Science and Technology Education in Asia and the Pacific
CEA	Conferencia Espacial de las Américas (Space Conference of the Americas)
CGWIC	China Great Wall Industry Corporation (under CASC)
CLEP	Chinese Lunar Exploration Programme
CLTC	China Satellite Launch and Tracking Control General
CMI	Civil–Military Integration
CMSA	China Manned Space Agency
CNES	Centre National D'Etudes Spatiales (French Space Agency)
CNP	Comprehensive National Power
CNSA	China National Space Administration
COSTIND	Committee on Science, Technology and Industry for National Defence

(continued)

M. Aliberti, *When China Goes to the Moon...*, Studies in Space Policy 11,
DOI 10.1007/978-3-319-19473-8

Acronym	Explanation
CPR	China's People Republic of China
CSS	Chinese Space Station
COSPAR	Committee on Space Research
CFSP	Common Foreign and Security Policy (EU)
CENC	China–Europe Navigation Satellite System Technical Training and Cooperation Centre
CFOSAT	Chinese-French Oceanographic Satellite
CSES	China Seismo-Electromagnetic Satellite
D	
DLR	Deutsches Zentrum für Luft- und Raumfahrt (German Space Agency)
E	
EAC	ESA Astronaut Centre
ECFR	European Council on Foreign Relations
EDA	European Defence Agency
EOR	Earth Orbit Rendezvous
ESA	European Space Agency
ESPI	European Space Policy Institute.
EU	European Union
EVA	Extra Vehicular Activity
G	
GAD	General Armaments Department
GDP	Gross Domestic Product
GER	Global Exploration Roadmap
GES	Global Exploration Strategy
GLONASS	Global Navigation Satellite System (Russia)
H	
HSPO	Human Spaceflight Project Office
HSTI	Human Spaceflight Technology Initiative
I	
IAF	International Astronautical Federation
IMF	International Monetary Fund
IGS	Information Gathering Satellite (Japan
IPR	Intellectual Property Rights
ISECG	International Space Exploration Coordination Group
ISEF	International Space Exploration Forum
ISRO	Indian Space Research Organisation
ISNET	Inter-Islamic Network of Space Technologies
ISS	International Space Station
ITAR	International Traffic in Arms Regulations
ITU	International Telecommunications Union
IMMARSAT	International Maritime Satellite Organisation
J	
JAXA	Japanese Aerospace Exploration Agency

(continued)

Acronym	Explanation
L	
LEO	Lower Earth Orbit
LEAG	Lunar Exploration Working Group
LLO	Low Lunar Orbit
LOR	Lunar Orbit Rendezvous
LSG	Leading Small Group
M	
MNE	Multi-National Enterprises
MOE	Ministry of Economy
MOST	Ministry of Science and Technology
MPCV	Multi-Purpose Crew Vehicle (Orion)
N	
NASA	National Aeronautics and Space Administration
NAS	National Academy of Science (United States)
NRC	National Research Council (United States)
O	
OECD	Organisation for Economic Cooperation and Development
OSTP	White House Office of Science and Technology Policy
P	
PLA	People's Liberation Army
PPP	Purchasing Power Parity
R	
R&D	Research & Development
RMB	*Renminbi*
S	
SAARC	South Asian Association for Regional Cooperation
SAST	Shanghai Academy of Spaceflight Technology (under CASC)
SASTIND	State Administration on Science, Technology, Industry for National Defence
SCOSA	Sub-Committee on Space Technologies and Applications (ASEAN)
S&T	Science & Technology
STEM	Science, Technology, Engineering, and Mathematics
T	
TNC	Trans National Corporations
TT&C	Telemetry, Tracking & Command
TCBMs	Transparency and Confidence-Building Measures
U	
UNCOPUOS	United Nations Committee on Peaceful Uses of Outer Space
UNOOSA	United Nations Office for Outer Space Affairs
V	
VSE	Vision for Space Exploration (NASA)
W	
WB	World Bank
WIPO	World Intellectual Property Organisation
WSLC	Wenchang Satellite Launch Center
WTO	World Trade Organization

Glossary of Chinese Terms

To transcript Chinese characters the *pinyin* system (without tonal distinction) has been utilised.

Chinese term	Characters	Explanation
BeiDou	北斗	The Big Dipper/China's navigation satellites' constellation
Chang Zheng	長征	Long March rockets
Chang'e	嫦娥	Moon Goddess/Chinese Lunar Exploration Programme
Danwei	单位	Work unit
Fuxing	复兴	National rejuvenation
Gaige kaifang	改革开放	Deng Xiaoping's policy of opening and reforms
Geming	革命	The Celestial Mandate Revocation
Guanxi	关系	Privileged relationships
He er Butong	和而不同	Harmony without uniformity
Hukou Bu	户口簿	Household Registration System
Kuafu	夸父	Hero of Chinese Mythology/China's space weather programme
Lingdao Xiaozu	领导小组	Leading Small Group
Nei luan, wai huan	内乱,外危险	Internal disorder, external danger
Shenzhou	神舟	Divine vessel: China's crewed space capsule.
Shijie Meng	世界梦	The World dream's policy of Xi Jinping
Tao guang yang hui	韬光养晦	"Hide the brightness, cherish obscurity" policy of Deng Xiaoping
Tiangong	天宫	Heavenly Palace: China's Space Laboratory/Chinese Space Station
Tianhe	天和	Harmony in Heaven/Core module of the CSS
Tianlian	天链	Sky link/China's Relay Satellite

(continued)

M. Aliberti, *When China Goes to the Moon...*, Studies in Space Policy 11,
DOI 10.1007/978-3-319-19473-8

Chinese term	Characters	Explanation
Tianming	天命	The Celestial Mandate
Tianxia	天下	Under the Heaven/the World
Tianzhou	天舟	Heavenly Vessel/cargo spaceship to the CSS
Wentian	问天	Great the Heaven (Experimental Module 1 of the CSS)
Xiaokang Shehui	小康社会	Moderately well-off society
Xitong	系统	Vertical systems
Xuntian	巡天	Cruise the Heaven (Experimental Module 2 of the CSS)
Yuan	元	Chinese currency
Yuan Wang	远望	Long View/China's Tracking Ships
Yuhangyuan	宇航员	Astronaut or Taikonaut
Yujun Yumin	寓军于民	Locating military potential in civilian capabilities (CMI)
Yutu	玉兔	The Jade Rabbit/China's first lunar rover
Zhong xue weiti xie xue weiyang	中学為體, 西学為用	Chinese knowledge as foundation, Western knowledge as function (Chinese external policy in the late nineteenth century)
Zhongguo	中国	The Central Kingdom
Zhongguo Meng	中国梦	The "Chinese dream" policy of Xi Jinping
Zizhu chuangxin	自主创新	Indigenous innovation

MIX
Papier aus verantwortungsvollen Quellen
Paper from responsible sources
FSC® C105338

If you have any concerns about our products,
you can contact us on
ProductSafety@springernature.com

In case Publisher is established outside the EU,
the EU authorized representative is:
Springer Nature Customer Service Center GmbH
Europaplatz 3, 69115 Heidelberg, Germany

Printed by Libri Plureos GmbH
in Hamburg, Germany